FUNDAMENTALS OF
STRUCTURAL DESIGN

The Intext Educational Publishers Series in Civil Engineering

Series Editor—Russell C. Brinker
                New Mexico State University

Brinker—*Elementary Surveying*, 5th edition
Clark, Viessman, and Hammer—*Water Supply and Pollution Control*, 2nd edition
Ghali and Neville—*Structural Analysis: A Unified Classical and Matrix Approach*
Hill—*Fundamentals of Structural Design: Steel, Concrete, and Timber*
Jumikis—*Foundation Engineering*
McCormac—*Structural Analysis*, 3rd edition
McCormac—*Structural Steel Design*, 2nd edition
Meyer—*Route Surveying and Design*, 4th edition
Moffitt—*Photogrammetry*, 2nd edition
Moffitt and Bouchard—*Surveying*, 6th edition
Salmon and Johnson—*Steel Structures: Design and Behavior*
Spangler and Handy—*Soil Engineering*, 3rd edition
Ural—*Finite Element Method: Basic Concepts and Applications*
Viessman, Harbaugh, and Knapp—*Introduction to Hydrology*
Wang—*Computer Methods in Advanced Structural Analysis*
Wang—*Matrix Methods of Structural Analysis*, 2nd edition
Wang and Salmon—*Reinforced Concrete Design*, 2nd edition
Winfrey—*Economic Analysis for Highways*

FUNDAMENTALS OF STRUCTURAL DESIGN:

Steel, Concrete, and Timber

Louis A. Hill Jr.
Arizona State University

Intext Educational Publishers
New York

Library of Congress Cataloging in Publication Data

Hill, Louis A   1927–
  Fundamentals of structures.

  (Intext Educational Publishers series in civil engineering)
  1. Structures, Theory of. 2. Structural design.
I. Title.
TA645.H48   624'.17   74–8428
ISBN 0–7002–2462–9

Intext Educational Publishers
666 Fifth Avenue
New York, New York 10019

This book is dedicated to
Dr. Jan. J. Tuma, P.E.
a teacher's teacher
and
Mr. Lee Hendrix, P.E., Consultant
an engineer's engineer

CONTENTS

PREFACE

This book is the outgrowth of my efforts to become a better engineer and a better teacher of engineering. It embodies the philosophy, concepts, and practices formulated as a design engineer, modified and organized by teaching. The content is based on the belief that competent, creative engineers combine two diverse attributes: (1) a critical mind searching for ever more complete understanding through comprehensive study of basic theory and (2) an agile mind seeking means for more efficient, economical solutions to very practical problems.

In keeping with this spirit, an integrated approach for structural design of concrete, steel, and timber members and connections is used. By contrast, a clear distinction is made between mathematical models, legally defined models as specified by codes, and the physical reality they represent. Careful delineation is also made between methodology applied to analysis as contrasted to that used for design. Theory, practice, and code development are closely related throughout, but the material is organized into subsections to provide instructors means for flexibility and variety of approach.

This book is written for students who have completed courses in statics and mechanics of solids to use as a textbook, and should also be a useful reference work for practicing engineers and architects. It is essentially self-contained, having some important redundancies with material from both elementary and advanced courses.

Chapters 1 through 8, 12, and Appendix A can be mastered with a minimum of mathematics. However, some developments and problems necessitate reasonable competency in calculus and linear differential equations with constant coefficients. (I have found the latter easily understood by architectural students with only two semesters of calculus.)

Level of mathematical maturity and available time govern the approach to be used with Chapters 9, 10, 11, and Appendixes B and C. Typically, I spend about three weeks lecturing on these subjects, using transparencies reproduced directly from this book. My objective is to provide "one-time" clear insight into the more theoretical developments, rather than long-term mastery. Such an approach has provided: (1) appreciation of code development and formulations, (2) transition to solution of more complex problems, and (3) reasonable competency for those students who have completed partial differential equations. Numerous students have reported enthusiastically their extensive use of this material in both graduate study and on the job.

Problems for student assignment are formulated so that a variety of input parameters can be used. This represents a compromise with my usual method of using a restricted number of comprehensive problems with parameters controlled by student team numbers in MOD functions. Both methods facilitate grading while enhancing variety.

I am grateful for the critique and encouragement of students, professional engineers, and professors, particularly to the following, listed alphabetically: Professor Russell C. Brinker, P.E.; Mr. William Carey, P.E., City of Phoenix; Mr. Bill Dye, P.E.; Professor Frederick M. Graham, Iowa State University; Professor Wilhelm Kubitza, University of Alabama in Huntsville; Professor Bogdan Kuzmanović, University of Kansas; Professor Harry Lundgren, Arizona State University (who tested manuscript in class); Mr. Ben Ong, P.E., Arizona Highway Department; Professor William Orthwein, Southern Illinois University; Professor Richard H. J. Pian, Arizona State University; Dr. W. C. Schoeller, P.E., Consulting Engineer, Dallas (who tested manuscript in class and initiated many valuable changes); Professor William J. Venuti, San José State University, and Mr. Jesse Wyatt, P.E., Portland Cement Association, Phoenix.

I am also appreciative of the continuing support and encouragement of Dean Lee P. Thompson and Associate Dean George Beakley, Jr. of Arizona State University, and of my wife, Jeanne.

FUNDAMENTALS OF
STRUCTURAL DESIGN

CHAPTER 1

CONCEPT OF DESIGN

A classical paradox of the learning process is that one must be able to see "the forest" in order to put "the trees" in their proper perspective—before one has had a chance to study "the trees."

Similarly, in the study of structural design an overall concept of design—"the forest"—is difficult to obtain until various miscellaneous details—"the trees"—are understood. Yet, without such a comprehensive framework of understanding, it is impossible to catalog and relate properly the miscellaneous details and build a strong foundation of learning.

This chapter, therefore, is written for those with an adequate background in mechanics and who are anxious to apply these fundamentals to real-world situations. Each of the following parts—The Engineer as a Designer, Basic Fundamentals, Tangible Foundations for Design Procedures, and Aids—has been chosen carefully to provide a frame of reference for specific developments of later subdivisions; that is, to show the "forest" into which the "trees" are to fit.

1-1 THE ENGINEER AS A DESIGNER

By the first definition of *Webster's Seventh New Collegiate Dictionary*, "design" (as a transitive verb) means "to conceive and plan out in the mind." As an intransitive verb, it means "to conceive or execute a plan; to draw, lay out, or prepare a design." Certainly all engineers are designers in both senses. But in this text *design* connotes the know-how and follow-through necessary to produce a useful real-world physical device or system which satisfies predetermined performance requirements.

1-1.1 Basis for Structural Design. A novice might question the need for any formalized study of structural design since "rules of thumb" were sufficient for

Egyptian pyramids, Greek temples, Roman aqueducts, and Gothic cathedrals. In the light of current knowledge, however, arbitrary rules cannot adequately fulfill three basic requirements of structural design, namely, safety, economy, and aesthetics. Although no formula exists for defining aesthetics, engineers usually appreciate and respond to beautiful structures and are often zealous in producing them. Study of structural design, therefore, relates primarily to the relationships between safety and economy. In general, these factors vary inversely so that a safe structure may not be economical and a structure of reasonable cost not necessarily safe. This inherent conflict governs most study of topics collectively called *structural design* in this text.†

1-1.2 Exact. Results of analysis and design of real-world structures are not precise—they are approximate. To clearly delineate this variance structural design will be related to the type of model being used.

A mathematical model is an equation or group of equations which may represent characteristics of the physical world. Solutions to such a prototype may be exact because mathematics is exact. But the mathematical model may not completely represent its real-world counterpart.

A legally defined model is represented by a set of characteristics, and for purposes of law may be exact. As an example, the expected live load on the floor of a schoolroom may be established by law as 40 pounds per square foot. When designing to satisfy a code containing such a specification, the value of the live load is exact.

Physical reality often can be replaced by a combination of mathematically and legally defined models. Solution of such mathematical formulations then gives results (often absolute) from which approximate conclusions can be drawn. The better the real world is represented by a mathematical model, the closer results of a computation will agree with reality. Judgment in proper representation of practicality obviously is critical if results must obtain proper balance between safety and economy.

1-1.3 Nonlinearities. A transfer from the physical world to a mathematical model provides no gain if the formulation cannot be solved. Because of this, many such models are linearized versions of the nonlinear real world. It is important to recognize three broad classes of nonlinearities, given below, because grossly unsafe and/or uneconomical structures can be postulated from mathematical solutions to grossly linearized versions of reality. Also, mathematical techniques are improving, and in conjunction with the electronic computer can yield solutions to complex formulations which in the recent past were considered unsolvable.

---

† Note that this definition differs from "considerations that enter into the formulation of a rational code or specification for the design of a member" as used on page 5 of *Advanced Mechanics of Materials* by Seely and Smith [1].

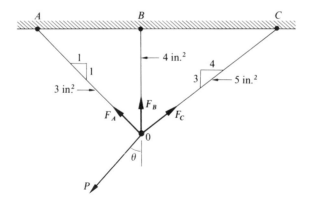

Fig. 1-1.1   Three-Member Truss

1. *Geometric nonlinearities* stem from the assumption that angles and member lengths remain constant between unloaded and loaded conditions. Consider the three-hinged truss of Fig. 1-1.1. Summing forces vertically, we have

$$F_A (\sqrt{2}/2) + F_B + F_C (3/5) = P \cos \theta.$$

Summing forces horizontally, we obtain

$$F_A (\sqrt{2}/2) + 0 - F_C (4/5) = -P \sin \theta.$$

By compatability, the vertical displacements must be identical, so using $\Delta = PL/AE$, we find that

$$\frac{F_A L_A}{3E} \left(\frac{\sqrt{2}}{2} \right) = \frac{F_B L_B}{4E} = \frac{F_C L_C}{5E} \left(\frac{3}{5} \right),$$

where $\Delta \equiv$ displacement, $P \equiv$ force, $L \equiv$ length, $A \equiv$ cross-sectional area, and $E \equiv$ modulus of elasticity.

From these equations, forces in the members can be obtained. In reality, however, as load is applied the members both elongate and rotate so that results are in error. By ineration and use of new values of angles and lengths for each calculation, true values eventually can be found. Even in so simple a case as this, serious errors may be obtained using only a linear solution, particularly if P is extremely large.

2. *Physical characteristics of materials* are often nonlinear, but in the traditional approach are often assumed to be linearly elastic. Figure 1-1.2 indicates a true stress versus true strain curve for concrete in compression as compared with its linear elastic representation. Obviously the error is not overly severe within a limited range, but such representations are being replaced slowly by a more realistic approximation so that overall factors of safety can be more uniform. Several examples are shown in this text.

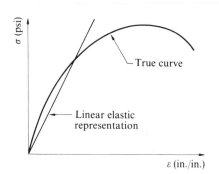

Fig. 1-1.2   Stress-Strain Diagram for Concrete

3. Loading configurations can lead readily to serious nonlinearities; as an example, consider the cantilever beam of Fig. 1-1.3. The load P is decomposed into vertical and horizontal components, P_V and P_H. If P_H is zero, then maximum bending moment is $P_V L$. But, if P_H is not zero, then maximum bending moment is increased by $P_H \Delta$. In addition, this horizontal component causes axial compression and as the ratio of P_H to P_V increases, the possibility of Euler buckling increases rapidly. In fact, if P_V is zero indicating a maximum bending moment of zero, the only obviously apparent stress is axial compression. Yet in such a case the cantilever is subject to failure by buckling before the compressive stress becomes significantly high—meaning a member with $P_V = 0$ can fail by bending even though the simple mathematical representation suggests that no bending can exist.

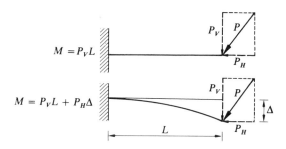

Fig. 1-1.3   Loading Configuration Inducing Nonlinearity

1-1.4 The Design Process. Although it is practically impossible to specify the steps most appropriate for all designs, it is useful to visualize four stages inherent in most. Based on complexity of the problem, coupled with experience and preference of engineers, the sequence may be either progressive through successive stages or iterative in application.

The first stage embodies a *definition of the problem* leading to formulation of a mathematical model. In almost all practical situations this requires (a) deletion of excess facts, (b) recognition of missing data (which often are never going to be available), and (c) a clear statement of assumptions used in setting up a mathematical model. Contrary to common beliefs held by students on the basis of academic experiences which place emphasis on example solutions, an engineer in practice finds that the most difficult aspect of design is in clearly stating the problem. A beginner in his first design course will find, if he analyzes the situation carefully, that most of his difficulty stems from trying to solve a problem before it is defined *clearly*.

Once a problem has been fully specified, a *solution* must be obtained for its mathematical model. In general the procedure is not overly difficult if all parameters are understood. Successive sections emphasize efficient methodology to provide more available time for critical evaluation.

The *solution* of the mathematical model must be *interpreted* into meaningful correlation with physical reality. It is not particularly unusual if the interpretation of results demands an alternate mathematical model.

Finally, *the results must be presented.* Typically, solutions are presented in the forms of (a) calculations, (b) sketches and drawings, and/or (c) specifications. A design is successful only if the presentation leads, without confusion or ambiguity, to production or construction of a physical or system entity.

In summary, the design process consists of: stage 1, *definition* of the problem; stage 2, *solution* of the problem as defined; stage 3, *interpretation* of the solution; and stage 4, *presentation of the results*. Note, however, that at stage 3, it might be necessary to return to stage 1 for a better problem definition.

1-1.5 Cost, Time, and Profit. Obviously, a structural engineer working for an aircraft company that builds ten thousand airplanes, all alike, spends much time on design. Great effort can be expended in assuring that all possibilities have been explored, and to lower the safety factor to a bare minimum. In such cases, design cost easily can exceed the cost of 20 of the airplanes.

On the other hand, consider the designer of a bridge or building, that costs $100,000. In this category, his total fee (structural design only) could be about $3,000. If the designer's methodology is so complicated and complex that his cost is $12,000, obviously he cannot remain in business. Conversely, if his work is slipshod, either he may be accused of overdesigning† the structure and refused additional work, or the structure may fail and his reputation is lost. In either case, his professional career is destroyed.

Since engineering fees usually represent a percentage of the actual cost of a structure as built, defining problems of and design procedures for

---

† This is often referred to colloquially as "loading the structure."

structures varies among jobs. As relative design fees decrease, factors increasing safety against unforeseen and uncomputed conditions proliferate and construction economy lessens. Yet, in every case, proper problem definition and choice of design technology rely to a very large extent on common basic fundamentals.

1-2   BASIC FUNDAMENTALS

Probably the two most important basic concepts are: (1) structures have only two "methods of failure" and (2) only two "types of stress" exist. Once fully understood, these insights lead to confidence in even the most esoteric design.

The combination of two "factors of uncertainty" in conjunction with three 'types of analysis," allows total design methodology to be presented in tabular form (see Table 1-2.1).

Finally, certain techniques universally appropriate to a wide range of applications are considered in some detail.

1-2.1 Methods of Failure. The failure of a structure usually means, to a layman, falling floors, tumbling walls, or crashing bridges. Actually, less spectacular failures can be equally devastating.

Fortunately, a structure can fail in only two primary ways—either by overstressing (insufficient strength) or by excessive deformation (insufficient stiffness). That is, a failure is due to one of two causes: (1) the material is of insufficient strength and actually crushes, tears, rips, or breaks, or (2) the structure is of insufficient stiffness and overdeflects, has excessive curvature, vibrates extensively, or buckles.

A member might fail by overstressing in compression, tension, bending, torsion, or by principal stresses such as "diagonal tension" in reinforced concrete. This type of failure is dependent primarily upon the various allowable stresses of each material. In this text, a measure of the ability of a member to resist overstressing is referred to as *strength of the member* and stress always means *force per unit area*.

Failures due to excessive deformation are those which produce excessive deflection, and depend primarily upon relative *moduli of elasticity* of the materials. Herein, resistance to overdeformation is termed *stiffness of the member*.

An example of excessive deformation occurred in a structure built with a roof supported by tension cables which, in turn, were carried by an arch (see Fig. 1-2.1). Because the cables looked too large relative to the rest of the structure, it was decided to replace them with cables of smaller diameter and much greater strength. The choice was made erroneously on the sole basis of strength. Axial stress was computed as P/A, and because load P was unchanged, area A was reduced inversely proportional to the ratio of the two allowable stresses.

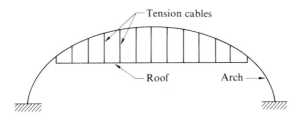

Fig. 1-2.1   Roof Hung from Tension Cables

No deflection check was made. When the structure was erected, the deflection actually increased in the same ratio that steel area decreased, because $\Delta = PL/AE$ and the load P, length L, and modulus of elasticity E, were identical for both cases. This was a costly failure involving overdeformation.

In another case, a steel column in a multistory building was replaced by one with a much higher allowable stress. Erroneously this decision was arrived at by using only a calculation based on the inverse ratio of stresses and areas. If the error had not been caught in time, a catastrophic buckling collapse would have occurred because stiffness was a function of the unchanged modulus of elasticity, not the material strengths.

It is important to realize that structures which are adequately designed for static loadings at normal temperatures might be completely unsafe for conditions that involve either or both of dynamic loads (including impact and fatigue) and immoderate temperatures (including brittle fracture in extreme cold and excessive deformation due to creep in extreme heat). However, even in these conditions, ultimate failure is due to overstressing (insufficient strength) or excessive deformation (insufficient stiffness).

1-2.2 Types of Stress. Only two types of stress exist: normal (axial) and tangential (shear). A moment's reflection will convince students that flexural stress is actually a normal stress *induced* by flexure. Likewise, a bearing stress is a normal stress *induced* by the bearing of the surface of one body on another. A torsional stress is a tangential stress *induced* by torsion. Clear understanding of the relationship between stress nomenclature and inducing force, coupled with creation of a mental picture of the normal or tangential stresses actually produced, prove to be invaluable. Not only is it enlightening but also it is rather appealing to know that nothing can happen to any structure except by axial and/or shearing stresses.

This concept of only two types of stress may seem oversimplified, especially if one has studied introductory Theory of Elasticity. Admittedly, both normal and shearing stresses have components and are most completely represented as tensors. Experience seems to indicate, however, that this simplistic view is particularly helpful in securing safe, economical designs in practice.

1-2.3 Factors of Uncertainty. To compensate for lack of precision in dealing with the physical world, engineers include a margin for error in every design. This margin quite correctly can be referred to as a *factor of uncertainty*, because its magnitude is proportional to the quantity of the involved "unknown." This factor, applied in divergent manners, leads to two distinct terms. A *factor of safety* is the numerical value of the factor of uncertainty. The reciprocal of the factor of uncertainty multiplied by the *ultimate* (or sometimes *yield*) stress yields an *allowable working stress* used with *actual* (that is, *working*) loads. A *load factor* is the numerical value of the factor of uncertainty which is multiplied by actual working loads to obtain *ultimate* (*failure* or *collapse*) *loads* used with *ultimate* (or sometimes *yield*) stresses.

1-2.4 Types of Analysis. In general, three types of analysis are recognized. (1) Historically the most important is based upon characteristics of an ideally linearly elastic material. (2) Of increasing importance is analysis based on the assumption of a perfectly elasto-plastic material. (3) In some applications, nonlinear analysis is the only way to obtain valid results. Stress-strain curves for these three analysis types are shown in Fig. 1-2.2.

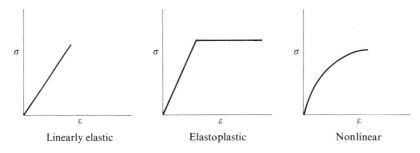

Fig. 1-2.2   Three Typical Stress-Strain Diagrams

A distinction is evident in the use of the word "analysis." To *analyze a frame* means to find the existing moments and forces imposed by loads. In such a case the structure is replaced by a mathematical model which represents members along their neutral axes by lines with lengths measured along their true centroidal axes. On the other hand, to *analyze a member* means to find the stresses in a particular plane (normally perpendicular to the centroidal axis). In such a case the structure is replaced by a mathematical model which represents a particular cross section by a plane. To distinguish between these two concepts, in points where the meaning may not be clear, the respective terms *frame analysis* and *member analysis* are employed.

Table 1-2.1 Design Methodology

Design Methodology	Factor of Uncertainty	Frame Analysis	Member Analysis
1. *Allowable Stress Design* in steel	Factor of Safety	Elastic	Elastic
2. *Working Stress Design (WSD)* of reinforced concrete			
3. *Elastic Design* in timber, prestress concrete, and metals			
1. *Strength Design (USD)* of reinforced concrete and prestressed concrete	Load Factor	Elastic	Plastic
1. *Plastic Design* in steel	Load Factor	Plastic	Plastic
2. Yield Line Theory of reinforced concrete			
1. Experimental analysis and design of metals	Load Factor	Nonlinear	Nonlinear
2. Some aerospace structures			
3. Experimental studies involving concrete			

1-2.5 Design Methodology. Design methodology can be specified by (a) factor of uncertainty, (b) frame analysis, and (c) member analysis (see Table 1-2.1). In the table, methodologies later developed in detail are italicized; other nonlinear examples also are developed. Except for strength design of reinforced concrete, note that frame analysis and member analysis are identical. Also, the factor of safety is used only with elastic analysis. Although other combinations are possible, they usually are not of major importance.

1-2.6 Universals. The title of this section was selected with utmost care because the four important entities listed here need to have proper recognition. Although often disguised or obscured in usage, they all tend to meet the definition of "universal" in *Webster's Seventh New Collegiate Dictionary*, "present or occurring everywhere ... existent or operative everywhere or under all conditions ... comprehensively broad and versatile."

The first universal is the equation

$$\sigma = P/A \pm Mc/I, \qquad (1\text{-}2.1)$$

where $\sigma \equiv$ normal (axial) stress, $P \equiv$ axial load, $A \equiv$ cross-sectional area of member, $M \equiv$ bending moment due to loading, $c \equiv$ distance from neutral axis to extreme fiber, and $I \equiv$ moment of inertia.

This equation, found in many disguised forms throughout the literature, often embodies modifications which relax the superposition requirement to account for buckling characteristics. A general form is developed in Chap. 9 according to the earlier work of Hardy Cross. Its fundamental nature is established in the subsequent text.

Mohr's circle is also widely used in the literature with symbols altered to fit particular circumstances. A brief development and usage of Mohr's circle is illustrated in sec. A-1.

A third universal is a table used to determine centroids, moments of inertia, and section moduli for a multitude of situations. This A, y, Ay, $Ay^2 + I_0$ *table* is discussed in sec. A-2 and used extensively throughout this book.

The fourth universal embodied in the equation $EIy'' = M$, with its relationships of deflection, slope, bending moment, shear, and load are discussed in sec. A-3.

1-3 TANGIBLE FOUNDATIONS FOR DESIGN PROCEDURES

To show that design concepts can be stated in simple terms, this section is devoted to consideration of methods of failure, of effects of different materials and shapes, and of different methods of loading. Although none of this subject matter is new to students of mechanics, it is hoped that these factors presented in a unified and nonmathematical way will be useful as a framework ("the forest") to which students may relate various details ("the trees") which are known or soon to be learned. Many concepts can be demonstrated personally with the use of simple models. Balsa wood is ideal for this purpose since it is inexpensive and easy to obtain.

1-3.1 The Part Materials Play. In the recent past the primary structural materials were steel, reinforced concrete, and many kinds of timber. When used in structures all of these were analyzed by elastic methods. Today, instead of one type of steel there are many, and design is either elastic and/or plastic. Instead of simply using the elastic theory of reinforced concrete, we use elastic, ultimate strength and yield line concepts in conjunction with both low- and high-strength masonry, reinforced concrete, and prestressed concrete. Lumber also is used in more ways, as conventional timber, laminated, in plywoods, and in sandwich panels. Aluminum, light gage steel, reinforced brick, and other materials are also bidding for a larger share of the structural work.

Based primarily on the historical preponderance of elastic design of reinforced concrete and steel only—and their divergent material characteristics—many texts and courses cover either one or the other but not both. Because the two materials are quite different, it is fortunate that research specialists have delved deeply into all aspects of only one or the other. However, a stand-

ardized approach to all design problems seems to have inherent advantages for both students and practicing engineers.

By means of a standard approach, possible causes of failure are considered in general, leaving only a conscious decision to eliminate the need for specific calculations based on knowledge of material characteristics. So, although not considered as a reason for independent design methods, differences in materials must be considered. For instance, steel is very strong in tension but many commercial shapes (due to inherently thin sections) tend to buckle rather readily in compression. On the other hand reinforced concrete, with its strength in compression, is extremely weak in tension but strong in shear (with a tendency to fail by a diagonal tensile stress when the shearing stress becomes too high).

Other material characteristics relate to general rather than specific differences. A linearly elastic assumption is validated by the stress-strain curve for steel, at least in its "elastic range," but a stress-strain curve for reinforced concrete is not straight. To make matters more difficult, reinforced concrete has an apparent modulus of elasticity which changes with time. Other structural materials also have nonlinear stress-strain diagrams, illustrated later in appropriate places. Obviously, then, the stress-strain characteristics of each material must be known fully to design adequately.

Various materials possess different relative *strengths*, as measured by their *allowable stresses*, and *stiffnesses*, as measured by their *moduli of elasticity*. To illustrate, a member composed of material S is three times as strong (has an ultimate allowable stress three times as high) as a member of material T, both of the same physical dimensions. Yet S can actually fail by buckling under lighter loads than those carried by a section made of T if the modulus of elasticity of T is greater than that of S. That is, S buckles sooner than T because it has less inherent stiffness, and buckling can occur before ultimate strength is reached.

These few paragraphs indicate the vital importance a material's characteristics have in structural design. Among other important material characteristics are: tendency toward brittle fracture, change of characteristics with time or under load, detrimental effects of repeated loads or deleterious environments, etc.

1-3.2 The Factor of Shapes.   Figure 1-3.1 shows six typical shapes used for cross sections of structural members. The simplest is the rectangle. Possibly the most common shape is the I or WF (I beam or wide-flange beam). The box, channel, angle, and T shapes also are frequently used, each in its own appropriate application.

Although other shapes also may be employed to make subsequent discussions realistic, only the rectangle and the WF are considered. Such an ap-

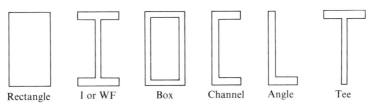

Fig. 1-3.1   Representative Shapes

proach is valid because combinations of rectangles and WFs can form essentially any shape. Once the rectangle and the WF beam are understood, the principles for almost all other shapes can be extrapolated rather easily.

To facilitate discussion, Fig. 1-3.2 shows the rectangle in expanded forms, first as a very thin, tall, vertical plate, second as a very flat, wide plate, and third as a square. The fourth shape is one that no longer looks like a rectangle; the uppermost part is rectangular, but the bottom part is made from small black circles; this represents the usual reinforced concrete (RC) beam. The last shape shown is a rectangle of more usual dimensions, its height about $2\frac{1}{2}$ times its width, and it is used as a basic shape for subsequent discussion as noted by the encircled one.

Figure 1-3.3 shows variations of the WF. First is a one-piece member and second is a one-piece section with plates added at top and bottom. The third and fourth sections, built up from other pieces, usually are called *plate girders*.

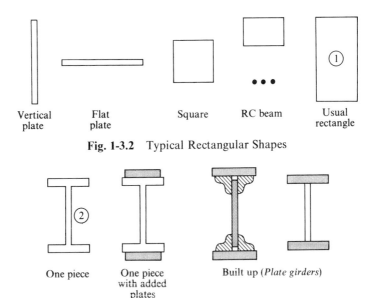

Fig. 1-3.2   Typical Rectangular Shapes

Fig. 1-3.3   Typical Wide-Flange Sections

The basic shape used for demonstration is the simple one-piece WF as noted by the encircled number two.

Students might find it worth-while to observe again that almost all cross sections can be made from these two basic shapes, the rectangle or wide flange. Furthermore, each possesses characteristics found in all others—right-angled corners, thin protruding flanges, and thin webs.

1-3.3 Studies in Loadings. The two basic shapes numbered 1 and 2 above will be employed in this section. The intent here is to isolate and delineate effects of loading and to relate clearly the details previously learned in mechanics courses. To do this, it is desirable for preliminary investigation to consider only axial forces. Yet, if one considers Mohr's circle, it is evident that axial forces also create shearing stresses. To overcome this problem, loading cases a through f *tacitly neglect combined stresses.*

Fig. 1-3.4   Case a, Tension Only, $f_a = P/A$

Case a. A very simple loading shown in Fig. 1-3.4 has a solid bar in tension due to two equal and opposite forces P. The force P is distributed over the entire end areas of the member. For this basic, simple case, only one type of stress (assuming no combined stress or else a material with high shear allowable) can cause failure—namely, tension. At failure the material will have either stretched too much (overdeformed) to meet serviceability requirements, or it will break before such excessive elongation (overstress). In the first case we might say the member was not stiff enough, and in the second that it lacked sufficient strength. Obviously, stress is equal to total force P divided by cross-section area A, or $f_a = P/A$, an old familiar formula.

To illustrate difficulties encountered by even this simple loading, one is reminded of early attempts at prestressing.† Allowable stresses were low and the modulus of elasticity approached 29×10^6 lb/in.^{2} The "stretch" to obtain "prestress" was, therefore, relatively short ($\Delta = PL/AE$). Due to time-related deformation called *losses,* members shortened as they aged, which reduced the original elongation of steel and consequently the prestressing force itself. Thus, after some months, all prestressing force was lost and the beams collapsed. Today, losses are still considered (sec. 7-3.1), but due to higher strength steel, prestressed area is reduced requiring longer initial stretching so that

---

† For a simple discussion of prestressing, refer to sec. 7-1.1.

Stress nearly uniform
(within 1%) at one beam
width from end of member

P

P

Fig. 1-3.5   Case b, Tension Only, WF Section

reasonable prestressing still remains under working loads after the member has shortened.

Case b. Figure 1-3.5 illustrates axial tensile forces acting on a WF beam, where again it is possible to consider that the load P is distributed uniformly over any cross section of the member. To make the problem more meaningful, however, load P acts at a concentrated location through the centroid. Although stress at such a point would be quite high, if the member itself is relatively long, uniform stress can safely be assumed at some distance from the ends (by St. Venant's principle). Further, it is probable that at the ends of this member, flange corners have no stress. Stresses could exist only if web and flanges were rigid enough to cause elongation of the flange itself parallel to the direction of the force P. A significant axiom of structural engineering is that stresses and strains almost always exist together. In general, if a member cannot strain (stretch or deform), it cannot stress, and vice versa.

An exception to this interdependence of stress and strain is evidenced by temperature changes. If the member of Fig. 1-3.5 were on a frictionless surface and the temperature increased, it would elongate without stress. On the other hand, if the member were rigidly restrained from elongation, stresses would build up without corresponding strains. Stresses due to temperature changes can be extremely large and must adequately be taken into consideration.

Case c. In Fig. 1-3.6 a member with rectangular section is loaded by colinear compressive forces P acting through the centroidal axis. In this example, failure could result from overstress in compression producing crushing, or excessive deformation causing too much shortening or buckling. Such a member buckles about the axis having the weakest moment of inertia. If it buckles before becoming overstressed, this too is a serious failure.

In every case in which compression occurs, buckling is a potential factor and must always be examined mentally and perhaps mathematically. (A bearing

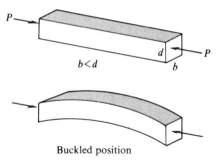

Buckled position

Fig. 1-3.6 Case c, Compression Only

plate between a pier and a girder will not buckle even though in compression.) Fortunately, types of buckling usually can be visualized, provided it is thought of in terms of simple member cross sections with pressure exerted as though to a column.

Case d. Figure 1.3-7 shows a rectangular member loaded by a tension force P *not* acting through the centroid. Displacement from the centroid to the line of action of force P is called the eccentricity. Since P acts a distance e from the centroid, it can be resolved into a couple and a force P' that acts through the centroid. Force P', due to this resolution, causes tension over the entire section and the couple can be replaced by a moment M.

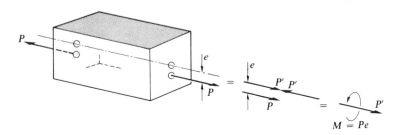

Fig. 1-3.7 Case d, Eccentric Tension, $f = P/A \pm Pey/I$

Effect of the couple, neglecting the concentrated load P' acting through the centroid, bends the member into a flat arch. Since plane sections are assumed to remain plane after bending, the top of this member is shortened and the bottom lengthened. Because stress and strain are compatible, however, shortening of the top signifies a compressive stress and concurrently the stretching of the bottom induces tension. Force at the bottom of this member is, therefore, always tensile caused by force P' acting through the centroid plus the moment couple.

This is not always true at the top of the member because if the eccentricity e is quite small, tension due to force P' acting through the centroid more than offsets compression due to bending, and top fibers are in tension. As eccentricity e increases compressive stress, due to bending, numerically becomes equal to tensile stress induced by P'. This represents the condition when P is at the lower kern point of the section and top fibers have zero stress. As e increases beyond the kern point, compressive force developed by bending becomes larger but tension force from the axial force P' remains constant. In summary, if P is within the kern, stress in the top remains tension; if P is lower than the bottom kern point, the top of the member is in compression and the possibility of buckling must be investigated.

Analytically, the previous discussion can be represented by Eq. (1-1.1), where moment M is represented by couple Pe. The stress at any point in this member then can be found by

$$\sigma = P/A \pm (Pe)y/I \qquad (1\text{-}3.1)$$

where $P \equiv$ axial load, $A \equiv$ cross-sectional area, $e \equiv$ eccentricity of axial load measured from centroid, $y \equiv$ distance from centroid where stress is computed, and $I \equiv$ moment of inertia of cross section.

From Eq. (1-3.1) stresses at any point in the member, including maximum values occurring at outermost fibers, can be determined. Specification limits are given for allowable compression and tension of various materials. If these limits are not exceeded, the member has sufficient strength.

When top fibers are in compression, buckling is possible because of lack of stiffness of the member. For a rectangular cross section, buckling occurs at right angles to the forces P (lateral buckling) so moment of inertia about vertical axis should be kept relatively high. Normally, this is accomplished by maintaining a relatively large ratio of width to the overall member length. Another technique is to reduce effective lengths to reduce possibilities of buckling.

Case e. Figure 1-3.8 shows a WF section loaded by axial tension P and with eccentricity e. As in the previous case, the bottom stress always will be tension, but the top flange may go from tension, to zero, to compression. So long as

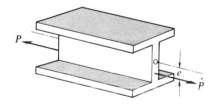

Fig. 1-3.8   Case e, Eccentric Tension, WF

the entire section is in tension, failure is limited to overstressing in tension (rupturing) or excessive deformation (stretching too far to be useful).

With the top flange in compression, failure is no longer restricted to buckling perpendicular to the web (lateral instability). This is evident because bending causes compression in the flange, which makes it act as a column, although it is stiffened by the web. Consider a situation in which this column acts alone with an axial compressive force; it is apparent that the plate will tend to buckle about the weak axis. This is precisely what can happen in wide-flange beams. Failure (flange buckling) in this mode means the top flange will crumple in accordion fashion. Thus buckling can be (1) general, where the member becomes laterally unstable and the top buckles sideways, or (2) local, where the flange is too thin and buckles locally. These phenomena are considered in detail in Chap. 11.

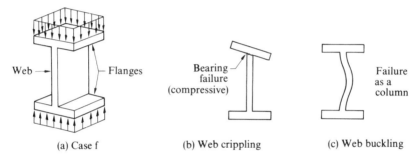

(a) Case f   (b) Web crippling   (c) Web buckling

Fig. 1-3.9   Simple Web Failure

Case f. The WF shape lends itself to the discussion of two common types of failures inherent in all members composed of relatively thin plates. Figure 1-3.9 depicts a wide-flange section used as a short column. The load imposed on top is assumed to act uniformly over the top flange and be supported by a similar force under the bottom flange. For the immediate discussion, top and bottom flanges are presumed infinitely rigid; that is, the flanges themselves cannot bend, break, or strain. With a rigid flange available, it is necessary to determine how the web (column) will react under the applied load. It may be seen clearly that the load bearing on this column (web) acts on a small area, and at the toe of the fillet the member might fail by crushing as shown in Fig. 1-3.9(b). The stress P/A was greater than the allowable in bearing, hence a compression failure occurred resulting in the flange tilting.

In Figure 1-3.9(c), the web does not fail by P/A (crushing); instead, it acts as a column of insufficient stiffness, and it buckles, a type of failure expected in wide-flange beams under heavy concentrated load.

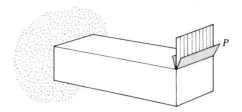

Fig. 1-3.10   Case g, Cantilever Beam

Case g. A cantilever with a concentrated load is shown in Fig. 1-3.10. At the point of load application, local compression failure is possible. If the local stress is too great, several methods may relieve the situation, such as using a bearing plate to increase the area over which the concentrated load acts.

Load *P* also tends to make the cantilever bend and deflect, causing the top fibers to stretch in tension and the bottom to shorten in compression.

Recall, at this point, that bending resistance in a beam is developed by axial compression and tension of its fibers. This means that bending moments actually are resisted by an infinite number of differential couples formed by tensile and compressive stresses.

Because of compressive stresses, a cantilever is subject to lateral instability. Lateral buckling generally can be prevented by use of a sufficiently wide member.

Cantilever beams also provide a good vehicle for discussing the problem of shearing force that can cause two types of failure, namely, by overstressing and by excessive deformation.

Deflection of a beam subjected to an applied load is produced by both bending moments and shearing forces. For most practical purposes, however, deflection produced by shearing force can be proven negligible by use of energy methods of frame analysis. Still, the member might fail by overstressing in shear itself, in which case a part of the cantilever would simply "slice off." Note that this shearing is completely different from normal stress produced by either bending and/or concentrated loads acting alone. Again, structures have only two stresses: axial (normal) or shearing (tangential). These types of stress are not additive but can be related using Mohr's circle (see sec. A-1). Remember, shear does not change the lengths of members; rather, it causes rectangles to distort into parallelograms.

Case h. Failure of a structure does not always occur by either pure shearing or normal stresses. Sometimes a critical combination of these two (principal stresses), as determined by Mohr's circle, is the cause.

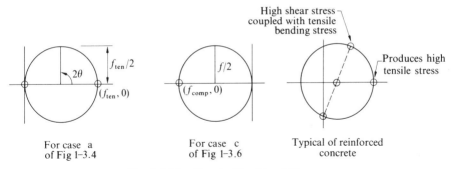

Fig. 1-3.11   Case h, Combined Stresses

   Thus in the member of Fig. 1-3.4 a tensile force induces shearing stress with a magnitude equal to one-half the tensile stress f_{ten} (Fig. 1-3.11). If shearing strength is less than one-half tensile strength, failure will be due to shear acting on a plane 45° counterclockwise from the neutral axis. A similar failure would result in the member of Fig. 1-3.6; see second Mohr's circle of Fig. 1-3.11. Other comparable situations readily come to mind. For instance, for a member fairly weak in tension, high shearing stress combined with bending stress can produce a maximum tensile principal stress that exceeds the allowable tensile strength of the material. Thus a "diagonal tension" failure (typical in re-inforced concrete) would result (also illustrated in Fig. 1-3.11). The opposite, of course, is also possible—a material quite weak in compression could be crushed by compressive stresses in the principal plane.

   Case i. In Fig. 1-3.12, all previously mentioned factors have been combined into a common situation. Practical possibilities of failure, starting from point of application of load P, are considered. First, the flange itself might fail, due to either shearing or bending as a cantilever beam. This means that the flange itself, if lacking sufficient strength or stiffness, could fail making the rest of the beam ineffective.

   A second possible type of failure is crushing of the web at the toe of the fillet. Excessive compressive stress resulting from the superimposed load P would cause the top flange to rotate as shown in Fig. 1-3.9(b) or the web to buckle as shown in Fig. 1-3.9(c).

   Shear and bending moment diagrams are shown in Fig. 1-3.12; bending force induces a compressive stress on the top flange. Due to this bending moment, overstress can result in either crushing or tearing. A bending failure might, however, come from excessive deformation in either of two forms: (1) The total beam might deflect more than considered allowable or desirable. If this beam is over a glass window, deflection might break the glass. As a

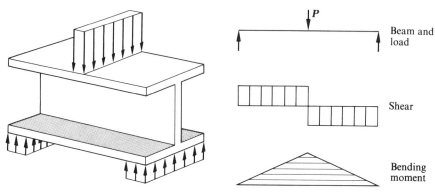

Fig. 1-3.12   Case i, Simple WF Beam

bridge girder, excessive vibration could result. (2) Failure could result from buckling of the top flange in a direction perpendicular to the web (lateral instability), or the flange might "accordian" (buckle locally). Any of these localized failures normally results in an overall structural failure.

Another problem concerns shearing stresses in centrally loaded beams. These shearing stresses must be limited to the allowable value for the material. Shape of the member has considerable effect upon shearing stresses as evidenced by the equation $v = VQ/Ib$, the derivation of which is to be found in sec. A-4.

Finally, failure might be due to principal stress combinations of shear and bending on some plane neither parallel nor perpendicular to the centroidal beam axis. For complete design, the load must be transfered to the reactions; these reactions act essentially as concentrated loads on the beam. For a design in which a very thin web is desirable for overall economy (welded plate girders), capacity of the web often is increased by adding stiffeners. "Bearing stiffeners" are designed as columns that assist the web in transmitting load throughout

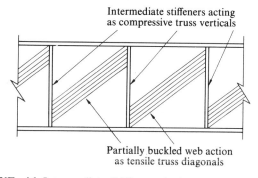

Fig. 1-3.13   WF with Intermediate Stiffeners Acting as a Truss at Near Failure

beam depth without web crippling. Intermediate stiffeners are used to overcome a tendency toward web buckling. At near-failure loads (Fig. 1-3.13) intermediate stiffeners serve as compression verticals of a truss while a portion of the web acts as a tension diagonal.

1-4   AIDS

Two basic aids available to the structural engineer are those imposed by law and others developed as shortcuts. In secs. 1-4.3 and 1-4.4 forms for presenting final calculations and "stress sheets" are given.

1-4.1 Legal and/or Official Aids.   These aids come under the general title of *specifications* and are classified normally as *structural-* or *material-oriented*. Typical specifications of the first type are the *Uniform Building Code* [2] (UBC), the *National Building Code* [3] (NBC), and *Standard Specifications for Highway Bridges* [4] (AASHO). In these references can be found complete sections embodying material-oriented specifications.

Material-oriented specifications, used extensively within this textbook, are *Specifications for the Design, Fabrication and Erection of Structural Steel for Buildings* [5] (AISC Code), *Building Code Requirements for Reinforced Concrete* [6] (ACI Code), and *National Design Specifications for Stress-Grade Lumber and Its Fastenings* [7] (Timber Code).

Every governmental authority, city, county, state, and federal bureau regularly adopts certain of the mentioned codes as their own, although some actually write their own codes. In such situations, specifications become part of the law and must be adhered to within the jurisdiction of the appropriate governmental bodies.

For purposes of these notes, the AISC Code, ACI Code, and Timber Code, will be considered as "legal" specifications. In practice, rigid adherence to these codes as "the law" is not required because most modern specifications allow use of new and innovative techniques as well.

Specifications given in "codes" are of great help in structural design. The codes show legal design procedures, allowable material stresses, and specify safe margins for engineers. Because of their mandatory nature, however, they often are either misunderstood or misinterpreted. This textbook will point out the origin of most "code formulas" to give the reader the ability to apply this knowledge astutely, rather than to simply follow the rules mechanically. The blind following of codes can lead to dire tragedies, particularly on new and unusual types of structures.

1-4.2 Shortcuts.   Most design involves redundant, that is, statically indeterminate, structures, which cannot be analyzed until the section properties are known. On the other hand, section properties cannot be determined without prior analysis. Thus, the "trial and error" process is inherent in most design. Further, it adds credence to the often heard: 90% of office time is spent to "find" the structure and 10% to "prove" it.

Based on this time ratio, an attitude used in most of this text is that short-cuts may be (in fact, should be) used in the trial design phase. Once this trial design phase is complete, however, virtually all design aids should be laid aside and the structure proved by an *independent, organized* analysis.

Design aids include books, graphs, tables, curves, etc. Appendix D presents shortcuts provided for design in reinforced concrete.

1-4.3 Final Calculations. In the design of any structure or structural component, the immediate aim of an engineer is to get the right answer—and promptly. A form for calculations designed to assist the engineer in accomplishing this goal is shown in Table 1-4.1. Before the functions that pertain to various headings in the table are discussed it is important to consider what final calculations are *not*. This negative approach is required since serious errors have often gone undetected because of poor "record" computations for permanent records.

Such calculations are *not* a history of all trials, errors, and tribulations, although such an approach easily becomes a habit for at least three reasons. (1) An engineer's pay is dependent on his fruitful output—the novice tends to confuse *fruitful* output with just output. (2) Because a novice tends to be unsure, he actually is afraid to dispose of anything for fear that yesterday's discards will be needed today. (3) Lacking experience, a beginner wants to make absolutely sure that any checker can follow his work, and possibly more to the point, that he can explain why certain steps were used, or from where certain figures came.

The most eminent peril to such an approach is bulk. While not necessarily a law of nature, it does seem that errors have definite periods of recurrence. Thus the more calculations made, the greater chance for error. At any rate, it is definitely more difficult to sort grain from chaff as the volume of computations increases. An even more critical danger from bulk is the probability of masking the physical structure behind a multitude of inconsequentials.

In summary, final calculations should comprise only the material needed to prove the chosen design. All other figures are merely parallel, and are neither included in final calculations nor kept in permanent files.

In practice, equations such as $f = P/A$ often are left unwritten. Instead, the symbol f is given so as to indicate calculations for axial stress, actual load replaces P, and a figure for actual area replaces A with the answer written on the same line. In analysis examples, such symbolic equations are written inside shaded boxes to indicate that their usage is not universal.

Table 1–4.1   Form for Final Calculations

1. **Loads** (Other Structural Considerations)
    a. Dead
    b. Live
    c. Dynamic
        1) Impact
        2) Wind
        3) Earthquake
        4) Blast

2. **Shear and Bending Moments** (Axial Forces and Torsional Moments)

3. **Section Properties** (Section and Allowable Stresses)

4. **Stresses**	1) Axial tension
	2) Axial compression
	a) Lateral stability
	b) Flange buckling
	3) Shearing and Torsion
	4) Principal stresses
	5) Shear flow (bond, shear connectors)
5. **Deflection**	6) Deflection
6. **Local Buckling**	7) Local buckling
	a) Web crippling
	b) Web buckling
7. **Special Considerations**	8) Fatigue failure
	9) Brittle fracture

In general, the Pythagorean theorem, binomial theorem, quadratic equations, and the like, should be stated on one line with the solution immediately following. Intermediate calculations are trivial and add bulk without increasing clarity.

The guiding outline of Table 1-4.1 is a procedural form for presentation of final solutions only. It is, therefore, appropriate to change estimates of loads to absolute values (within legitimate accuracy). In general, Dead (D), Live (L), and Impact (I) loads are treated separately, and their origins clearly (and typically) indicated as:

$$
\begin{aligned}
&\text{D: RC Beam } 1' \times 1' \times 150 && = 150 \,\#/\text{ft} \\
&\text{L: Office Floor (8' strip) } 8 \times 80 && = \underline{640} \\
&\text{Total} && = \overline{790 \,\#/\text{ft}}
\end{aligned}
$$

In the practical use of Table 1-4.1, other structural considerations should be noted in parentheses beside the first heading.

Loads (Simple 40′ span, A36 Steel, Laterally Supported)

In common practice, shear and bending moments due to dead and live loads are separated except in cases where it is absolutely clear that this is unnecessary. (Note also that this is the place to show the reactions for later use with connections.) Only maximum values of shear and bending moments (M&V or BM&V, typically) are tabulated. Except in unusual circumstances this is sufficient data, but occasionally shear and bending moment diagrams must be given for completeness.

For section properties, it is appropriate to show the actual section along with allowable stresses. Other data may be required as well. That is, sometimes r, the radius of gyration, or I for computing deflections, etc., may be needed information.

Actual stresses are computed for the five cases shown in Table 1-4.1. Note that in each case, necessary data should be taken directly from M&V and section properties. If a particular stress check does not apply, it can be dispensed with mentally or by writing "OK."

Deflections are figured, often for dead and live loads separately, and occasionally for short- and long-time loadings. Usually, they are limited by span length (L) ratios to $L/360$ for plastered ceilings and to $L/800$ for bridges. Actual and allowable deflections should always be compared.

Local buckling must be considered in almost all cases, but the only written calculation may be "OK by inspection."

Special considerations include fatigue and brittle fracture. Others might be extremely high or low temperatures or deleterious environment (caustic gases, ground water high in acids, etc.).

The form of Table 1-4.1 should become "second nature" to the reader as its components are used throughout this text.

1.4-4 Stress Sheets. To verify the safety and economy of a structure, the stress sheet can be an important aid. Primarily its purposes are for use by another engineer as the basis for his check, for review by a superior for overall reasonableness, and for a detailer to follow in preparing engineering drawings. A stress sheet should contain the absolute minimum number of facts necessary to satisfy these three primary purposes. It must include the component or assemblage name, project identification, date, and the original design engineer's initials. Almost always it is presented by a sketch, often with different vertical and horizontal scales. It has sufficient detail to be precise, but *no* double dimensioning. It shows cutoff points and similar configurations. It is not uncommon to include curves which relate allowable and actual bending moments, shears, etc. Design loads as well as controlling and material specifications are mandatory.

1-5   THE GOAL—WITH INFINITE SUMMITS

The title of this section is intended to suggest the dynamic nature of an engineering education. As each new bit of material is learned by engineers, as each goal is reached, new and brighter horizons and higher summits of effort appear. Solution to one problem develops questions to a dozen never before conceived. This is the quality that makes engineering a variant and vitally alive profession. And because knowledge is ever-changing, it seems advisable to show how "the goals" can be constant—but with new summits on each ensuing day.

1-5.1 Ability to Read the Literature. A student must be able to read the literature of his field and recognize the inherent principles learned, even when the principles are used in different contexts. Obviously a student must learn a variety of methods, but he must also learn that these methods are related, and how they are related.

In addition, a student must realized that methods are not bad simply because they are old. Lack of present use does not preclude future return to popularity. Quite often new breakthroughs are obtained by using long-forgotten ways of doing things, as exemplified by the "tangent modulus" concept developed and employed in Chaps. 3 and 10. To be able to read and understand the literature, a student must know the old and the new, plus basic principles *and* the way they are interrelated.

1-5.2 Ability to Relate. It is not uncommon to hear an engineer who has been practicing for many years say to a new engineer in rather irritable tones, "Can't you show a little 'common sense'?" Frequently, the method which this older engineer shows the inexperienced beginner is simpler than that being used, yet it is based upon well understood material which is merely presented in a different way. Common sense is the ability to relate knowledge gained in one sphere to problems in another, a faculty not easy to attain. Common sense probably should be called experience, and the ability to relate is heavily dependent upon a person's length of experience. Nevertheless, students should have this as a goal, and be consciously, continuously attempting to relate all knowledge acquired in the past to current investigations.

1-5.3 Ability to "Feel." In addition to competence in reading the literature and relating various aspects of design, another goal for students is getting *feel* for structures being designed. This is not easy to do. Feel can best be gained by visiting construction sites, keeping one's eyes open, relating sizes to loads and shapes seen each day. Although everyone drives across bridges and goes into buildings, only the astute engineer sees and develops a feel for their structural features.

PROBLEMS

Note to Students. Your instructor will assign you one value of those available for each unknown. In addition, he will specify which parts of *find* you are required to do.

Note to Instructors. Arrangement of possible values as *given* makes it possible to have many different problems from one problem definition. For example, with L fixed at 10, $a = 0.1$, $b = 0.8$, variation of θ and P alone provides nine different problems.

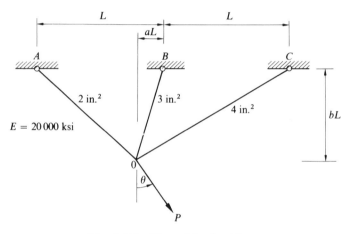

Fig. 1-P.1   Three-Member Truss

1-1 given: Three-member truss of Fig. 1-P.1 with $L =$ _____ (10, 12, or 14 ft), $a =$ _____ (0.1, 0.2, or 0.3), $b =$ _____ (0.8, 0.9, or 1.0), $\theta =$ _____ (10, 20, or 30°), $P =$ _____ (120, 140, or 160 kips).

find: a) Stresses in members and displacement of point zero considering fixed geometry.
b) Same as (a) except let dimensions change through _____ (2, 3, or 4) cycles.
c) Obtain a solution using a computer program.

1-2 given: Cantilever (weightless) of Fig. 1-P.2 with $I =$ _____ (300, 320, or 340 in.⁴), $L =$ _____ (24, 20, or 16 ft), $P =$ _____ (4, 5, or 6 kips), $\theta =$ _____ (30, 45, or 60°). Member does not buckle.

find: a) Primary bending moment, axial force, and deflection.
b) Find increase in bending moment and deflection due to axial force.
c) Iterate _____ (2, 3, or 4 times).

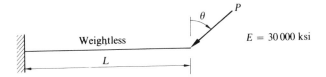

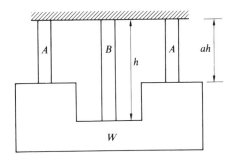

Fig. 1-P.2   Cantilever with Load at an Angle

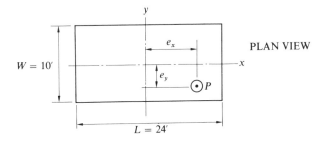

Fig. 1-P.3   Hanging Weight

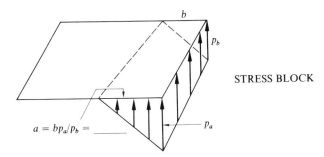

Fig. 1-P.4   Footing—High Eccentricity

1-3 given: Structure of Fig. 1-P.3 with
Material A with $E = 20{,}000$ ksi;
area = _____ (2, 3, or 4 in.^{2}).
Material B with $E = 30{,}000$ ksi;
area = _____ (1.5, 2.5, or 3.5 in.^{2}).
$h =$ _____ (20, 22, or 24 ft), $a =$ _____ (0.7, 0.8, or 0.9).
$W =$ _____ (180, 200, or 220 kips).

find: a) Displacement of weight W.
b) Displacement of weight W if all areas are reduced by one-half.

1-4 given: Footing of Fig. 1-P.4 which can be solved by statics, where $b =$ _____ (9, 10, or 11 ft), $p_b =$ _____ (5, 6, or 7 k/ft^2), $p_a =$ _____ (9, 10, or 11 k/ft^2).

find: P(kips), e_x(ft) and e_y (ft).

REFERENCES†

1. Seely, Fred B. and James O. Smith, *Advanced Mechanics of Materials,* 2d ed., Wiley, New York (1962).

2. *Uniform Building Code*, International Conference of Building Officials, 50 South Los Robles, Pasadena, California 91101 (1970).

3. *National Building Code*, The National Board of Fire Underwriters, New York, NY.

4. *Standard Specifications for Highway Bridges*, 10th ed., The American Association of State Highway Officials, 341 National Press Building, Washington, D.C. 20004, (1969).

5. *Specification for the Design, Fabrication and Erection of Structural Steel for Buildings*, American Institute of Steel Construction, 101 Park Avenue, New York, N.Y. 10017 (February 12, 1969).

6. *Building Code Requirements for Reinforced Concrete* (ACI 318-71), American Concrete Institute, P.O. Box 4754, Redford Station, Detroit, Michigan 48219 (1971).

7. *National Design Specification for Stress-Grade Lumber and Its Fastenings,* National Forest Products Association, 1619 Massachusetts Avenue NW, Washington, D.C. 20036 (1968).

8. *Manual of Steel Construction*, 7th ed., American Institute of Steel Construction, Inc., 101 Park Avenue, New York, N.Y. 10017 (1970).

† Complete addresses are given to specifications so that they can be ordered if desired.

CHAPTER 2

FLEXURE

In the real world bending of beams is a complex phenomenon. This chapter therefore, is divided into three main parts, Part A deals with bending of ideal beams as represented by mathematical models. Next, the nomenclature and techniques imposed by specifications are introduced so that analysis and design is accomplished for legally defined models. Last, with the idealized conditions understood, considerations of practical reality are introduced to highlight the difficulties inherent in valid interpretation of computations.

PART A
MATHEMATICAL MODELS IN BENDING

In this part consideration is given, in turn, to stress-strain relations as applied to bending, to the couple method, to the flexure formula, and to idealized elastoplastic analysis.

2-1 STRESS-STRAIN RELATIONS AS APPLIED TO FLEXURE

For any given material the stress-strain diagram is obtained by a relatively simple tension and/or compression test. A tensile test is performed by measuring the initial length and cross-sectional area (usually assumed constant) of the test specimen. Then the specimen is put in a testing machine, loaded by a specified load P, and remeasured. The nominal stress is then determined as $\sigma = P/A$ and the nominal strain as "the stretched length minus the original length" divided by "the original length," or $\epsilon = \Delta/L$. When the load is incremented a sufficient number of times before failure, the locus of points having successive ordinates (σ, ϵ) defines the stress-strain curve.

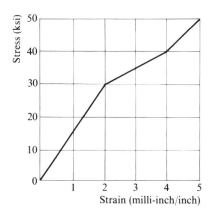

Fig. 2-1.1 Stress-Strain Curve (Tension *or* Compression)

A "perfect" stress-strain curve is shown for a fictitious material in Fig. 2-1.1. From this curve, it is evident that when strain is 0.002 in./in., corresponding stress is 30 ksi. Or, if stress is 40 ksi, corresponding strain is 0.004 in./in.

2-1.1 Stress-Strain Curve Applied to Beams. In order for such a curve to be applicable to the flexure problem it is necessary to recall the classical statement that plane sections before bending remain plane sections after bending. Based on this assumption, one with a very wide range of reliability, it is possible to visualize strains as proportional to their distance from the neutral axis, that is, strain increases linearly as its distance from the neutral axis increases. Strain also is proportional to the loads applied.

In the idealized stress-strain curve, a 1:1 relationship exists between stress and strain, so stress is a function of the value of strain at every point in any member, and under every loading condition. Said in another way, the occurrence of a stress implies an accompanying strain, and vice versa (temperature considerations excepted).

Figure 2-1.2 shows a beam with a general loading that increases linearly in magnitude with time. At a certain magnitude, the maximum strain in the

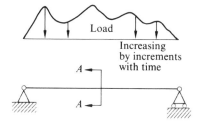

Fig. 2-1.2 Beam with General Load, Magnitude Increasing Linearly with Time

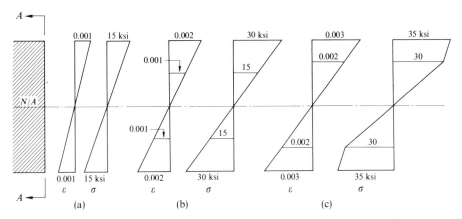

Fig. 2-1.3   Variations of Stress-Strain on Beam Section with Loading

extreme fibers of the rectangular beam is 0.001 in./in. at section *A–A*. If the material is represented by the stress-strain diagram of Fig. 2-1.1, then the corresponding maximum stress is 15 ksi. This is shown in Fig. 2-1.3(a). Note that both the strain and stress vary linearly from the neutral axis in corresdence with the pertinent part of the stress-strain diagram. As the magnitude of the general load increases, an outer fiber strain of 0.002 in./in. is reached and the maximum stress varies linearly to become 30 ksi, as shown in (b). As the load continues to increase, the strain remains proportional to its distance from the neutral axis but the stresses do not (c). This change in pattern is evident from the stress-strain diagram of Fig. 2-1.1 which shows a linear stress variation only up to a strain of 0.002 in./in. That is, the linear variation of stress differs in slope between 0.002 and 0.004 in./in., having a value of 35 ksi at 0.003 in./in.

The upper halves of the symmetrical strain and stress diagrams are shown in Fig. 2-1.4 as the load continues to increase. In (e), the outer fiber is at failure and the stress diagram has the shape of Fig. 2-1.1, with the axes rotated. Note, however, that the strain diagram is still linear.

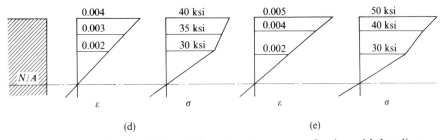

Fig. 2-1.4   Further Variations of Stress-Strain on Beam Section with Loading

Thus, for an ideal material, the stress and strain relationship induced by bending can be determined directly from the stress-strain curve obtained from either tension or compression tests. Of course, this is just another way of saying that bending stresses are normal stresses induced by bending.

2-1.2 Stress Block. With the magnitude of stress at any point and the cross-sectional dimensions of the member known, it is possible to visualize all incremental stresses acting together in a "stress block" and replaced by a resultant force. A stress block is equivalent numerically to the product of its cross-sectional area times the average stress; therefore, it has units of force.

To find a statically equivalent resultant, in addition to determination of the magnitude, the correct position of this force with respect to the neutral axis must be found. Both magnitude and position of the resultant of the stress block are determined by elementary statics, where $T \equiv$ the resultant of a tensile stress block, $C \equiv$ the resultant of a compressive stress block, and $jd \equiv$ distance between T and C.

The resisting moment of the cross section is computed as the moment produced by a couple composed of the two equal, but oppositely directed, resultants.

The stress-strain diagram of Fig. 2-1.1 used in conjunction with the WF beam of Fig. 2-1.5 indicates how such calculations may be performed. Note that the beam is at maximum load because it fails when strains exceed 0.005 in./in. In Table 2-1.1, the magnitude and position of the resultant force C is computed. To facilitate the computations, the stress block is divided into volumes which are simple to compute and which have readily defined centroids. Thus the portion of the stress block acting on the flange is divided into parts

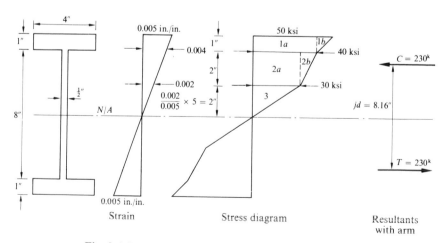

Fig. 2-1.5   Resolution of Stress Block into Resultants

Table 2-1.1 Typical Stress Block Computation

Block	Average Stress (ksi)	PL Area (in.²)	Stress vol. (k)	Distance to Centroid Stress Block from N/A (in.)	First Moment about N/A (in.-k)	
1a	40	4	160	4.50	720.0	
1b	5	4	20	4.67	93.3	
2a	30	1	30	3.00	90.0	
2b	5	1	5	3.33	16.7	
3	15	1	15	1.33	20.0	
Total Volume		230		4.08	940.0	Total for First Moment

1a (parallelepiped 4 in. × 1 in. × 40 ksi) and 1b (a triangular-based prism acting on a 4 in. × 1 in. area with an *average* stress of 5 ksi). The centroid to 1a is at the center of the flange, 4 in. + $\frac{1}{2}$ in. above the neutral axis, whereas the centroid of 1b is one-third of the way down from the top of the flange so that the "arm" is 4 in. + $\frac{2}{3}$ in. above the neutral axis. Note the parallel of this computation to that of the A, y, Ay tabulation of sec. A-2.

Because $C = T$ and the stress blocks are symmetrical, the moment couple that expresses the maximum bending moment that the section can carry without failure is

$$M = \frac{(230)(4.08 + 4.08)}{12} = 156.5 \text{ ft-kips,}$$

where 230 represent either C or T in kips and 12 is used to convert inches to feet.

2-2 COUPLE METHOD

The couple method, developed in this section, provides solutions to generalized flexural problems.

To use the couple method, position of the neutral axis must be known and can be determined by (1) visual inspection, (2) trial and error, or (3) calculation based on $C = T$.

2-2.1 Procedure.

1. Obtain the compressive resultant C and its location.

2. Obtain tensile resultant T and its location.

3. Resulting couple is magnitude of resisting moment.

If two or more materials are used in the same cross section, the problem is not significantly more difficult because (1) plane sections remain plane after bending, and (2) the magnitude of stress is dependent upon the magnitude of strain—the exact nature of this function depends upon the particular material under consideration.

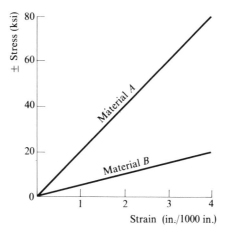

Fig. 2-2.1   Two Linearly Elastic Stress-Strain Diagrams

2-2.2 Example Using Two Materials: Neutral Axis Known.   Figure 2-2.1 shows the linearly elastic stress-strain diagrams for two materials A and B. Their moduli of elasticity are

$$E_A = 80,000/0.004 = 20 \times 10^6 \text{ psi}$$

and

$$E_B = 20,000/0.004 = 5 \times 10^6 \text{ psi}.$$

In Fig. 2-2.2, a section is shown with pertinent dimensions that include the position of the neutral axis. For maximum permissible bending moment, strain must be 0.004 in./in. at least at one point, but cannot exceed 0.004 in./in. at any point. Because the extreme fiber of material A is farther from the neutral axis than the extreme fiber of material B, at failure B is not highly strained.

The corresponding stress diagram is divided into parts for easy computations of volume and centroid. Stresses are determined from Fig. 2-2.1 to correspond to the strains determined by ratio and proportion. The solution is presented in Table 2-2.1. The resultants and distances to centroids (arms) are sketched on Fig. 2-2.2. The allowable bending moment is the volume of the stress block, 200 kips, times the couple arm (2.93 in. + 4.80 in. = 7.73 in.)

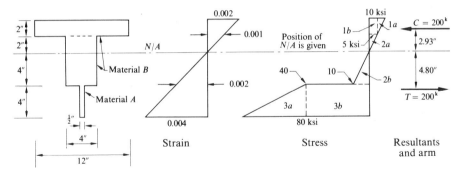

Fig. 2-2.2 Couple Method for Two Materials

Table 2-2.1 Couple Method Computation with Two Materials

Block	Average Stress (ksi)	Plate Area (in.²)	Stress Volume (k)	Centroid to N/A (in.)	First Moment about N/A (in.-kips)
Above N/A					
1a	2.5	24	60	$3\frac{1}{3}$	200
1b	5.0	24	120	3	360
2a	2.5	8	20	$1\frac{1}{3}$	27
			200	2.93 in.	587
Below N/A					
2b	5.0	16	80	$2\frac{2}{3}$	213
3a	20.0	2	40	$6\frac{2}{3}$	267
3b	40.0	2	80	6	480
			200	4.80 in.	960

converted to feet or $M = 200(7.73)/12 = 129$ ft-kips. Note that the neutral axis was given correctly because C does equal $T = 200$ kips.

2-2.3 Example Using Two Materials: Neutral Axis Unknown (Special).

The traditional approach used in analysis of reinforced concrete beams is reflected by this example. Figure 2-2.3 indicates the *idealized* linearly elastic stress-strain diagrams for steel S and concrete C.

In Fig. 2-2.4, the cross section is shown with symbols typical of those found in the literature of reinforced concrete.

Figure 2-2.5 illustrates the idealized beam with concrete in tension *completely disregarded*. It is assumed that concrete and steel are both stressed to

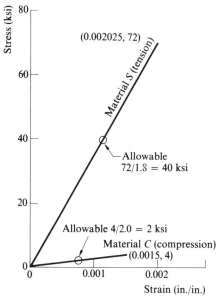

Fig. 2-2.3   Idealized Linearly Elastic Reinforced Concrete

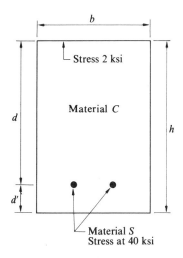

Fig. 2-2.4   Typical Reinforced Concrete Beam

their maximum allowable stresses (in reality this is seldom true), with $\sigma_S =$ 40 ksi (factor of safety $= 1.8$) and $\sigma_c = 2$ ksi (factor of safety $= 2.0$).

Using the stress-strain diagrams, $\epsilon_c = (2/4)(0.0015) = 0.00075$ in./in. and $\epsilon_s = (40/72)(0.002025) = 0.001125$ in./in. Then, by relating geometrically the

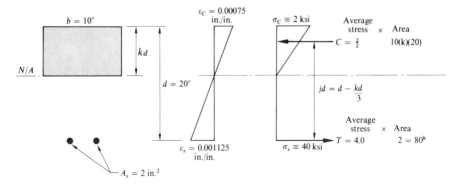

Fig. 2-2.5   Idealized Beam and Stress Block

strain diagram and the cross-sectional dimensions

$$\frac{\epsilon_C}{\epsilon_C + \epsilon_S} = \frac{0.00075}{0.001875} = \frac{kd}{20},$$

so that $kd = 8$ in.

A more common procedure to locate the position of the neutral axis kd in. from the top of the beam is to equate C with T. Then

$$\frac{2}{2}(10)(k)(20) = 40(2)$$

so

$$k = 0.4 \quad \text{and} \quad kd = 0.4(20) = 8 \text{ in.}$$

Equally, either method is valid.

With position of the neutral axis established by kd, the moment arm (commonly referred to as jd) is computed as $d - kd/3 = 20 - 8/3 = 17.33$ in. Since $T = C$ is known to be 80 kips, the allowable bending moment is

$$80(17.33)/12 = 115 \text{ ft-kips.}$$

The couple method is directly applicable to bending problems without regard to nonlinearities, but as parameters become more complex, location of the neutral axis becomes more difficult.

2-3   FLEXURE FORMULA

The flexure formula is derived and related terms are defined in sec. A-3. This procedure is used in a high percentage of bending problems, but to be completely valid the following criteria must be met.

1. The material must be linearly elastic and the members relatively straight.

2. The load must be static and applied through the shear center.

3. The load must not have a significant axial component.

4. Plane sections must remain plane after bending.

5. The flexure formula $\sigma = My/I$ must be applied with respect to a principal axis.

An extension of the flexure formula called *transformed area* has been widely used in conjunction with both reinforced concrete (steel "transformed" to concrete) and composite design (concrete "transformed" to steel).

The example cross section is identical to that solved previously by the couple method. The stress-strain diagrams are depicted in Fig. 2-2.1 and the cross section in Fig. 2-2.2.

Modular ratio n signifies the ratio between two moduli of elasticity. Modular ratios n_1 and n_2 will be defined as (A is higher strength material)

$$n_1 = E_A/E_B \tag{2-3.1}$$

and

$$n_2 = E_B/E_A. \tag{2-3.2}$$

An example of the transformation of areas and the modular ratio n is shown in Fig. 2-3.1, where $n_1 = 4$ and $n_2 = 1/4$.

The cross section of Fig. 2-2.2 is repeated in the center of Fig. 2-3.1; to its left is an equivalent beam made only of material B; to its right is an equivalent beam composed of material A only. The transformation is made by using the modular ratio n to relate stresses for a given strain in the two materials. When modular ratio n is greater than unity, a higher strength material is converted to

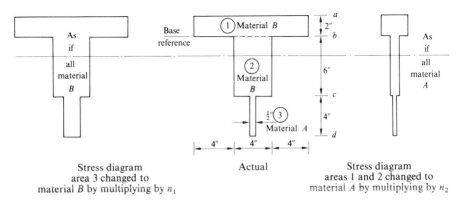

Fig. 2-3.1   Transformed Sections

a lower strength material by increasing its *width n* times. The equivalent beam, if composed of only the weaker material, must have more section.

To visualize the necessary transformation, try to obtain by the use of differing maximum stresses identical stress block volumes without changing the position of the resultant. Clearly, this can be accomplished only by altering cross-sectional width. Any other approach changes the magnitude and/or position of the resultant.

Table 2-3.1   Section Property Calculations

Part	All Material B A	y	Ay	$Ay^2 + I_0$	All Material A A	y	Ay	$Ay^2 + I_0$
1	$(2)(12) = 24$	1	24	24	$(2)(12)(\frac{1}{4}) = 6$	1	6	6
				8	$\underbrace{\quad}_{n_2}$			2
2	$(6)(4) = 24$	-3	-72	216	$(6)(4)(\frac{1}{4}) = 6$	-3	-18	54
				72	$\underbrace{\quad}_{n_2}$			18
3	$(4)(\frac{1}{2})(4) = 8$	-8	-64	512	$(4)(\frac{1}{2}) = 2$	-8	-16	128
	$\underbrace{\quad}_{n_1}$			10.7				2.7
	56	-2 in.	-112	842.7	14	-2 in.	-28	210.7
				-224.0				-56.0
				$618.7 \times \frac{1}{4}$	$=$	$\longrightarrow$		154.7

Computations for two different transformations are shown in Table 2-3.1. Note from Fig. 2-3.1 and the table that base reference is taken from the flange bottom. For a discussion of this tabular procedure, refer to sec. A-2.

Of particular importance is use of the modular ratios n to change dimension. Note also that the moments of inertia are proportional to their modular ratios. True stresses may be obtained by transforming back to the original material. To convert from "all material B," the computations are:

$$\sigma_a = \frac{12(129)(4)}{618.7} = 10.0 \text{ ksi}, \qquad \sigma_b = \frac{(12)(129)(2)}{618.7} = 5.0 \text{ ksi},$$

$$\sigma_c = \frac{12(129)(4)}{618.7} = 10.0 \text{ ksi}, \qquad \sigma_d = \frac{(12)(129)(8)}{618.7} \times \underbrace{4}_{n_1} = 80.0 \text{ ksi}.$$

To transform from "all material A" (the stronger material):

$$\sigma_a = \frac{12(129)(4)}{154.7} \times \underbrace{\frac{1}{4}}_{n_2} = 10.0 \text{ ksi}, \qquad \sigma_b = \frac{12(129)(2)}{154.7} \times \underbrace{\frac{1}{4}}_{n_2} = 5.0 \text{ ksi},$$

$$\sigma_d = \frac{12(129)(8)}{154.7} = 80.0 \text{ ksi}.$$

2-4   IDEALIZED ELASTO-PLASTIC ANALYSIS

Ultimate failure load of a steel frame cannot be predicted accurately by use of elastic analysis because steel can yield to about one-fourth of its length before rupturing. Plastic analysis was introduced to overcome this difficulty and thus obtain a more realistic factor of safety against collapse. What started as an analysis technique soon became a rationale for design.

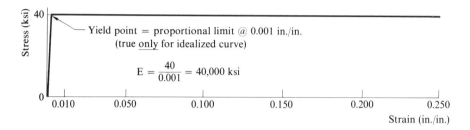

Fig. 2-4.1   "Perfect" Elasto-Plastic Stress-Strain Curve for Mild Steel

The stress-strain diagram of Fig. 2-4.1 is an idealization of the true curve for mild steel, with physical world modifications neglected so that the salient points may be clearly expounded.

2-4.1 Shape Factor and Plastic Section Modulus.   Shape factor sF is defined as

$$sF = Z_p/Z_e, \tag{2-4.1}$$

where $Z_e \equiv$ elastic section modulus; that is, in the flexure formula $\sigma = My/I$, if y_{max} is called c, then $Z_e = I/c$ so $\sigma = M/Z_e$; and where Z_p, the value of the plastic section modulus, is defined such that

$$\sigma_p = M_p/Z_p, \tag{2-4.2}$$

where $\sigma_p \equiv$ stress in *extreme* fiber and $M_p \equiv$ the plastic moment, that is, the maximum moment which the section can take without excessive rotation.

The rectangular cross section of Fig. 2-4.2 is used to clarify these concepts. Throughout the elasto-plastic range, strains are linear and, up to the proportional limit, stresses also are linear. However, as the strain exceeds 0.001 in./in., the stress becomes constant at 40 ksi. With increasing load, the member continues to rotate and this 40 ksi stress extends downward until it approaches the neutral axis. Without appreciable error, the maximum resisting bending moment may be computed by assuming that the 40 ksi stress does reach the neutral axis. This yields two rectangular stress blocks, typically used to represent the plastic stress curve.

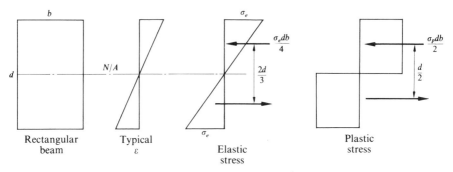

Fig. 2-4.2   Shape Factor of Rectangular Section

Two distinct resulting stress blocks lead to two equations which represent their moment couples:

$$M_e = \left(\frac{\sigma_e}{2} b \frac{d}{2}\right) \frac{2d}{3}, \quad M_p = \left(\sigma_p b \frac{d}{2}\right) \frac{d}{2},$$

$$M_e = \frac{\sigma_e bd^2}{6}, \quad M_p = \frac{\sigma_p bd^2}{4}. \tag{2-4.3}$$

Division of both sides of each equation by allowable stresses yields the section moduli (for a rectangle only):

$$Z_e = \frac{M_e}{\sigma_e} = \frac{bd^2}{6}, \quad Z_p = \frac{M_p}{\sigma_p} = \frac{bd^2}{4}. \tag{2-4.4}$$

Shape factor is defined as the ratio of the plastic to elastic section modulus, so for any rectangle

$$sF = \frac{bd^2/4}{bd^2/6} = 1.5. \tag{2-4.5}$$

More direct means† are normally employed to compute elastic section modulus. A more direct approach to plastic section modulus is obtained if allowable plastic stress σ_p is set at unity, because $M_p = Z_p \sigma_p$ then implies that $M_p = Z_p$.

As an example, consider computations leading to the shape factor for a circular cross section such as that shown in Fig. 2-4.3. The elastic section modulus may be computed by standard means or taken from *Steel Construction* as $Z_e = 0.785R^3$. With σ_p set at unity, $Z_p = M_p$, so

$$Z_p = \left(\frac{\pi R^2}{2}\right) \left(\frac{8R}{3\pi}\right) = \frac{4R^3}{3}. \tag{2-4.6}$$

---

† Note that by definition $Z_e = I/c = (bd^3/12)/(d/2) = bd^2/6$.

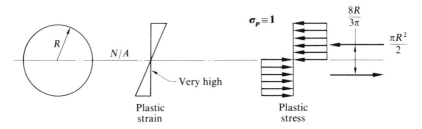

Fig. 2-4.3 Shape Factor of Circular Cross Section

Shape factor for a circle is the quotient of Z_e and Z_p, or

$$sF = \frac{1.333R^3}{0.785R^3} = 1.7. \tag{2-4.7}$$

The shape factor shows much more clearly the true factor of safety against ultimate failure, and actually is used to obtain elastic allowable stresses. As an example, the allowable elastic flexural stress for a WF beam is $0.60\,F_y$[†] assuming a shape factor of about 1.13. The allowable elastic flexural stress for a pin (round and therefore with a shape factor of 1.7) is given in *Steel Construction* as 0.9 F_y. Division of both allowable stresses by their respective shape factors yields $0.60\,F_y/1.13 = 0.53\,F_y$ and $0.9\,F_y/1.7 = 0.53\,F_y$. Obviously, these different allowable elastic stresses have taken into account the shape factor quite uniformly.

2-4.2 Plastic Hinge. A concept of plastic hinge is essential for calculations involving elasto-plastic members loaded into plastic range. In Fig. 2-4.4, strain and stress diagrams corresponding to increasing load (moment) are shown as the member yields closer and closer to its neutral axis.

The cross section is a rectangle 10 in. wide and 20 in. deep. The stress-strain curve is that shown in Fig. 2-4.1. At strains equal to or greater than 0.001 in./in., the stress is 40 ksi. For lower strains, stresses vary proportionally. Under each strain and corresponding stress diagram is shown the allowable bending moment (in foot-kips) computed by the stress block method.

When the strain at the extreme fiber reaches 0.250 in./in., the member will have yielded completely except for the innermost 0.08 in. In such a case, the extreme fiber would have extended horizontally through a distance of 1.25 in. Concurrently, the bending moment will have reached a maximum value of 3333 ft-kips or 1.5 (the shape factor) times the maximum allowable elastic bending moment of 2222. Theoretically at this point, the member would fail.

[†] F_y is defined as yield stress of the steel, that is, (1) the specified minimum yield point or (2) a specified minimum yield strength when a yield point does not otherwise exist.

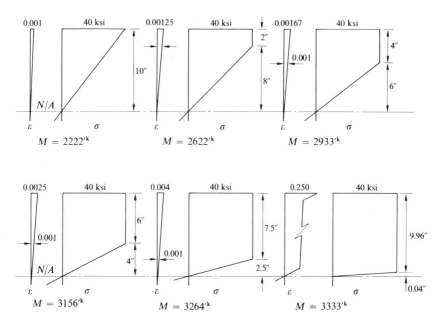

Fig. 2-4.4   Plastic Hinge Development for 10 in. Wide by 20 in. Deep Rectangle

In current practice, the assumption is made that a member has a constant moment capacity from the time it reaches yield until it fails. This means that the "plastic hinge" simply rotates without carrying higher moments. The validity of such an assumption is illustrated in Fig. 2-4.5, plotted from the data of Fig. 2-4.4 (ϕ is the angle of rotation in radians, that is, maximum strain divided by 10 in.). The assumed curve shows a plastic hinge that forms suddenly at a concentrated point, and then continues to rotate without a change in moment-carrying capacity. As pointed out by Kuzmanović [1], this theory is subject to some question; it is re-examined in Part C of this chapter.

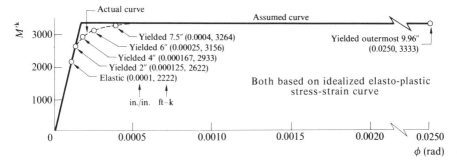

Fig. 2-4.5 $M-\phi$ Curve

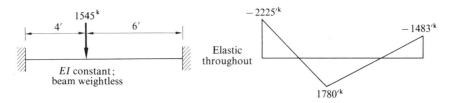

Fig. 2-4.6   Moment Redistribution—Stage 1

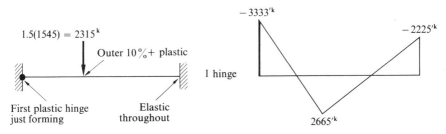

Fig. 2-4.7   Moment Redistribution—Stage 2

2-4.3 Moment Redistribution. An increasing load on a member concurrent with a specified maximum moment at a particular location causes either beam failure or moment redistribution. This process of moment redistribution is treated in the next few paragraphs and the accompanying figures.

Figure 2-4.6 shows a fixed end, weightless 10 in. × 20 in. beam with a constant EI. The concentrated load of 1545 kips produces a maximum moment of −2225 ft-kips at the left end. Thus the full depth of the beam is stressed in only the elastic range (*Note.* Calculations are made to slide-rule accuracy.)

As the load increases upward by a factor of 1.5, the maximum moment reaches −3333 ft-kips and a plastic hinge forms at the left support as shown in Fig. 2-4.7. Throughout time while the load increased from 1545 to 2315 kips, the magnitude of the bending moment at every point was directly proportional to the load. Under the full concentrated load, the outermost 10% of the beam has yielded, but the remainder is still elastic (see curve of Fig. 2-4.5). At the right support, the beam is completely elastic but the extreme fibers have just reached the yield point.

As the load increases still further, the plastic hinge can rotate, but cannot carry more moment than −3333 ft-kips. Therefore, the *shape* of the bending moment diagram begins to change. This change can be visualized as the moment on a true propped beam carrying the increase in load shown in Fig. 2-4.8. At a load of 388 kips, such a propped beam would have a maximum positive moment of 668 ft-kips and a maximum negative moment of −650 ft-kips. Super-

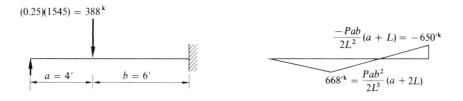

Fig. 2-4.8 Auxiliary Bending Moment Diagram

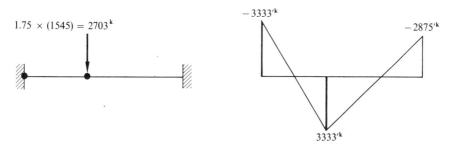

Fig. 2-4.9 Moment Redistribution—Stage 3

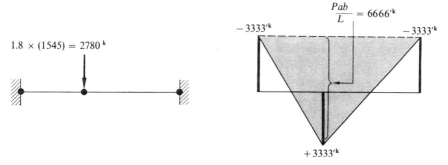

Fig. 2-4.10 Moment Redistribution—Final Stage

position of the auxiliary bending moment condition on that shown in Fig. 2-4.7 results in Stage 3 of the moment redistribution (see Fig. 2-4.9). Two plastic hinges have now formed.

As the load is increased still more, the third plastic hinge forms; the beam becomes a mechanism and collapses. This condition is illustrated in Fig. 2-4.10. Of course, the shaded area represents the simple beam bending moment diagram with a maximum value of $2780(4)(7)/10 = 6666$ ft-kips. Because the plastic hinge moments *at failure* are all equal, this final diagram can be constructed easily with no need to proceed through each successive stage to obtain the mechanism. (*Note.* The moments have been redistributed, that is, the elastic and plastic composite bending moment diagrams are *not* similar.)

As illustrated in Fig. 2-4.11, only two plastic hinges are required to produce a mechanism because the left reaction is a simple support. The simple beam moment (SBM) is Pab/L or $40(12)(8)/20 = 192$ ft-kips. The plastic moment, M_p, is unknown. Comparison of the bending moment drawn by parts in Fig. 2-4.11(a) and the composite bending moment diagram of Fig. 2-4.11(b) suggests equating bending moments at the point of load. Thus

$$-0.6\,M_p + 192 = M_p \qquad \text{or} \qquad M_p = 120 \text{ ft-kips.}$$

Additional "plastic" considerations are introduced in appropriate discussions throughout this text. In Chap. 8, plastic design and analysis is formalized and applied to continuous beams with and without variable moments of inertia.

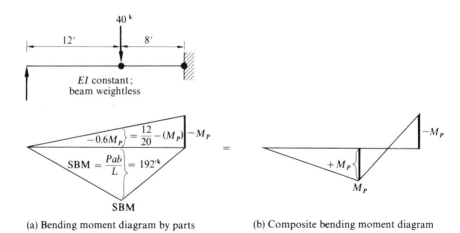

(a) Bending moment diagram by parts   (b) Composite bending moment diagram

Fig. 2-4.11   Direct Computation of Redistributed Moments

PART B
LEGALLY DEFINED MODELS

Discussion in this part is limited to simple bending of reinforced concrete and of structural steel members. In this context of design, as legally formulated, the real loads, allowable stresses, and material characteristics are replaced by *mandatory* mathematical models.

2-5   BENDING OF REINFORCED CONCRETE

A continuous reinforced concrete frame is generally considered to be linearly elastic under assumed working loads. This assumption is quite valid for working loads; and, in fact, there is little moment redistribution in the normal frame as ultimate loading of the structure is approached. In the determination of bending moments in a structure, therefore, it is common to assume a linear response whether the structure is to be designed according to the working stress concept or the ultimate strength approach.

The two methods of design shall be referred to as WSD (*working stress design*) and USD (*ultimate strength design* or simply *strength design*).† Briefly, in WSD it is assumed that reasonable live and dead loads act on the structure and that concrete and steel both respond elastically. Members are proportioned so that maximum allowable stresses are not exceeded. In USD, strength design, loads that were used for WSD are multiplied by appropriate load factors. The members are then proportioned so that they would be near failure under these modified loads. A limitation is placed on the amount of steel that can be used in USD to ensure yielding of the steel prior to crushing of the concrete. This limitation results in a *fail-safe* structure, one which will not exhibit a sudden failure.

Working stress design was used almost exclusively prior to 1960, but ultimate strength design has gained in popularity and is replacing the older method. In general, USD results in a more uniform actual factor of safety for all members than does WSD. Both methods are included in this text.

2-5.1 General Discussion and Definitions.   In the development of techniques for member analysis and design, the student is expected to distinguish between preliminary and final calculations. Preliminary calculations may be performed with all possible charts, curves, tables, etc. Once the final design is selected, the preliminary calculations become simply waste paper. Final calculations should be performed directly without the aid of shortcuts and in a standard form as discussed in sec. 1-4.3. This variation in the approach to analysis and design provides a powerful check on the accuracy of answers to problems without excessive loss of time.

Standard symbols employed in conjunction with reinforced concrete should become familiar as they facilitate both comprehension and efficiency. The most common symbols follow.

$b \equiv$ width of beam (inches).

---

† ACI-71 uses the term "strength design." Both that and USD are used in this book because of the extensive use of the latter by practicing engineers.

$d \equiv$ effective depth of beam (inches); that is, the distance from the compression face of concrete to the centroid of the tensile steel.

$h \equiv$ overall beam depth (inches).

$d' \equiv$ distance from centroid of tensile steel to tensile face of concrete (inches).

$f_s \equiv (\sigma_s)$ working stress of steel (ksi); that is, the allowable steel stress which includes a factor of safety.

$f_y \equiv (\sigma_y)$ yield stress of steel (ksi).

$f_c \equiv (\sigma_c)$ working stress of concrete (psi); that is, the allowable concrete stress which includes a factor of safety.

$f'_c \equiv$ ultimate concrete strength (psi), that is, strength of standard test cylinder which is 28 days old.

$E_c \equiv$ modulus of elasticity of concrete $= w^{1.5} 33 \sqrt{f'_c} (\text{psi})$, where $w \equiv$ weight of concrete (pounds per cubic foot) 145 lb/ft^3.

$E_s \equiv$ modulus of elasticity of steel,
   $= 29 \times 10^6$ psi.

$n \equiv$ modular ratio $= E_s/E_c$ (nearest whole number).

$kd \equiv$ distance to neutral axis from compressive face of concrete.(Some engineers use kd only with WSD and use c with strength design.)

$k \equiv$ coefficient of kd. (Note. kd often is considered as either a single or a composite symbol.)

$jd \equiv$ distance between resultant compression and tensile forces. (Some engineers use jd only with WSD and use another symbol, possibly r, for USD.)

$j \equiv$ coefficient of jd.

$F \equiv bd^2/12,000$.

$p \equiv A_s/bd$ (Note. d is *not* ·entire depth h!), where $A_s \equiv$ area of tensile steel.

One symbol has alternate and unrelated meanings for WSD and USD.

$a = f_s j/12,000$.   (for WSD)

$a =$ distance from extreme compression fiber to opposite   (for USD)
   end of *Whitney's stress block* (discussed below)
   and *not* equal to kd.

The following special symbols are used for WSD:

$K \equiv \frac{1}{2} f_c k j$.

$k \equiv nf_c/(nf_c + f_s)$. This is used for design only because balanced design is assumed; that is, materials are at maximum allowable working stress. Section dimensions are unknown and desired.

$k \equiv \sqrt{2np + (np)^2} - np$. This is used for analysis only because $p = A_s/bd$, meaning that the dimension of the section and area of the steel are known; actual working stresses need to be determined.

$j \equiv 1 - (k/3)$.

Three special symbols are used for USD:

$$q \equiv \frac{A_s}{bd}\left(\frac{f_y}{f'_c}\right) = \frac{pf_y}{f'_c} \text{ (sometimes called } \omega).\dagger$$

$K_u \equiv \phi q f'_c(1 - 0.59q)$, where $\phi \equiv$ capacity reduction factor. For flexure $\phi = 0.9$, for shear and torsion $\phi = 0.85$, for bearing on concrete $\phi = 0.7$, for spirally reinforced columns $\phi = 0.75$, and for other columns $\phi = 0.70$.

Reinforcing bar sizes are specified as #6, #8, etc. For bar sizes #3 through #8, the nominal diameter of the bar is formed by dividing the bar size by 8. Thus nominal diameter of a #7 bar is $\frac{7}{8}$ in. Approximate values for the nominal diameters of larger bars also can be determined in this way.

Bars in slabs are usually designated by size of a particular bar to be used at a specified spacing: typically, #3 @ 6 or #4 @ 3.

Bars in beams are usually designated by the number of bars of a particular bar size: typically, $6-\#6$ or $4-\#8$.

Cover refers to the clear distance between the reinforcement and the nearest concrete surface:

$d' = \text{cover} + (\text{bar diameter})/2 + \text{stirrup diameter}$

or   (2-5.1)

$d' = \text{cover} + (\text{bar size number})/16 \text{ (up to \#11 bars)} + \text{stirrup diameter}$.

As a preliminary estimate, d' is often chosen as 1 in. for slabs and 2 in. for beams to correspond to respective cover requirements of $\frac{3}{4}$ in. and $1\frac{1}{2}$ in. These values are actually correct only when #4 bars are used in slabs and #8 bars in beams without stirrups.

2-5.2 Validation of Whitney's Stress Block. Early efforts‡ in the design of reinforced concrete considered the nonlinear stress-strain relationship of concrete. However, the resulting complex design procedure never was adopted

---

† q is used here because it is used less frequently than ω as a symbol in related subject matter.

‡ Interesting reading and valuable insight concerning the stances of proponents of plastic versus elastic theory can be found in the references cited at the end of this chapter, particularly references 2, 3, and 4.

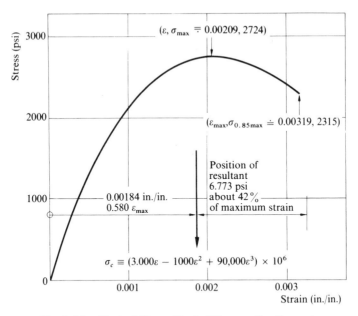

Fig. 2-5.1 Typical Stress-Strain Diagram for Concrete

formally in this country. Instead, the method used was an approach based upon linearly elastic considerations, or the method known as WSD. Over a period of time, a simplified design method was developed which reflected the nonlinearity of concrete at ultimate loads. The newer method, USD, is now in the process of replacing WSD as the major design procedure, as previously noted.

The logic of the design concept in USD can be understood by considering the action of a reinforced concrete beam as the applied bending moment is increased. Recall that the amount of reinforcing steel is intentionally limited to a "small" quantity. This allows the steel to reach the yield point well before the concrete has been heavily stressed. From that point on, steel, since it is elasto-plastic, carries the same tensile force. An increased bending moment is accommodated by a shift of the neutral axis and the resultant compression force and higher stresses and strains in the concrete. In the limit the concrete crushes and the beam can carry no more moment. It is important to note that the resultant compression and tension forces are constant after the steel yields, although steel strain increases.

Figure 2-5.1 shows a typical stress-strain curve for concrete, which can be closely approximated by the equation†

$$\sigma_c = (3.000\epsilon - 1000\epsilon^2 + 90,000\epsilon^3) \times 10^6 \qquad (2\text{-}5.2)$$

where σ_c has units of psi and ϵ in./in.

---

† This equation was developed by the author to provide a simple, yet practical, formulation similar to actual test curves.

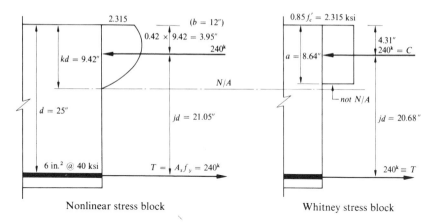

Fig. 2-5.2   Comparison of Stress Blocks

The compressive stress block for a beam can be developed from a stress-strain curve by means of an affine† transformation in which the strain axis is transformed to the depth of the compression block, with zero strain at the neutral axis and ultimate strain at the compression face.

Area under the stress-strain diagram and location of the centroid can be determined by either direct or numerical integration. Multiplication of this area by the actual depth of the stress block of a beam yields the magnitude of the compressive force per unit beam width. This resultant, also may be located from the centroid by direct proportion between the stress-strain diagram and the actual beam kd.

Consider the compressive stress block for a rectangular concrete beam, using the stress-strain curve of Fig. 2-5.1. The depth of the stress block is determined by equating the tensile force in the steel bars ($T = f_y A_s$) to the compression force in the stress block for concrete (C).

Assume that a beam 12 in. wide has an effective depth d of 25 in. and 6 in.^{2} of reinforcing steel with a yield strength of 40 ksi. The tensile resultant T is $(6)(40) = 240$ kips. The volume of the compressive stress block is $C = (A_{SSD})(b)(kd)/\epsilon_{max}$ where $A_{SSD} \equiv$ area under stress-strain diagram in psi (6.773 psi from Fig. 2-5.1), $b \equiv$ beam width (12 in. by assumption), $kd \equiv$ unknown distance to neutral axis, and $\epsilon_{max} \equiv$ maximum strain at failure (0.00319 in./in. from Fig. 2-5.1). Thus

$$C = (6.773)(12)(kd)/0.00319 = T = 240,000 \text{ lb.}$$

When C is equated to T, $kd = 9.42$ in. This resultant is 42% of kd from the extreme fiber (Fig. 2-5.1), so $jd = 25 - 0.42(9.42) = 21.05$ in. The resulting

---

† A transformation that transforms straight lines into straight lines, and parallel lines into parallel lines, but may alter distances between points and angles between lines.

moment is $240(21.05)/12 = 421$ ft-kips. These relationships are shown in Fig. 2-5.2.

Whitney proposed to replace the curved stress block by an equivalent rectangular one (see Fig. 2-5.2) with a uniform stress of $0.85 f_c'$. With the stress block of this shape, resultant compressive force is $C = (2.315)(12)a$. Again, C equated to T results in $a = 8.64$ in. Therefore, an approximate value of jd is $25 - (8.64/2) = 20.68$ in. The resisting moment is $240(20.68)/12 = 414$ ft-kips.

The small error, 421 ft-kips compared to 414 ft-kips, involved in the assumption of a rectangular stress block is typical for concrete beams. It is interesting to observe that the simplicity inherent in Whitney's stress block depends upon tacitly assuming that the stress block does not extend to the neutral axis.

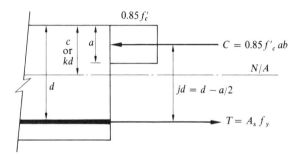

Fig. 2-5.3   USD Equations

2-5.3 Stress Method Equations.   USD flexural equations are developed by the couple method, as shown in Fig. 2-5.3. Equating horizontal forces C and T, we have

$$a = A_s f_y / 0.85 f_c' b. \tag{2-5.3}$$

The ultimate bending moment, expressed by the couple Tjd, is

$$M = A_s f_y (d - a/2) \tag{2-5.4}$$

where a can be substituted from Eq. (2-5.3) to yield

$$M = A_s f_y (d - 0.59 A_s f_y / f_c' b). \tag{2-5.5}$$

This equation can be simplified by factoring d and noting that $A_s f_y / f_c' b d = q$ by previous definitions; therefore

$$M = A_s f_y d (1 - 0.59q). \tag{2-5.6}$$

An alternate form, obtained by multiplying both numerator and denominator by bdf'_c, reduces Eq. (2-5.6) to

$$M = bd^2 f'_c q(1 - 0.59q).$$ (2-5.7)

This converts to ACI format when a capacity reduction factor is added:

$$M(\text{in.-lb}) = bd^2 \phi f'_c q(1 - 0.59q).$$ (2-5.8)

Equation (2-5.8) is converted to ft-kips by dividing by 12,000, so

$$M(\text{ft-kips}) = (bd^2/12,000)[\phi f'_c q(1 - 0.59q)].$$ (2-5.9)

By use of previous definitions, this may be restated:

$$M(\text{ft-kips}) = FK_u.$$ (2-5.10)

Values of F are tabulated in restricted† form in Fig. D-1; values of K_u are plotted in Fig. D-5. Note that the selection of q indirectly specifies p, from which the area of steel may be determined:

$$q = pf_y/f'_c = A_s f_y/bdf'_c,$$ (2-5.11)

or

$$A_S = qbdf'_c/f_y = pbd.$$ (2-5.12)

2-5.4 A Design Example. The procedure below illustrates a sequence used for trial calculations. These computations are in accordance with ACI Standard 318-71. This requires use of the capacity reduction factor ϕ in accordance with ACI 9.2.

1. *Select materials to be used.* Say, for instance, $f'_c = 3000$ psi and $f_y = 36$ ksi.

2. *Determine approximate design moment.* Assume a 20-ft simple span with 1 k/ft dead load and 1.1 k/ft live load. The dead load does not include the beam itself although designs can be shortened by good estimates based on experience. Live loads are defined precisely by the code and dead loads are reasonable approximations. Multiply by *load factors* as appropriate from ACI 9.3.1: dead load by 1.4, and live load by 1.7. Thus, the total load is $1 \times 1.4 + 1.1 \times 1.7 = 3.3$ k/ft. $M_u = 3.3(20)^2/8 = 165$ ft-kips.

3. *Assume q.* When q is less than 0.18, experience has shown that deflection seldom is a consideration. So, for the present, use a trial q of 0.18. (A discussion of this parameter is presented later.)

4. *Determine K_u* based on q from Fig. D-5. With $q = 0.18$ and $f'_c = 3000$, then $K_u = 434.4$.

5. *Calculate F.* $F = M/K_u = 165/434.4 = 0.38$.

---

† Charts D-1, D-2, and D-3 are reduced forms of those used in practice and are to be used by students to minimize design parameter space.

6. *Choose b and d* based on *F* from Fig. D-1 (restricted values). (An un-restricted *F* table is available in supplementary publications; however, until the fundamentals are mastered, only the restricted table will be used.) The smallest *F* greater than 0.38 is 0.423, which provides $b = 11\frac{1}{2}$ in. and $d = 21$ in. Assume that d' equals about 2 in. so that *h* is 23 in.

7. *Compute beam dead load.* Beam dead load $= (11.5)(23)(145)/144 = 0.264$ k/ft. Although 145 lb/ft was used as the weight of reinforced concrete, many engineers use 144 or 150 lb/ft³.

8. *Recompute the moment for additional dead load of beam.* $M = [3.3 + (0.264)(1.4)] 20^2/8 = 184$ ft-kips. (Note that the 1.4 load factor is included.)

9. *Recompute F.* $F = M/K_u = 184/434.4 = 0.424$. Therefore, according to Fig. D-1, $b = 11\frac{1}{2}$ in. and $h = 23$ in. with $d \doteq 21$ in. is still reasonable. If, at this point in the calculations, it is determined that a larger beam is required, return to step 6 and repeat through step 9.

10. *Choose steel.* Use Eq. (2-5.12) to determine the area of steel required. The simplest procedure is to continue to assume that $q = 0.18$ to obtain $A_S = 0.18(11.5)(21)(3)/36 = 3.62$ in.².

Any time the actual beam is of a value *F* very different from that required (or if the matter is in doubt), then compute a more precise value of *q* by (1) $K_u = M_u/F$, and (2) use this computed value of K_u in conjunction with Fig. D-5 to obtain a better value. In this case, however, $K_u = 184/0.423 = 435.0$ and *q* is only slightly higher than 0.18.

Within the restricted bar list of Fig. D-3, 3.62 in.² of steel would indicate use of, say, #8 bars. Since 5-#8 are suitable for 3.95 in.², but the minimum web width is 13 in., larger bars should be tried:

4-#9   4.00 in.² $b \geq 12$ in. (These will not fit into the beam.)
3-#10 3.81 in.² $b \geq 10\frac{1}{2}$ in. (OK.)
3-#11 4.68 in.² $b \geq 11$ in. (These fit, but provide an excessive amount of steel.)

Thus, use 3-#10 for 3.81 in.² which fits well within the width of $11\frac{1}{2}$ in.

Design problems in which the restricted tables D-1 and D-3 of Appendix D are used have shown that it is not uncommon that a larger beam results to accommodate the steel. With the use of unrestricted tables such occurrences are less frequent.

2-5.5 Analysis. Analysis by the stress method of the beam designed in the previous section is shown in Fig. 2-5.4. The distinction between the preceding design procedure and the following analysis is characterized by these points:

1. Loads and load factors have been determined.

2. Critical shear and bending moment have been determined.

Loads   (20′ Simple span)

DL: 1.00 k/ft

Beam $11\frac{1}{2}$ × 23 × $\dfrac{145}{144}$ $\underline{0.27}$

                                                 1.27 × 1.4 = 2.03

LL: 1.10 × 1.7 = $\underline{1.70}$

                                                          $\overline{3.73\,\text{k/ft}}$

V & BM

$V = (\frac{1}{2})(3.73)(20) = 37.3$ kips (reaction)

$M = (3.73)\,20^2/8 = 186.5$ ft-kips ($\text{\textcentoldstyle}$ of beam)

Section properties $(f'_c = 3000 \text{ psi}, f_y = 36 \text{ ksi})$

$\phi = 0.9$ for flexure, 0.85 for shear

$q = A_s f_y/bdf'_c = \dfrac{(3.81)(36)}{11.5(20.5)3} = 0.194$

$v_1 = 2\sqrt{f'_c} = 109$ psi (without stirrups)

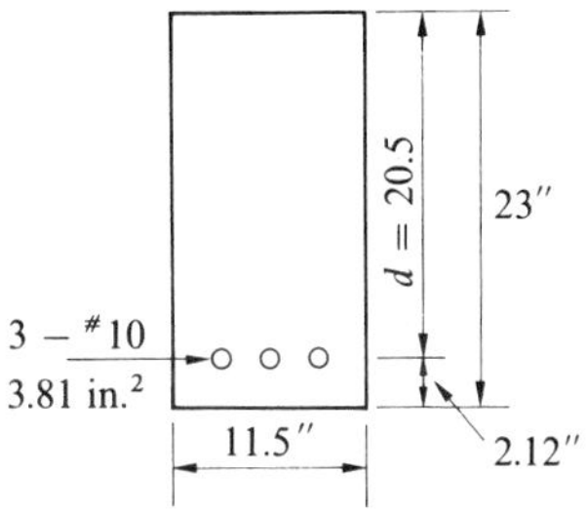

23″

$d = 20.5$

3 − #10
3.81 in.²

11.5″   2.12″

Check

$M_u = \phi bd^2 f'_c q(1 - 0.59q) = \dfrac{0.9(11.5)(20.5)^2(3)(0.194)(0.886)}{12} = 186.8^{\prime k} > 186.5$ OK

$v_u = \dfrac{V_u}{\phi bd} = 37300/0.85(11.5)(20.5) = 186$ psi > 109 ∴ use stirrups

Defl. check $h = 23″ > 12L/16 > 12(20)/16 = 15″$

Fig. **2-5.4**   Analysis by the Stress Method

3. Section properties and material properties have been selected and the allowable stresses computed; $d = h -$ (bar diameter)/16 − cover − stirrup diameter; or assuming #3 stirrups, $d = 23 - (10/16) - 1.5 - 0.375 = 20.5$ in.

4. *Actual d* is used, rather than assumed d;

5. Computation of allowable M_u is by formula *without* the use of aids given in Appendix D.

In addition to these comparative characteristics, the analysis also includes a check on shear by $v = V/\phi bd$† according to ACI 11.2.1. Allowable shear

---

† *Note*. v is shearing stress; V is shearing force.

stress without stirrups is computed under section properties according to ACI 11.4.1 as $2\sqrt{f'_c}$. Shear as a measure of diagonal tension and stirrups is discussed in sec. 6-2. Deflection is checked indirectly by ACI table 9.5(a) which requires (h and L both in inches):

$h \geq L/16$ for simple span beams;
$h \geq L/21$ for continuous spans—both ends continuous;
$h \geq L/18.5$ for continuous spans—only one end continuous;
$h \geq L/8$ for cantilevers.

When members exceed these criteria, deflection must be checked in accordance with Chap. 6, Part D.

2-5.6 Discussion of q Ratio used in Strength Design. For over fifty years, the WSD method of calculating for reinforced concrete led to generally safe, if not always economical, results. As discussed below, *balanced design* was the goal. That is, steel and concrete each was made to work as nearly as possible to their maximum allowable working stresses. Under these conditions, years of experience indicated that deflection calculations were unnecessary. For such balanced sections, q is approximately 0.18 [5]. For this reason, the ACI Code of 1963 stated that only members with q greater than 0.18 need be checked for deflection. With increased experience, a higher figure might well become acceptable, but in the meantime, deflections must be considered in relation to ACI 9.5. (See Chap. 6, Part D).

Too much steel used in a beam will cause the concrete to crush before the steel begins to yield. Failure of such a beam under a nonrelaxing load is explosively sudden and occurs without warning. Flexural tests of beams have shown that when a sufficient amount of steel is used so that it does not yield, the maximum value of the ultimate bending moment is approximately $(M_u)_{max} = \frac{1}{3}f'_c bd^2$, its limit being the crushing of the concrete [2]. If the amount of steel were selected so that it would yield at this moment the section would be "balanced" according to USD. This amount of steel is $p = 0.456f'_c/f_y$, so $q = 0.456$, and $a/d = 0.537$. This is the basis for the code's limitation of q to 0.30 (which includes a factor of safety) in ACI 18.8.1.

To emphasize the need for limitations on q, the failure modes are summarized alternately.

1. When a beam is under-reinforced and the load becomes excessive, the steel begins to yield although the concrete remains well understressed. As the load continues to increase, the steel further elongates and the member deflects quite noticeably. The large and visible deflection warns that the load must be decreased before excessive damage is done. If, instead, the load is increased, the steel will continue to yield, tension cracks will increase in size and number, and ultimate complete failure will occur as the concrete becomes overstressed in compression.

2. When a beam is over-reinforced, the steel does not reach yield. Deflections, therefore, are very small and not detectable by the naked eye. Suddenly, the concrete fails, leading to complete collapse, but the occupants have absolutely no warning. Obviously, this type of failure is much more serious than that of an under-reinforced member.

2-5.7 WSD Equations† The WSD equations are also determined by the couple method, using Fig. 2-5.5. Thus, for a beam of rectangular cross section,

$$M_c = \tfrac{1}{2}f_c kjbd^2 \quad \text{or} \quad f_c = 2M/kjbd^2 \tag{2-5.13}$$

and

$$M_s = A_s f_s jd \quad \text{or} \quad f_s = M/A_s jd,$$

where moments are in inch-pounds and stresses in pounds per square inch. A more useful equation can be formulated by dividing both sides of the first of Eqs. (2-5.13) by 12,000 and writing it as

$$M \text{ (ft-kips)} = KF, \tag{2-5.14}$$

where $K = \tfrac{1}{2}f_c kj$ and $F = bd^2/12,000$. F is identical to that used in USD, and K is comparable to K_u.

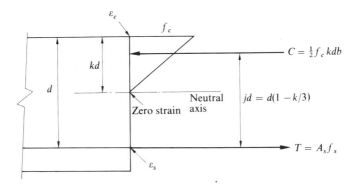

Fig. 2-5.5 WSD Equations

In an analysis approach, that is when the section proportions are known or assumed, the location of the neutral axis can be found readily by the methods of sec. 2-2. Another approach is to relate stresses and strains by the modular ratio $n = E_s/E_c$ and Hooke's law $E = f/\epsilon$. Thus

$$\epsilon_c/\epsilon_s = nf_c/f_s. \tag{2-5.15}$$

† Allowable stresses are presented in Fig. D-4 taken from ACI-63, page 49, because ACI-71 considers WSD as an alternate method with little detail.

But, by geometry (see Fig. 2-5.5)

$$\epsilon_c/\epsilon_s = kd/(d - kd) = k/(1 - k). \tag{2-5.16}$$

Equating (2-5.15) and (2-5.16),

$$f_s/f_c = n(1 - k)/k. \tag{2-5.17}$$

But, using the statical relation, $T = C$ yields

$$\tfrac{1}{2} f_c kdb = A_s f_s \tag{2-5.18}$$

or

$$f_s/f_c = k/2p \qquad (p = A_s/bd). \tag{2-5.19}$$

Equations (2-5.17) and (2.5.19) show that

$$n(1 - k)/k = k/2p$$

or

$$k = \sqrt{2np + (np)^2} - np. \tag{2-5.20}$$

This equation is presented as a curve in Fig. D-7. It is obviously useful only if section proportions are known or have been assumed.

By contrast, when the section is unknown, a balanced design approach is usually considered ideal. That is, it is assumed that concrete and steel are both stressed exactly to their respective allowable values. Based on this assumption, and reordering Eq. (2-5.17),

$$k = nf_c/(nf_c + f_s). \tag{2-5.21}$$

This equation is to be used only when *balanced design is assumed*, a condition which is often approached but seldom actually achieved. With k determined as a function of $f_c.f_s$, and n, the constant K equals $\tfrac{1}{2}f_c kj$. Representative values are tabulated in Fig. D-4.

2-5.8 Design by WSD. As in sec. 2-5.4, assume that the dead load is 1 k/ft and the live load is 1.1 k/ft.

1. *Select materials*, $f_c = 1350$ and $f_s = 24$ ksi.
2. *Determine design moment* (simple span is 20 ft). Design moment is 2.1 ✕ $20^2/8 = 105$ ft-kips.
3. *Determine K* $= 204$ from Fig. D-4.
4. *Calculate required F* $= 105/204 = 0.51$ (M/K).
5. *Choose b and d* from Fig. D-1. $b = 12$ in., $d = 22$ in. for which $F = 0.484$ (slightly less than 0.51). Assume d' is equal to 2 in. $h = 24$ in.
6. *Compute additional dead load* of beam as 12 ✕ 24 ✕ $(144/144)/1000 = 0.288$ k/ft (weight of concrete 144 lb/ft^3).

7. *Recompute bending moment.* $M = 105 + 0.288(20^2)/8 = 119.4$ ft-kips.

8. *Recompute required* $F = 119.4/204 = 0.585$ (M/K).

9. *Choose larger beam.* $b = 13$ in., $d = 25$ in., $h = 27$ in. (should try $d = 24$ in.)

10. *Recompute bending moment.* $M = 105 + (13)(27)20^2/8000 = 122.5$ ft-kips.

11. *Recompute required* $F = 122.5/204 = 0.60$. $b = 13$ in., $d = 25$ in., $h = 27$ in., therefore it is OK.

12. *Choose steel.* $A_s = M/f_s jd = 12(122.5)/(24)(0.88)(25) = 2.78$ in.^{2}. Note j is normally assumed as 0.88.

$$4\text{-}\#8 = 3.16 \text{ in.}^2 \qquad b \geq 11 \text{ in.}$$

$$3\text{-}\#9 = 3.00 \text{ in.}^2 \qquad b \geq 9\tfrac{1}{2} \text{ in.} \leftarrow \text{USE}$$

$$2\text{-}\#11 = 3.12 \text{ in.}^2 \qquad b \geq 8 \text{ in.}$$

Loads (20′ simple span)

DL		1.00 k/ft
	Beam 13″ × 27″	0.35
LL		1.10
		2.45 k/ft

V & BM $V = \tfrac{1}{2}(2.45)(20) = 24.5$ kips (reaction)

$M = 2.45(20)^2/8 = 122.5$ ft-kips (mid-span)

Section properties $(f_c = 1350 \text{ psi}, f_s = 24 \text{ ksi}, n = 9)$

Fig. D-4

$$p = A_s/bd = \frac{3.00}{13(24.56)} = 0.0094$$

$$np = 0.084$$

Fig. D-7 or Eq.(2-5.20)

$k = 0.335$ $kd = 8.23''$

$j = 0.890$ $jd = 21.8''$

$f_v = 60$ psi   **Fig. D-4**

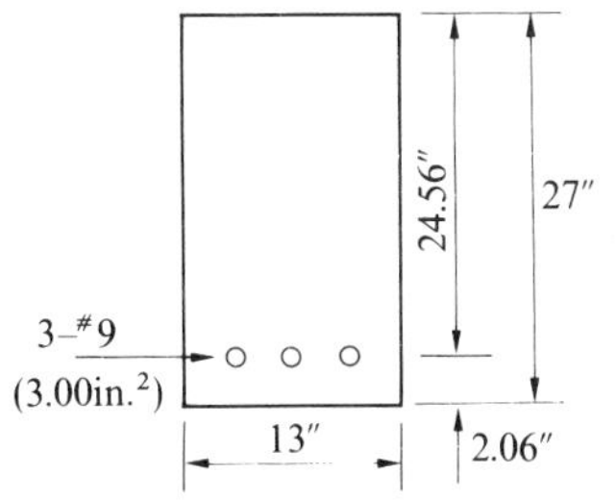

3-$\#9$ (3.00in.^{2})

24.56″   27″   13″   2.06″

d: 27 h
$-1\tfrac{1}{2}$ cover
$-\tfrac{9}{16}\ \tfrac{1}{2}$ bar ϕ
$-\tfrac{3}{8}$ stir ϕ
24.56″

Stress check   convert to in.-k

$f_c = 2M/kjbd^2 = (2)(12)(122.5)/(0.335)(0.890)(13)(24.56)^2 = 1.26$ ksi < 1.35 (OK)

$f_s = M/A_s jd = 12(122.5)/(3.00)(21.8) = 22.5$ ksi < 24 (OK)

$f_v = V/bd = 24\,500/(13)(24.56) = 77$ psi > 60

use stirrups

Fig. 2-5.6 WSD Analysis

2-5.9 Analysis by WSD. Analysis for this beam is shown in Fig. 2-5.6. As usual, an improved value of d is used. Also, both steel and concrete flexural stresses are checked. The shear stress is checked against the allowable stress given in Fig. D-4. Note stress comparisons indicating 13 x 26 would have been better choice.

An alternate direct analysis formulation method is possible by means of transformed area. By this method, after computing kd, the transformed moment of inertia is calculated. Stresses, appropriately modified by the modular ratio n, are then determined by the flexure formula.

2-5.10 Steel Design; M, b, and d Given. In the design of continuous reinforced concrete members, beam size and bending moment are often fixed so only steel has to be chosen. For ultimate strength design the procedure is quite simple. With F and M known, K_u is obtained by division. K_u corresponds to a precise q from Fig. D-5 and $A_s = qbdf'_c/f_y$ from Eq. (2-5.12).

When using working stress design, an iterative procedure is often helpful. In the formula $A_s = M/f_s jd$, M, f_s, and h are known. Assume j at 0.9, $d = h - 2$ inches, and solve for A_s. Then choose steel. Now, recompute d (include diameter of stirrups if used) and read j accurately from Fig. D-7. Adjust as necessary.

Analysis for both procedures is direct. The computed value of M_u is compared to that required, in USD. Using WSD, concrete and steel stresses are calculated and compared with allowable stresses.

2-5.11 Limitations on p. ACI formula 10-1 specifies that

$$p_{min} = \frac{200}{f_y} \tag{2-5.22}$$

unless reinforcing is one-third in excess of that required at *every* section. This may be expressed in terms of q as

$$q_{min} = \frac{200}{f_y} \frac{f_y}{f'_c} = \frac{200}{f'_c}. \tag{2-5.23}$$

Respective values of q_{min} for f'_c of 3000, 4000, 5000, and 6000 are 0.067, 0.05, 0.04, and 0.033.

ACI 10.3.2 limits reinforcing to $0.75 p_b$. The following section defines p_b for a balanced condition when maximum concrete strain is 0.003 in./in. and reinforcing reaches f_y. Assuming linear strain distribution:

$$\frac{\text{strain top fiber of concrete}}{\text{distance top fiber to neutral axis}}$$
$$= \frac{\text{strain in reinforcing}}{\text{distance from centroid of reinforcing to } N/A}. \tag{2-5.24}$$

With reinforcing strain expressed by Hooke's law as f_y/E_s, where $E_s \equiv$ 29,000,000 psi and $\epsilon_c \equiv 0.003$, Eq. (2-5.24) yields

$$\frac{0.003}{kd} = \frac{f_y/29,000,000}{d - kd} \tag{2-5.25}$$

from which

$$kd = d\left(\frac{87,000}{87,000 + f_y}\right) \tag{2.5-26}$$

Resultant of the tensile stress block $T = A_s f_y = p_b b d f_y$ (see Fig. 2-5.2), where p_b is steel ratio for balanced design. Resultant of the compressive stress block is $0.85 f_c' ab$, or $0.85 f_c'(\beta_1 kd)b = 0.85 \beta_1 f_c' bd \, 87,000/(87,000 + f_y)$.† Equate T with C, and p_b for balanced condition is obtained as

$$p_b = \frac{\beta_1 (0.85)f_c'}{f_y} \frac{87,000}{87,000 + f_y} \tag{2-5.27}$$

where $\beta_1 \equiv 0.85 - 0.05 \langle f_c' - 4000\rangle/1000$ (ACI 10.2.7), and $\langle f_c' - 4000\rangle = 0$ when negative.

Expressed in terms of q

$$q_b = \frac{\beta_1 (0.85)(87,000)}{(87,000 + f_y)}. \tag{2-5.28}$$

Several representative values are shown in Table 2-5.1. Compare with discussion of sec. 2-5.6.

Table 2-5.1   Values of q_b for Various Combinations of f_y and f_c' ($0.75\, q_b$ shown in parentheses)

	$f_c' \leq 4000$	$f_c' = 5000$	$f_c' = 6000$
$f_y = 40,000$	0.495 (0.371)	0.466 (0.349)	0.437 (0.328)
$f_y = 50,000$	0.459 (0.344)	0.432 (0.324)	0.405 (0.304)
$f_y = 60,000$	0.428 (0.321)	0.402 (0.302)	0.377 (0.283)

2-5.12 Practical Considerations.   Cost of Grade 60 reinforcing ($f_y = 60$ ksi) is not much more than Grade 40 ($f_y = 40$ ksi). In many design offices, therefore, reinforcement and concrete strength used is higher than that shown in previous examples. A typical combination is $f_c' = 4000$ psi and $f_y = 60,000$ psi ($f_c' = 5000$ or more for columns). Such parameters reduce beam sizes and can lead to increased deflections and difficulties in steel placement.

---

† β_1 is the ratio between a and kd, that is, $\beta_1 = a/kd$. This ratio changes downward from 0.85 as concrete stress f_c' exceeds 4000 psi at an ACI-defined rate of 0.5 per 1000 psi change in f_c'.

To obtain overall economy, which equates material, labor, and space costs, it is desirable to put an upper limit on p_b. If this is not done, it may be extremely difficult to pour concrete successfully, particularly at beam and column junctions. In addition, ACI 8.6 allows redistribution of moments in continuous flexure members only when p is less than $0.5\ p_b$. Some engineers strive for this ratio and believe it leads to optimum cost structures.

For these reasons, Fig. D-5

1. stops bottom of curves at minimum q, implying minimum p and A_s;
2. indicates upper limit of curves as controlled by concrete and steel strength $p \equiv 0.75\ p_b$;
3. shows $q = 0.18$ which represents a lower range of q for maximum economy; and
4. shows points where $p = 0.5\ p_b$ for various grades of steel and strengths of concrete.

With these aids, it is possible to restrict designs to permissible values of p, and possibly to obtain true economy.

Figure D-6 is similar to D-5, except $f'_c \equiv 4000$ psi, $f_y \equiv 60,000$ psi and the horizontal scale p is used rather than q. Specialized curves such as this are easy to construct and are of practical value in design offices.

2-6   ROLLED STEEL BEAMS

Continuous structural steel frames can be analyzed elastically, as in reinforced concrete, or plastically. The type of analysis is associated with the design method employed. Working stresses are used with expected loads in allowable stress design, whereas yield strengths are used with collapse loads for plastic analysis. Ultimate strength of steel members, identified by general yielding that results in excessive deformation rather than fracture or crushing as in concrete, is usually referred to as *plastic design*. Plastic design is becoming increasingly popular.

As noted previously, concepts associated with plastic design have influenced allowable stresses used in the allowable stress design method. Another example is that allowable flexural stresses for a "compact section"† are higher than for one which is not. A member is considered compact if, without buckling its compression flange, it can bend sufficiently to form a plastic hinge.

Regardless of design method, the procedure employed to select a beam section is the same. Initially, it is sized for flexure and then checked for shear and/or deflection.

---

† A detailed discussion of compact section is presented in sec. 11-2.2.

2-6.1 Flexural Design with Lateral Stability Assured. A beam is laterally stable when a sufficiently stiff floor or properly spaced diaphragms or cross beams keep the compression flange from displacing sideways. To facilitate design of such members, the flexure formula may be rearranged as

$$Z = 12M/F_b, \qquad (2\text{-}6.1)$$

where $Z \equiv$ required section modulus (in.^{3}), $12 \equiv$ coefficient to convert bending moment to inch units, $M \equiv$ bending moment in ft-kips, and $F_b \equiv$ allowable stress in ksi (for compact sections using working stress design, $0.66\,F_y$; for plastic design, F_y).

For ASTM A36 steel (with a yield strength of 36 ksi and in general referred to as A36), Eq. (2-6.1) reduces to

$$Z_e = 12\,M/24 \quad \text{or} \quad Z_e = M'^k/2 \text{ (elastic)} \qquad (2\text{-}6.2)$$

and

$$Z_p = 12M/36 \quad \text{or} \quad Z_p = M'^k/3 \text{ (plastic)}, \qquad (2\text{-}6.3)$$

where Z_e = required elastic section modulus (called S_x in *Steel Construction*), and Z_p = required plastic section modulus (called Z_x in *Steel Construction*).

1. To choose a trial beam for an elastic bending moment of 200 ft-kips, refer to page 2–9 of *Steel Construction*.† The required section modulus is $Z_e = 200/2 = 100$. The table lists steel sections in order of decreasing elastic section moduli in column 4, titled S_x. The smallest $Z_e(S_x)$ which is still greater than 100 is 103 for a W 14 X 68. However, note that beams are all set in groups with bold-face type to delineate the first entry of a section. The top of this group, which includes the W 14 X 68, is a W 21 X 55 with a section modulus of 110. This latter member is lighter than any other section below it which has a section modulus greater than 100, as can be observed. Thus, for a bending moment of 200 ft-kips, select a W 21 x 55 with a section modulus of 110 and weight of 55 lb/ft. Alternately, when the steel has a yield stress of 36 ksi, the most economical beam can be found by using the bending moment, called M_R, in the rightmost column.

2. In some cases, the depth of the beam must be limited. Again assume a bending moment of 200 ft-kips, but with a depth limited to 13 in., choose the W 12 X 79 as the lightest beam adequate in flexure.

3. For a plastic bending moment of 300 ft-kips, the required section modulus is $300/3 = 100$ in.^{3}. Refer to page 2–18 of *Steel Construction* and choose a

---

† The American Institute of Steel Construction *Manual of Steel Construction*, 7th ed., is used unless otherwise specifically noted. W 16 X 68 indicates type of cross section: W means a wide-flange beam, 16 means it is about 16 in. deep and 68 means it weighs about 68 lb/ft. When computing costs, steel is sold by pound and the weight tabulated is used as exact.

W 21 X 49 with a plastic section modulus of 108. The plastic section modulus is listed in column 1 as Z_x and the plastic moment for A36 in the seventh column as M_p.

4. As a final example, consider an elastic bending moment of 320 ft-kips to be carried by a rolled steel section of *42 ksi yield strength*. The allowable working stress is 28 ksi; 0.66 X 42 rounded off. Thus $Z_e = 12(320)/28 = 137$ in.³. Select a W 21 X 68 having a section modulus of 140 in.³ from page 2–8.

Notice that the tables for allowable stress design and plastic design are ordered differently. This acts as a perpetual reminder of the danger inherent in using the incorrect one.

2-6.2 Design Modification. From a knowledge of mechanics, it is known that when shear is a function of length to the nth power, bending moment is a function of length to the $(n + 1)$st power and deflection is a function of length to the $(n + 3)$d power. Therefore for relatively short spans, shearing stress can control design. Conversely for relatively long spans, deflection often controls.

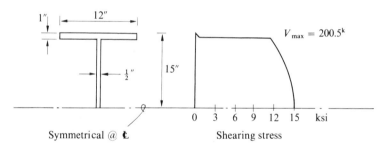

Fig. 2-6.1   Shearing Stress Distribution in a WF Beam

1. Design controlled by shear. Shearing stress in a beam can be computed as $v = VQ/Ib$ (see sec. A-4). The results of applying this formula to a WF beam are shown in Fig. 2-6.1.† When shear force of 200.5 kips is divided by total web area (30 in. X $\frac{1}{2}$ in.), an average shearing stress of 13.4 ksi is obtained. Comparison of this 13.4 average to the 15 ksi maximum of Fig. 2-6.1 shows that actual maximum shearing stress is 112% of average. To simplify shearing stress calculations, the AISC Specification has modified allowable shear stress so that average shear stress is directly compatible with maximum computations involving VQ/Ib.

As an example, consider the beam of Fig. 2-6.2. Page 2-10 of *Steel Construction* gives the most economical A36 rolled shape as W 16 X 31 with a section

---

† The stresses in the vicinity of the juncture of the web and flange are, of course, more complex than can be represented by this formula.

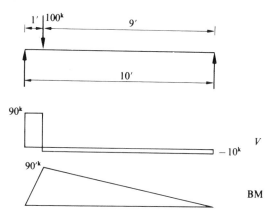

Fig. 2-6.2 Shear Controls Design

Table 2-6.1 Search For Optimum Rolled Steel Beam When Shear Controls

Trial Section	Page in *Steel Construction*	Z_e	d_w	t_w	A_w	Comment
W 16 × 58	1–32	94.4	15.86	0.407	6.45	This is OK, but weighs 58lb/ft. It shows up in the fourth grouping on page 2–9 of *Steel Construction*.
W 18 × 50	1–32	89.1	18.00	0.358	6.44	Try this as a reasonable choice; it works and weight is down to 50 lb/ft.
W 18 × 45	1–32	79.0	17.86	0.335	5.98	NG; W 18 × 50 still best with W 16 and 18 exhausted.
W 21 × 44	1–30	81.6	20.66	0.348	7.19	OK, and better at 44 lb/ft. Can it be best? Nothing else on page 2–9 of *Steel Construction* will do.
W 14 × 43	1–36	62.7	13.68	0.308	4.21	First possibility on page 2–10 of *Steel Construction* NG; no W 14 beats W 21 × 44.

Mental arithmetic can eliminate W 12 × 40 and W 10 × 39, so use W 21 × 44.

Loads   (10′ span — laterally stable — simply supported)

LL : Concentrated 1′ from left support 100^k

DL :   (say) $50^{\#}$/ft

V&BM V(Lt. support) = 90 + 5(0.050) = 90.25 kips

M(@ load)   = 100(1)(9)/10   = $90'^k$

DLM neglected

Section properties	<u>W21 × 44 A36</u>

$A_w = (0.348)20.66 = 7.19 \text{ in.}^2$ $F_v = 14.5 \text{ ksi}$

$Z_e = 81.6 \text{ in}^3$ $F_b = 24.0 \text{ ksi}$

Check $\sigma_L = 12\,M/Z = 12(90)/81.6 = 13.2 \text{ ksi} < 24.0$

$\sigma_v = V/A_w = 90.25/7.19 = 12.6 \text{ ksi} < 14.5$

Fig. 2-6.3   Analysis of Short WF Beam

modulus of 47.2 in.³. The full depth web area is (15.84 **X** 0.275) 4.36 in.² (*Steel Construction*, page 1–32). The allowable shear stress (*Steel Construction*, page 5–64) is 14.5 ksi. But, 90/4.36 = 20.6 ksi > 14.5 so that the W 16 × 31 is inadequate in shear strength.

To be adequate, the full depth web area must be (90/14.5) at least 6.2 in.² and the elastic section modulus must exceed 45.

Therefore, use the Allowable Stress Design Selection Table as a guide and check the web area of successive sections. This is shown in Table 2-6.1 where required properties are $Z = 45$ in.³ and $A_w = 6.2$ in.².

Shear in rolled sections is considered in more detail in sec. 6–1. However, a simple proof for the selected W 21 **x** 44 of Table 2-6.1 is shown in Fig. 2-6.3. Notice that the increment of bending moment due to dead load (DLM) is neglected purposefully.

2. Design controlled by deflection.   When the allowable deflection is exceeded by the first trial choice, it usually is easiest to invert the deflection equation and solve for the required moment of inertia. Then, select the lightest beam having an adequate Z_e (S_x) and moment of inertia (I) simultaneously.

For plastered walls, deflections are normally limited to $L/360$ with other criteria often specified for special situations. It is not uncommon to combine (1) some deflection calculation with (2) reasonable visual inspection to prove a section, rather than to spend an excessive amount of computation time. It also is advantageous to recall that for simply supported, constant EI spans, the mid-span deflection cannot vary by more than 2.54% from the maximum deflection.

2-6.3 Flexural Design in which Lateral Stability must be Considered. When the compression flange of a beam is without continuous lateral support, the allowable compressive stress is reduced to preclude lateral buckling. The pertinent equations are specified in AISC Specification 1.5.1.4 and summarized here:

1. Continuous lateral support is assumed for compact members if the distance between points of lateral support

$$\ell' \leqq 76.0 \, b_f / \sqrt{F_y} \qquad (2\text{-}6.4)$$

and also

$$\ell' \leq 20,000/(d/A_f)F_y. \qquad (2\text{-}6.5)$$

(d/A_f can be read directly from *Steel Construction* "Properties for Designing.") If these conditions are satisfied, $F_b = 0.66 \, F_y$.

2. Stress for all other cases may not exceed $0.60 \, F_y$.

3. Where the conditions of 1 are not satisfied, the allowable stress is taken as the larger of the following expressions that does not exceed $0.60 \, F_y$.

AISC formula 1.5-7 includes lateral bending and St. Venant torsional resistance:

$$F_b = \frac{12 \times 10^3 C_b}{Ld/A_f} \qquad (2\text{-}6.6)$$

AISC formulas 1.5-6a and 1.5-6b provide stiffness against lateral displacement of the compression flange. If

$$L/r_T \geqq \sqrt{\frac{102 \times 10^3 C_b}{F_y}},$$

$F_b = 0.60 \, F_y$ without further check. If

$$\sqrt{\frac{102 \times 10^3 C_b}{F_y}} \leqq \frac{L}{r_T} \leqq \sqrt{\frac{510 \times 10^3 C_b}{F_y}}, \qquad (2\text{-}6.7)$$

then

$$F_b = \left(\frac{2}{3} - \frac{F_y(L/r_T)^2}{1,530 \times 10^3 C_b} \right) F_y, \qquad (2\text{-}6.8)$$

but if

$$\frac{L}{r_T} \geq \sqrt{\frac{510 \times 10^3 C_b}{F_y}},$$

then

$$F_b = \frac{170 \times 10^3 C_b}{(L/r_T)^2} \tag{2-6.9}$$

where $b_f \equiv$ flange width (in.), $F_y \equiv$ yield strength (psi), $A_f \equiv$ area of flange, $d \equiv$ depth of beam, $L \equiv$ actual unbraced length of compression flange, $r_T \equiv$ radius of gyration of a section composed of the compression flange and one-third of the compression web area, computed about an axis in the plane of the web (in.), and $C_b \equiv$ a coefficient which takes into account that bending moment may not be constant along the span. Conservatively, its value is set at one (1); C_b is discussed more fully at the end of this section.

Substitution of Eq. (2-6.7) into Eq. (2-6.8) shows that this representation of F_b lies between $0.60\,F_y$ and $\frac{1}{3}F_y$. Equation (2-6.9) is a typical Euler curve which limits F_b to a maximum value of $\frac{1}{3}F_y$. A complete development of these equations is given in Chap. 11.

For a specific structural member, the only variable in these equations is the unbraced length. Thus it is relatively easy to plot curves for each member showing the allowable bending moments (allowable stress times section modulus) that correspond to various unbraced lengths. Such curves are shown in *Steel Construction* starting on page 2-87.

For example, the maximum unbraced length for a W 16 × 40 (A36, with $b_f = 7.000$ in., $S = 64.6$ in.^{3}, $r_T = 1.84$ in., and $d/A_f = 4.54$) is $L_c = 7.4$ ft. This is determined by use of Eqs. (2-6.4) and (2-6.5) as the minimum of

[by Eq. (2-6.4)] $\dfrac{76b_f}{\sqrt{F_y}} = \dfrac{76(7.000)}{\sqrt{36}} = 88.67$ in. or 7.39 ft,

or

[by Eq. (2-6.5)] $\dfrac{20,000}{(d/A_f)F_y} = \dfrac{20,000}{(4.54)(36)} = 122.36$ in. or 10.19 ft.

This point is shown on page 2-93 of *Steel Construction* by the solid black circle, at an allowable moment of $(S)(F_b)/12 = (64.6)(24)/12 = 129.2$ ft-kips, where $F_b = 0.66\,F_y$.

A vertical drop then occurs that terminates on a horizontal line for $F_b = 0.60\,F_y = 22$ ksi, or an allowable moment of $(64.6)(22)/12 = 118.4$ ft-kips. Located at the right extremity of this horizontal line is an open circle, representing the maximum unbraced length of the compression flange for this condition.

To find the abscissa of this point, it is necessary to solve for the length by the use of Eqs. (2-6.6) and (2-6.8). The allowable bending moment of $0.6\,F_y$ is now used as 21.6 rather than 22.0 ksi. (This anomaly accounts for the short vertical line below the open circle.) From Eq. (2-6.6)

$$21.6 = \frac{12 \times 10^3(1)}{L(4.54)},$$

L is 122.3 in. or 10.19 ft. This illustrates the derivation of Eq. (2-6.5) as previously used to compute L_c. From Eq. (2-6.8)

$$21.6 = \left(\frac{2}{3} - \frac{36(L/r_T)^2}{1530 \times 10^3(1)} \right) 36,$$

$L/r_T = 53.23$. More simply, this minimum magnitude of L/r_T can be found from Eq. (2-6.7) as $\sqrt{102 \times 10^3(1)/36}$. Since $r_T = 1.84$ is known, $L = 97.94$ in. or 8.16 ft. Using the *larger* figure, $L_u = 10.2$ ft. at an allowable moment of 118.4 ft-kips. Note that L_u is computed with $F_b = 0.6 F_y = 21.6$ ksi, but the allowable moment is calculated with F_b rounded to 22.0 ksi.

The values of L_c and L_u are also listed in the Allowable Stress Design Selection Tables.† A more direct procedure for calculating them is to use only Eqs. (2-6.4), (2-6.5), and (2-6.8) with L/r_T in the latter equation set by the first of Eqs. (2-6.7). For A36 steel, with $C_b = 1$, Eq. (2-6.7) may be represented as:

$$(53.23)\ 53 \leqq \frac{L}{r_T} \leqq 119\ (119.09).‡ \qquad (2\text{-}6.10)$$

To complete the plot, the maximum value of Eq. (2-6.6) or Eq. (2-6.8) is used until L/r_T exceeds 119; after which Eq. (2-6.6) is compared with Eq. (2-6.9). It could be fairly easy to discern which equation is critical from the plots, because Eq. (2-6.6) represents a concave curve, Eq. (2-6.8) curves downward (for W 16 x 40 from $M \doteq 104$ ft-kips at 11.4 ft to $M = 64.1$ ft-kips at 18.25 ft), and Eq. (2-6.9) is concave upward (for W 16 x 40, below 64.1 ft-kips), with the last equation effective only below one-half the maximum allowable moment $(\frac{2}{3} F_y \div \frac{1}{3} F_y)$. However, as the curves were inked, smoothing masked the break points.

As used in selecting trial sections, all members above and right of a point, found by the intersection of the bending moment in foot-kips and unbraced length, are adequate. In general, a broken line indicates a more economical section exists, farther up or to the right, which is shown by a solid line.

Assume for example, an A36 beam spanning 41 ft with diaphragms in 13.1 ft from each end. Design for an unbraced length of $41 - 2 \times 13.1 = 14.8$ ft and a bending moment of 730 ft-kips. From *Steel Construction*, page 2–89, two possible choices are W24 x 145 or W27 x 145. Both beams are represented by broken

---

† *Steel Construction*, p. 2-20.
‡ More precise values shown in parentheses would be excessively accurate for this purpose.

lines in this area, however, so proceed upward and to the right from the point (14.8, 730) to the first solid line to find that the most economical choice is W36 x 135. Of course, the first solid line may not be adequate in cases of (1) high shear, (2) large deflections, or (3) limited beam depth.

In final calculations for members without full lateral stability show, under the "Section Properties" caption, computations that lead to allowable compressive flexural stresses.

The value of C_b is equal to one (1) when bending moment at any place between two lateral supports exceeds the value found at both support points. If the moment at either support points is greater than any intermediate value (AISC 1.5.1.4.6a),

$$C_b = 1.75 + 1.05(M_1/M_2) + 0.3(M_1/M_2)^2, \qquad (2.6-11)$$

where M_1 is the larger of the moments at points of lateral support and M_2 is of smaller magnitude (use Frame Sign Convention; see sec. E-2.2), that is, single curvature is negative. [*Note.* The symbols and sign convention are reversed for reinforced concrete in Eq. (3-5.1.).]

Practice in applied computations to arrive at solutions to mathematical and legally defined models is necessary before proficiency in design can be attained. It is also very important for a designer to be sensitive to variations between models and the physical reality which they attempt to represent.

PART C
PHYSICAL REALITY IN BENDING

Since the advent of high-speed calculators and electronic computers, some engineers seem to have lost sight of the physical structure and the limitations of accuracy which it imposes. Unfortunately, it is not uncommon to see answers in eight significant figures when the calculations are based on data for which no more than three are reliable. This inherently dangerous procedure can lead to serious errors in judgment. Engineering requires acumen, not simply manipulation of mathematics. When mathematics, process, or method inhibits clear insight, then either internal mental processes and/or external procedures must be modified.

2-7   PHYSICAL VARIANTS

As a general rule three significant figures suffice for structural design. Rarely do special circumstances necessitate more. Three-place accuracy is predicated upon several factors: loads, tolerances, material characteristics, time and temperature variations, and assumptions.

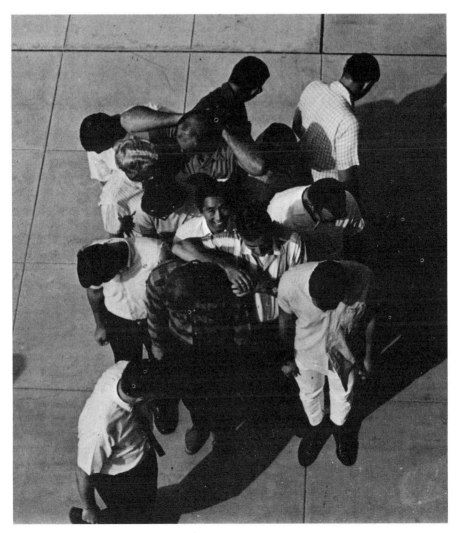

Fig. 2-7.1 Illustration of 100 lb/ft^2 Live Load (Note crowding and that some people are hanging over edges.)

2-7.1 Loads and Tolerances. To insure safety, and to eliminate conflict, "codes" legally define live loads for most occupancies (see *Steel Construction*, page 5–221 ff). Live loads are treated as exact values when specified under jurisdiction of such rules, even though they represent estimations based on experience and judgment without much substantiation by research. Even then, care must be exercised because usage of the structure can change with time.

The author is reminded of an overloaded warehouse which was at the point of failure because the contents differed dangerously from those for which specified adequate design loading was calculated some twenty years before.

In another case, catastrophe was averted when occupants of a building telephoned for consultation about excessive deflections. The cause was attributed to outdated files stored on a floor originally designed for *no* live load. The space had been "found" when the building was modified architecturally to correspond to its neighbors by adding dormers in the previously closed sloping roof.

Some feeling for live load specifications can be obtained by summing the weight of every member in a class and having them stand within areas producing an average load of say 40 and 100 pounds per square foot, respectively. Figure 2-7.1 illustrates this experiment for 100 pounds per square foot. Extrapolation to loads observed on various floors at heavily loaded times gives credence to code-specified live loads.

First thoughts might lead one to believe that dead loads can be determined accurately. This is not true because physical characteristics of members vary within the same nominal size. A frequent misconception is that a W 36 x 150 weighs exactly 150 pounds per foot because that *is* the weight used in costing and pricing. Actual weight varies by $\pm 2.5\%$ although the member still is within specified tolerances. As weights vary, so do moments of intertia, section moduli, etc., so that values specified in *Steel Construction* are only approximate.

To build with reinforced concrete the discrepancy to be taken into account is even greater because concrete and reinforcing steel are not generally placed within close tolerances. To complicate matters, section properties, may be based on gross or transformed section depending upon circumstances. Weight fluctuates according to several factors, including the amount of steel used; the nominal design weight of 145 lb/ft^3 can be in error by upward of 7%.

In addition to this, many truly dynamic loads are treated as quasi-static. Such a procedure overlooks the part which the structure itself plays in dynamic analysis, the random nature of typical vibrations, and the alteration of material characteristics.

Because many of these factors are present in most design problems (regardless of material), it is evident why the slide rule has been the engineer's primary method of calculation. The standard ten-inch slide rule† provides built-in accuracy that approaches the difference between a physical structure being designed and its mathematical model. Excessively precise calculations performed on inaccurate mathematical models represent poor engineering when safety considerations are hidden and/or calculation time is increased.

---

† The author is enthusiastic about the increased use of mini-electronic calculators, and anticipates that the slide rule will soon be a museum piece, but this only emphasizes the need for judgment in utilization of significant figures.

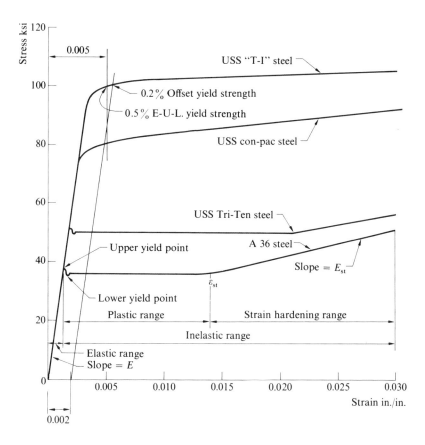

Fig. 2-7.2 Typical Initial Stress-Strain Curves for Structural Steels with Specified Minimum Tensile Properties [8, page 3]

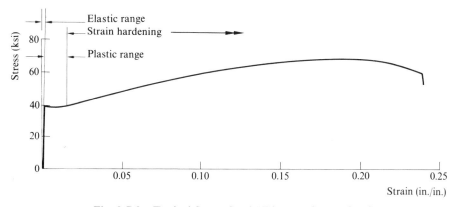

Fig. 2-7.3 Typical Stress-Strain Diagram for A7 Steel

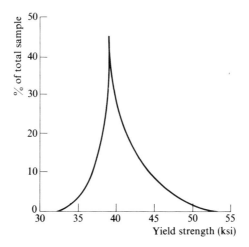

Fig. 2-7.4   Smoothed Curve for 33,000 Tons of A7 Steel (adapted from reference 6)

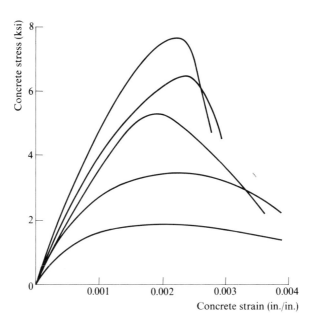

Fig. 2-7.5   Typical Compressive Stress-Strain Curves for Concrete

2-7.2 Material Characteristics. Figure 2-7.2 shows some typical initial stress-strain curves for various metals, and Fig. 2-7.3 shows a more complete stress-strain curve for A7 steel.

The contrast between the two curves is rather spectacular. Figure 2-7.3 indicates that steel has a strain capacity actually more than eight times greater than suggested by Fig. 2-7.2.

But, this curve is *typical*, and not *the* stress-strain curve for mild steel, as evidence by the data displayed in Fig. 2-7.4. Shown is a smoothed curve of yield strengths obtained from some 33,000 tons of A7 steel (3,974 tests on nine projects between 1938 and 1951) with most values well above the 33 ksi specified. In presenting this data, O. G. Julian [6] points out that results may not be conclusive for three reasons: (1) Tests are taken by the drop of the beam or halt of the gauge method to obtain the upper limit of the yield point, producing an error estimated at between 5% and 10%. (2) The high time-rate at which the tests are conducted causes an estimated error of another 10%. (3) The coupons from which the tests are made are normally cut from the webs. Because webs are thinner than flanges, the process tends to produce additional error of 5% to 10%. Because these errors are cumulative, actual yield points are 20% to 30% below those given by mill tests.

Typical stress-strain curves for concrete in compression are shown in Fig. 2-7.5. As the mix varies, the stress-strain curves show marked differences. Further, according to ACI 4.3.3, one in three tests on specimens taken from the same pour on the same job may fall 500 psi below f'_c so long as their average is greater than f'_c. In fact, modulus of elasticity changes with type of aggregate, type and even brand of cement, method of placing and curing, and elapsed time between mixing and testing. In an attempt to resolve these difficulties, Adrian Pauw[7] developed an empirical formula which correlates well with test data for both normal and lightweight concrete. This equation was adopted and is specified in ACI 8.3.1 as

$$E = w^{1.5}\, 33\sqrt{f'_c}, \qquad (2\text{-}7.1)$$

where $E \equiv$ modulus of elasticity (28 days) in psi, $w \equiv$ weight of concrete in pounds per cubic foot (90 to 155), and $f'_c \equiv$ 28-day ultimate compressive strength of concrete in psi.

Timber structures are designed by considering strengths and moduli as specified by type of wood, grade, and classification (see Chap. 4). Actual parameters vary from those specified because of differences in (1) characteristics of geographic growth areas, (2) grain and cutting methodologies, (3) growth patterns, (4) amounts of and fluctuations in moisture content, and (5) the duration of applied loads.

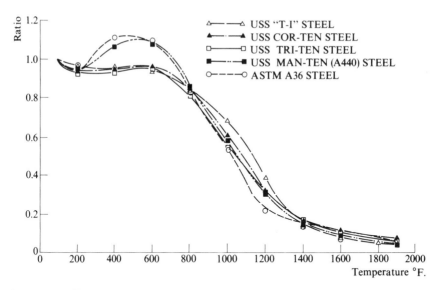

Fig. 2-7.6 Effect of Temperature on the Ratio Between Elevated-Temperature and Room-Temperature Tensile Strength [8, page 13]

2-7.3 Time and Temperature Variations. Related changes in reinforced concrete stress and strain with respect to time are given by Eqs. (2-7.2) and (2-7.3).†

$$\sigma_t = f_c' \left(\frac{1 - e^{-t/28}}{1 - e^{-1}} \right), \tag{2-7.2}$$

where $\sigma_t \equiv$ ultimate compressive strength of concrete in psi at t days, $f_c' \equiv$ ultimate compressive strength of concrete in psi (28 days), $e \equiv 2.7183$, and $t \equiv$ time in days.

$$\epsilon_t = \epsilon_e + \left[\epsilon_{sh} + \epsilon_{pl} \right] \left[1 - e^{-t/200} \right], \tag{2-7.3}$$

where $\epsilon_t \equiv$ strain at time t (in./in.), $\epsilon_e \equiv$ elastic strain (in./in.), $\epsilon_{sh} \equiv$ maximum shrinkage $\equiv 3 \times \epsilon_{e(\max)}$, and $\epsilon_{pl} \equiv$ plastic flow $\equiv \epsilon_{e(\max)}$. These equations are used in Sec. 2-8.1 to demonstrate time-related deflection characteristics of reinforced concrete. Although the office techniques discussed in Chap. 7 do not use them, they prove to be quite helpful as background material.

Time does not have a marked effect on steel except at higher temperatures. Figures 2-7.6, 2-7.7, and 2-7.8, from *Steel Design Manual* [8], show that as temperatures increase, yield stress, tensile stress and modulus of elasticity all

---

† Equations (2-7.2) and (2-7.3) have been found to be reasonable closed form equations to which students can relate. Although they seem to describe the phenomena rather well, they have not been rigorously verified.

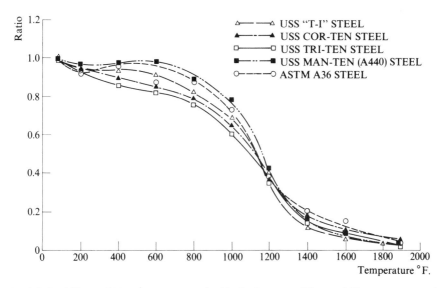

Fig. 2-7.7   Effect of Temperature on the Ratio Between Elevated-Temperature and Room-Temperature Yield Strength [8, page 13]

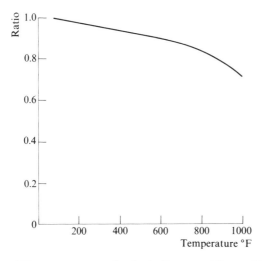

Fig. 2-7.8 Effect of Temperature on the Ratio Between Elevated-Temperature and Room-Temperature Modulus of Elasticity for Structural Steels [8, page 14]

decrease (except for limited transitional ranges). The converse is true as temperatures drop.

At elevated temperatures, steel is subject to creep also. Figure 2-7.9 shows limitations placed on stress for a mild steel (a) to resist rupture at a specified

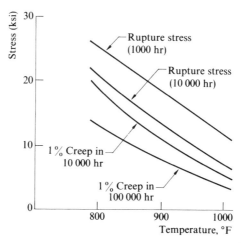

Fig. 2-7.9   Carbon Steel Creep at Elevated Temperatures [8, page 33]

temperature maintained for 1,000 or 10,000 hours (42 days or 1.14 years), and (b) to restrict creep to 1% at a specified temperature maintained for 10,000 or 100,000 hours (1.14 years or 11.4 years).

Ductility, the ability of a material to undergo large deformation without fracture, decreases as temperatures decrease below room temperature. (A decrease also occurs in the temperature range between 300° and 700° Fahrenheit). At lower temperatures, structures are subject to failure by brittle fracture, as evidenced by photos in a 1951 *Engineering News Record*[10]. *Toughness*, ability of a member to absorb large amounts of energy, is desirable in applications subject to such nonductile failures. For interesting, comprehensive reading, which provided the background for this summary, see references 8, 9, 10, and 11.

Shear fractures, associated with ductile behavior, present a gray, silky-appearing fracture surface. *Cleavage fractures* occur with little or no plastic deformation (hence *brittle*) and appear bright and granular. Figure 2-7.10 illustrates a fracture transition curve related to the percent of maximum shear strength for a mild steel specimen. The temperature corresponding to a 50% shear is often called the *fracture transition temperature*; the lower this value, the better the resistance to cleavage (brittle) fracture.

In general, brittle fracture is resisted by well designed structures with excellent details and craftsmanship and composed of correctly selected materials. Triaxial states of stress (with resulting low shearing stress) exist at sudden changes in geometry, notches, punched holes, rough edges, etc. and lead to brittle fracture under adverse atmospheric conditions (primarily

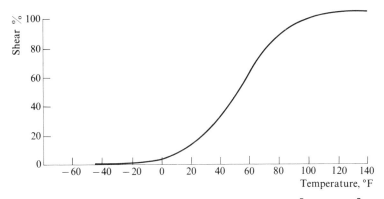

Fig. 2-7.10   Fracture Transition Curve for Mild Steel [9, page 19]

temperature). When designing for an extreme temperature environment, reference works, such as 12 and 13, are helpful for the selection of proper materials.

2-7.4 Variations to Assumptions.   Differences sometimes exist between physical world entities and their mathematical models because of assumptions necessary to make solutions tractable. Some linearizing techniques were discussed in Chap. 1. Other discrepancies are caused when (1) plane sections do not remain plane, as when beams become relatively deep, (2) shearing deflections cannot be ignored, and (3) superposition does not hold as, for example, members loaded by combined axial forces and bending moments when axial stress is relatively large.

Professor Kuzmanović [1] gives mathematical verification to the invalidity of the assumptions of an idealized elasto-plastic stress-strain curve. He states:

... if the formation of plastic hinges is possible at all, strain hardening must occur and the expected equalization of different moments on critical sections cannot be fully achieved in any manner. The magnitude of the error that is made is dependent on the elasticity conditions and the loading programmes. If the difference to be equalized between the moments is great and other conditions are favourable, the error in the calculation based on plastic hinge design is also great and vice versa.

As technology progresses, the type and nature of assumptions will change, but it appears that reasonable assumptions will be required in the foreseeable future when translating the physical world into suitable mathematical models.

2-7.5 Repeated Loads—Fatigue. Although fatigue failure due to repetition of loads may be crucial in design of aerospace structures, the effect is less pronounced in bridges and even less in buildings.

AISC Code, Appendix B, classifies four loading conditions, based on a 25-year time span: (1) from 20,000 to 100,000 repetitions (2 to 10 times per day), (2) from 100,000 to 500,000 repetitions (10 to 50 times per day), (3) from 500,000 to 2,000,000 repetitions (50 to 200 times per day), (4) over 2,000,000 repetitions (over 200 times per day).

AASHO 1.7.3 [14] use two classifications with three loading conditions each: (1) For freeways, expressways and major highways and streets, 2,000,000 repetitions for load spanning, $0 \longrightarrow 14$ feet (H Loading); 500,000 repetitions for load spanning, $14 \rightarrow 44$ feet (HS Loading); and 100,000 repetitions for load spanning more than 44 feet (Lane Loading). (2) For all other roads, 500,000 repetitions for loads spanning $0 \rightarrow 14$ feet (H Loading); and 100,000 repetitions for loads spanning more than 14 feet (HS and Lane Loading). In these classifications, an H Loading is only one two-axle truck on full bridge length (in one lane), an HS Loading is only one three-axle tractor-truck on full bridge length (in one lane), and a Lane Loading is an equivalent uniform load plus concentrated load(s) which represent a truck-train.

Using these criteria, allowable stresses are modified by both codes with regard to the particular structural component involved. In buildings, crane runway girders and machinery supports are most likely to have stress reduction to preclude fatigue failures.

2-8   EFFECT OF PHYSICAL VARIANTS

In this section it will be shown how physical variants affect the development and interpretation of mathematical models.

2-8.1 Modulus of Elasticity of Concrete. Figure 2-8.1 (an extension of Fig. 2-5.1) shows a typical stress-strain curve for concrete with various modulae of elasticity shown.

Initial tangent modulus is defined by slope to the stress-strain curve at the origin. For this idealized case,

$$\sigma_c = (3.000 \epsilon - 1000 \epsilon^2 + 90,000 \epsilon^3) \times 10^6, \qquad [2\text{-}5.2]$$

where $\sigma \equiv$ stress (psi) and $\epsilon \equiv$ strain (in./in.). The slope of this curve is

$$d\sigma/d\epsilon = (3.000 - 2000 \epsilon + 270,000 \epsilon^2) \times 10^6 \qquad (2\text{-}8.1)$$

which, when evaluated at $\epsilon = 0$, defines initial tangent modulus at 3×10^6 psi.

Secant modulus is not frequently used and is defined as the slope of a line drawn from the origin through the coordinates where the stress is a specified

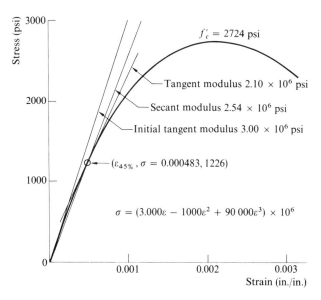

Fig. 2-8.1 Various Definitions of Moduli of Elasticity

percentage of f'_c. For this example, secant modulus is computed by dividing a stress of $0.45f'_c$ (1226 psi) by the corresponding strain of 0.000483 in./in. to yield 2.54 X 10⁶ psi.

Tangent modulus is defined as the slope of the line tangent to the stress-strain curve at a point specified by a percentage of f'_c. For this example, the specified stress is $0.45 \, f'_c$ which corresponds to a strain of 0.000483 in./in. Substitution of the value of strain into Eq. (2-8.1) yields a tangent modulus of 2.10×10^6 psi.

Apparent modulus of elasticity, alternately called *reduced* or *long-time* modulus of elasticity, is computed as the quotient of stress/strain at some particular time. It represents a secant modulus value which can be used in computation of long-time deflections under sustained loads.

As an example, the stress and strain for a 504 day old concrete is figured using Eqs. (2-7.2) and (2-7.3). It is also assumed that elastic strain due to live load only is 0.0003 in./in., with a corresponding stress of 812 psi. Therefore

$$\sigma_t = f'_c \ \frac{1 - e^{-t/28}}{1 - e^{-1}} \qquad\qquad [2\text{-}7.2]$$

implies that if the actual 28-day stress is used $\sigma_{504} = 812(1 - 0.000000015)/(1 - .368)$ or $\sigma_{504} = 1285$ psi.

At 504 days, $\epsilon_{sh} = 3$ X $0.0003 = 0.0009$ and $\epsilon_{pl} = 0.0003$, so, by use of Eq. (2-7.3), we have $\epsilon_{504} = 0.0003 + (0.0009 + 0.0003)(1 - 0.0805)$ or $\epsilon_{504} = 0.0014$ in./in.

Thus the apparent modulus of elasticity is $E_{504} = \sigma_{504}/\epsilon_{504} = 1285/0.0014 = 0.92 \times 10^6$ psi, or 36% of the elastic secant modulus. The word "apparent" calls attention to the fact that changes in length due to plastic flow and shrinkage are used. Actually, the true modulus of elasticity increases as the strength increases. For this example at 504 days, true long-time $E = 1285/0.0003 = 4.28 \times 10^6$ psi.

These results can be applied to a beam which deflects 1 in. due to dead load and 1 in. due to live load at 28 days. If the live load is intermittent, the resulting deflection at 504 days is:

$$DL\Delta_{504} = \Delta_{D,28} \times E_{28}/E_{app\,504} = 1 \times 2.54/0.92 = 2.8 \text{ in.}$$
$$LL\Delta_{504} = \Delta_{L,28} \times E_{28}/E_{true504} = 1 \times 2.54/4.28 = \underline{0.6}$$
$$\Delta_{504} = 3.4 \text{ in.}$$

Here $DL\Delta_{504} \equiv$ deflection due to dead load at 504 days, $\Delta_{D,28} \equiv$ deflection due to dead load at 28 days, $E_{28} \equiv$ secant modulus at 28 days, $E_{app504} \equiv$ apparent modulus at 504 days, $LL\Delta_{504} \equiv$ deflection due to live load at 504 days, $\Delta_{L,28} \equiv$ deflection due to live load at 28 days, and $E_{true504} \equiv$ true modulus at 504 days.

Of course, these calculations are reliable only if questions such as the following can be answered with assurance:

1. Is this the true stress-strain curve for the beam?

2. Is a valid quantity used for plastic flow?

3. Is a valid quantity used for shrinkage?

4. Is the moment of inertia known with reasonable accuracy?

2-8.2 Moment of Inertia of Reinforced Concrete Beams. A singly reinforced concrete beam is shown in Fig. 2-8.2. It is assumed that the concrete is cracked ch from the tensile face and carries tension for a distance th. Primary variables used to compute variations in moment of inertia, with their corresponding upper and lower limits are:

h	$\equiv$ overall depth (in.)	19.7	to	20.3
b	$\equiv$ width (in.)	11.7	to	12.3
c	$\equiv$ crack depth	0	to	$h - kd$
A_s	$\equiv$ area of steel (in.²)	2.97	to	3.03
wt	$\equiv$ weight per ft³	139	to	151
f'_c	$\equiv$ 28-day ultimate strength (psi)	2800	to	4000

These variables imply corresponding secondary variables:

$d \equiv h - d'$, where $d' \equiv 2$ in. $t \equiv (h - ch - kd)/h$, and $n \equiv 29 \times 10^6/(wt)^{1.5}\,33\sqrt{f'_c}$.

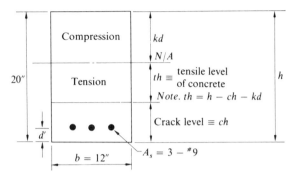

Fig. 2-8.2   Singly Reinforced Concrete Beam

By taking moments of areas about the neutral axis, the following expression is obtained:

$$\frac{b(kd)^2}{2} = \frac{b(h - ch - kd)^2}{2} + nA_s(d - kd). \tag{2-8.2}$$

From this equation, kd is determined as

$$kd = \frac{bh^2 - 2bch^2 + bc^2h^2 + 2A_snd}{2(bh - bch + A_sn)} = \frac{bh^2(1 - c)^2 + 2A_snd}{2[bh(1 - c) + A_sn]}. \tag{2-8.2}$$

Using this expression for kd, the transformed moment of inertia is determined from the sum of:

part due to compressive concrete $= b(kd)^3/3$

part due to tensile concrete $= b(h - ch - kd)^3/3$

part due to tensile steel $= nA_s(d - kd)^2.$

With the use of this formulation, a computer solution gives moments of inertia for all combinations of varying parameters. Typical results are discussed below.

Figure 2-8.3 shows the variation of I as h varies and all other dimensions are held to the fixed values of Fig. 2-8.2; the concrete is completely cracked to

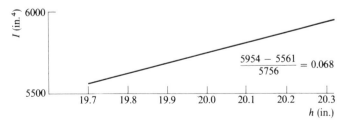

Fig. 2-8.3   Variation of I with h ($t \equiv 0$)

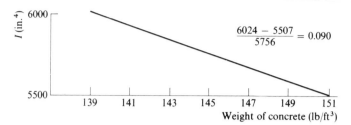

Fig. 2-8.4   Variation of I with Weight ($t \equiv 0$)

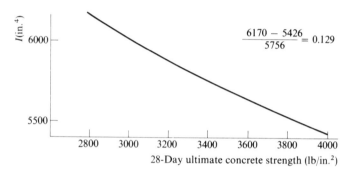

Fig. 2-8.5   Variation of I with Strength of Concrete ($t \equiv 0$)

the neutral axis. The maximum error, taken with respect to $h = 20$ in., between $h = 19.7$ in. and $h = 20.3$ in. is 6.8%. Thus a deviation from $h = 20$ in. of $\pm\frac{1}{4}$in. causes an error of about 3.4%.

As the width varies between 11.7 in. and 12.3 in., the moment of inertia varies from 5715 in.^{4} to 5795 in.^{4}. Between these ranges, the error (based on a 12-in. width) is $100(5795 - 5715)/5756 = 1.4\%$.

As the area of reinforcing steel varies from 2.97 in.^{2} to 3.03 in.^{2}, the moment of inertia varies from 5714 in.^{4} to 5797, the difference being about 1.4%.

As the weight of concrete increases, the modulus of elasticity goes up and the modular ratio n goes down. These values are shown plotted in Fig. 2-8.4 with a maximum variation of 9%. Note, however, that as I goes down, E goes up so that EI actually shows an overall increase of about 3.8%.

A similar phenomenon occurs as the f_c' goes from 2800 psi to 4000 psi, shown in Fig. 2-8.5. Although I decreases by 12.9%, EI actually increased by about 4.9%.

Considerations of crack depth produce very significant variations *if* the concrete actually acts in tension. Such a plot is shown in Fig. 2-8.6 where the difference, figured against the gross moment of inertia, is 57.5%.

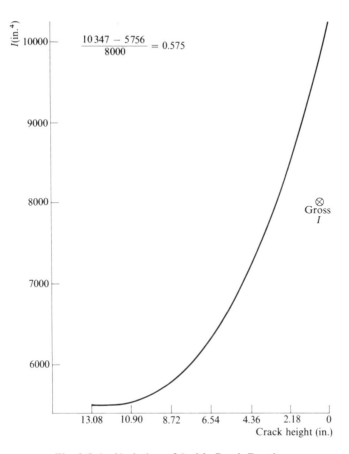

Fig. 2-8.6 Variation of *I* with Crack Depth

When the material properties are constant, that is, when f'_c and weight do not vary, the modulus of elasticity and the modular ratio both remain constant. Two curves are shown in Fig. 2-8.7 where the beam dimensions all change from minimum to maximum values concurrently through seven steps.

In summary, as the weight and strength, dimensions, and crack depth vary from the prescribed dimensions, corresponding maximum errors would appear to be in the neighborhood of

weight and strength	$EI \pm 4\%$
dimensions	$I \pm 5\%$
crack depth	$I \pm 15\%$.

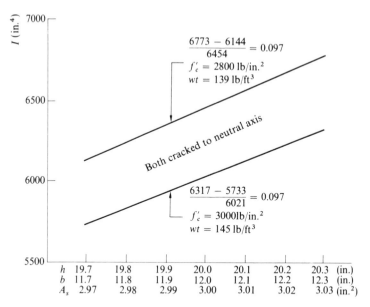

Fig. 2-8.7   Variation of I with h, b, and A_s ($t \equiv 0$)

In reality, many of these terms tend to offset each other so that the total errors are not so pronounced as they might appear. Required tolerances are stated in ACI 7.3.2.

2-8.3 Analysis of Reinforced Concrete Frames.   Variations in loading, in E, and in I also affect the results of analysis. Loads are usually designated such that variations between a mathematical model and its real-world counterpart are on the safe side.

To analyze redundant frames, stiffness as a function of EI must be used. Normal practice is to compute I based on gross concrete section. This procedure does not account for variations in moment of inertia caused by the amount and the position of steel reinforcing, form placement, material strength, or the unknown depth and position of cracks. A refinement of analysis based on transformed area is not too helpful, because cracking is not only discontinuous but is also relatively random. Nonetheless, safe and economical structures continue to be built. Some further ramifications of analysis techniques are discussed briefly in Chap. 10.

2-8.4 Effect of Dimensional Variations on Steel Sections.   A W 36 × 160 may vary in overall depth (according to page 1–127 of *Steel Construction*), measured through the web, by $\pm\frac{1}{8}$in. Approximately 78% of its moment of inertia comes from the flanges, transfer moment of inertia Ay^2, and 22% from the remaining web.

Nominal web depth, 36 in. $-$ 2(1.02) in., is 33.96 in. varying from 34.085 in. to 33.835 in. Maximum error involved in cubing this number is $\pm$ 1.11%.† The arm for transfer moment of inertia of the flanges varies between 17.5525 in. and 17.4275 in. Maximum error involved in squaring these numbers is $\pm 0.71\%$.

Total possible error from just this one cause, for a W 36 × 160 is, therefore,

$$(0.78)(0.71) + (0.22)(1.11) = 0.8\%.$$

In *Steel construction* I is 9760. But, 0.008 × 9760 = 78 in.[4], which gives some indication of the accuracy of the published value.

The division between mathematical models, legally defined models, and the real world should now be evident. These categories are used explicitly or implicitly in the following text to assist in a clear formulation of the hypotheses. In essence, engineering judgment is enhanced greatly by mathematical calculation, but mathematics is not synonymous with engineering.

PROBLEMS

2-1 *given*: Stress-strain diagram _____ (**1, 2, 3,** or **4** of Fig. 2-P.1). Cross-section _____ [(a), (b), (c), (d), or (e) of Fig. 2-P.2)] . $d = 20$, $b = 10$, $w = 1$, and $t = 2$.

    find: (a) Resultant of stress block at maximum allowable strain and its position. (b) Maximum allowable bending moment using the couple method. (c) Check maximum stress using flexure formula.

2-2 *given*: Stress-strain diagrams **1** and **3** of Fig. 2-P.1. Cross section _____ [(a), (b), (c), (d), or (e) of Fig. 2-P.2)]. Material **1** occupies the mid-portion of cross section. Material **3** occupies the portion from extreme fibers inward by a distance of $c = b/2$. $d = 20$, $b = 10$, $w = 1$ and $t = 2$.

    find: Maximum allowable moment using (a) couple method; (b) transformed area and flexure formula.

---

† $(34.085^3/33.96^3 - 1)100\ = 1.11$
$(17.5525^2/17.49^2 - 1)100 = 0.71$

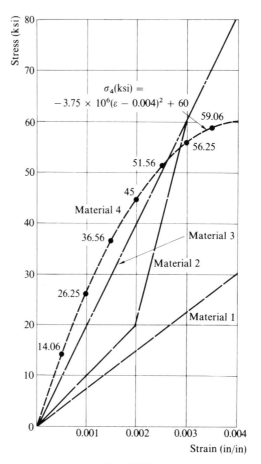

Fig. 2-P.1

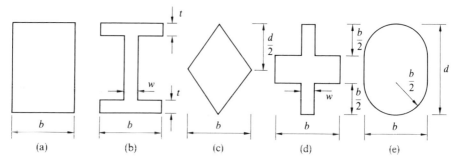

Fig. 2-P.2

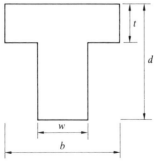

Fig. 2-P.3

2-3 given: Stress-strain diagrams **1** and **3** of Fig. 2-P.1. Cross section of Fig. 2-P.3. $d = 20$, $b = 10$, $w = 6$, $t = 4$. Material **1** extends $c = 0.4d$ downward from top. Material **3** extends $e = 0.6d$ upward from bottom.

   find: (a) neutral axis; (b) maximum allowable moment by couple method; (c) maximum allowable moment by transformed area and flexure formula.

2-4 given: Section and stress-strain parameters of 2-P.4. $b = 12$, $d = 18$, $h = 21$, $A = 3$. $f_c = $ _____ (2, 3, 4, or 5 ksi). $f_s = $ _____ (20, 22, 24, 30, or 36 ksi).

   find: (a) Additional parameters (1) maximum strains corresponding to maximum stresses; (2) position of neutral axis. (b) Maximum allowable bending moment (1) by couple method; (2) by transforming S to C and using flexure formula.

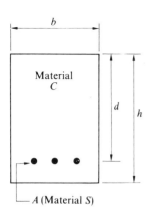

STRESS-STRAIN PARAMETERS

	Modulus of elasticity (E ksi)	Maximum allowable stress (f ksi)
Material C*	$E_c = 3\,000$	f_c
Material S	$E_s = 30\,000$	f_s

*Good for compression *only*

Fig. 2-P.4

2-5 given: Cross section ————— [(a), (b), (c), (d), or (e) of Fig. 2-P.2]
with $d = 20, b = 10, w = 1$, and $t = 2$. Or given one of the following:
(*Hint.* See *Steel Construction*, pages 6-26 thru 6-29.)
    1. Two parabolas common at their bases;
    2. Four compliments of half-parabolas, back to back;
    3. Ellipse;
    4. Two triangles common at their bases;
    5. Four parabolic fillets, back to back;
    6. Four elliptic compliments, back to back;
    7. Single triangle.

   find: (a) Elastic section modulus; (b) plastic section modulus; (c) shape
factor.

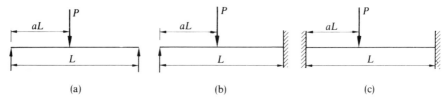

(a)   (b)   (c)

Fig. 2-P.5

2-6 given: Stress-strain curve of Fig. 2-4.1. Cross section ————— [(a), (b),
(c), (d), or (e) of Fig. 2-P.2]. $d = 20$, $b = 10$, $w = 1$, $t = 2$.
Beam ————— [(a), (b), or (c) of Fig. 2-P.5]. $L = 30$ ft, $a = 0.3$,
so $aL = 9$ ft.

   find: (a) Maximum P_e if beam just reaches yield at a critical point(s);
(b) maximum P_p if beam just reaches mechanism; (c) ratio of
P_p to P_e.

2-7 given: Figure 2-5.1.

  prove: Magnitude and position of resultant for area under stress-strain
diagram by (a) integration; (b) using an A, y, Ay table (see sec. A-2)
with area broken into ————— (5, 6, 8, 10) vertical strips of equal
width.

Note. In problems 2-8 through 2-11, beam may have to be oversized to accommodate steel because of restricted tables. Also, in computing d, if stirrups are
required, assume they are #3 bars.

2-8 given: Strength method with $f_y =$ ————— (40, 50, or 60 ksi), $f'_c =$
————— (3000, 4000, or 5000 psi), $M_u =$ ————— (60, 180, or
270 ft-kips) including load factors.

find: Minimum section using Table D-1 *only* with $0.18 < q < 0.20$; minimum steel using Table D-3 *only*.

2-9 given: Strength method with $f_y = $ _____ (40, 50, or 60 ksi), $f_c' = $ _____ (3000, 4000, or 5000 psi). Beam _____ [(a), (b), or (c) of Fig. 2-P.5], with $L = $ 20 ft, $aL = $ 12 ft, $P_{DEAD} = $ 7 kips and $P_{LIVE} = $ 8 kips, load factors *not* included.

design: Smallest beam for maximum moment *only* if (1) beam dead load is _____ (included, excluded); (2) beam is to be selected from Table D-1; (3) steel is to be selected from Table D-3; and (4) q is to be (a) less than 0.18; (b) less than 0.20; (c) less than $p_b/2$. (5) Choose steel for moments at other relative maxima.

present: Final analysis sheet.

2-10 given: Working stress method with $j \equiv 0.9$ and $f_s = $ _____ (20, 22, or 24 ksi), $f_c = $ _____ (1350, 1800, or 2250 psi) $M_u = $ _____ (40, 120, or 180 ft-kips).

find: Minimum section using Table D-1 *only*, and minimum steel using Table D-3 *only*.

2-11 given: Working stress method with $j \equiv 0.9$ and $f_s = $ _____ (20, 22, or 24 ksi) $f_c = $ _____ (1350, 1800, or 2250 psi) Beam _____ [(a), (b), or (c) of Fig. 2-P.5] with $L = $ 20 ft, $aL = $ 12 ft, $P = $ 15 kips.

design: Smallest beam for maximum moment *only* with beam dead load _____ (included, excluded) and using only Tables D-1 and D-3 for section and steel selection.

present: Final analysis sheet (a) using formulas (2-5.20) and second set of (2-5.13), or (b) using transformed area and flexure formula.

2-12 given: Strength method with $f_y = $ _____ (40, 50, or 60 ksi), $f_c' = $ _____ (3000, 4000, or 5000 psi), $b = $ _____ (8, 13, or 19 in.), $d = $ _____ (13, 22, or 27 in.).

find: (a) Maximum allowable moment if $p \leq 3p_b/4$; (b) maximum working stress dead, live, and total moments if moment consists of 4/10 dead and 6/10 live load. (c) Choose steel to fit within b and compute h rounded upward to nearest whole inch assuming #3 stirrups for _____ (1, 0.9, 0.8, 0.7, 0.6, 0.5, 0.4, 0.3, 0.2, 0.1) times M_{max}.

2-13 given: Working stress method with $f_s = $ _____ (20, 22, 24 ksi), $f_c = $ _____ (1350, 1800, 2250 psi), $b = $ _____ (8, 13, or 19 in.), $d = $ _____ (13, 22, or 27 in.).

find: (a) Maximum allowable moment. (b) Choose steel to fit within b and compute h rounded up to nearest whole inch assuming #3 stirrups for ———— (1, 0.9, 0.8, 0.7, 0.6, 0.5, 0.4, 0.3, 0.2, 0.1) times M_{max}.

2-14 given: Same as in problem 2-9.

design: (a) Smallest beam for maximum moment only with beam dead load ———— (included, excluded) and so that deflection criteria are satisfied without need for actual deflection calculations. Beam cross section and steel selections unrestricted. (b) Steel for moments at other relative maxima.

present: Final analysis sheet.

2-15 given: Beam ———— [(a), (b), or (3) of Fig. 2-P.5], supported laterally throughout with parameters set (as below) ———— (1, 2, 3, 4, 5, 6, 7).

Parameter Set	L (feet)	a	P (kips)
1	40	0.5	24
2	60	0.3	17
3	32	0.1	30
4	80	0.5	12
5	80	0.1	12
6	12	0.1	80
7	40	0.4	24

Use ———— (A36, A440 at f_y = 50 ksi) standard rolled sections. Limit deflection to $L/360$ and do not exceed allowable shear stress. Beam dead load is to be ———— (included, excluded).

design: Most economical beam using allowable stress design.

present: Final analysis sheet.

2-16 given: Beam ———— [(a), (b), or (c) of Fig. 2P.5] laterally supported (a) at point of load and both sides of load, midway between point of load and reaction or (b) at 1/4 points; with L = ———— (40, 60, or 32 ft), a = ———— (0.5, 0.3, or 0.1), P = ———— (24, 17, or 30 kips). Use ———— (A36, A440 at f_y = 50 ksi) steel. Deflection is to be ———— (limited to $L/360$, unlimited). Beam dead load is to be ———— (included, excluded).

design: Most economical beam using allowable stress design. (Caution. Can critical section be for a bending moment less than maximum occurring in a long unsupported length?)

present: Final analysis sheet.

2-17 given: $f'_c =$ _____ (3000, 4000, or 5000 psi), $t =$ _____ (60, 120, 180, 240, 300, or 360 days), $\epsilon_{sh} =$ _____ (2, 2.3, 2.6, 2.9, 3.2) times $\epsilon_{e(\max)}$, $\epsilon_{p\ell} =$ _____ (0.9, 1, 1.1) times $\epsilon_{e(\max)}$, $\Delta_{D,28} =$ _____ (0.9, 1, 1.1 in.), $\Delta_{L,28} =$ _____ (0.7, 1, 1.3 in.), ϵ_{sh} and corresponding $f'_c =$ _____ and _____ (0.00029 in./in. and 900 psi, 0.003 in./in. and 950 psi, or 0.00031 in./in. and 1000 psi).

find: Using Eqs. (2-7.2) and (2-7.3), for time t, find deflection due to (a) dead load, (b) live load, and (c) total load.

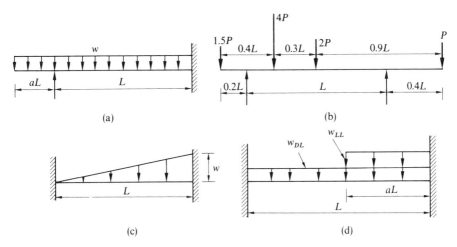

Fig. 2-P.6 General Beams

2-18 given: Beam _____ [(a), (b), (c), or (d) of Fig. 2-P.6], lateral support provided throughout, with $L = 30$ ft, $a = 0.4$, $w = w_{DL} = w_{LL} = 1.2$ kips/ft, $P = 10$ kips (undesignated loads are 40% dead and 60% live). $f_y =$ _____ (40, 50, or 60 ksi), $f'_c =$ _____ (3000, 4000, or 5000 psi).

design: Using strength method.

present: Final analysis sheet.

2-19 given: Same as in problem 2-18 except $f_s =$ _____ (20, 22, or 24 ksi) and $f_c =$ _____ (1350, 1800, or 2250 psi).

design: Using WSD.

present: Final analysis sheet.

2-20 given: Same as in problem 2-18 except $f_y =$ _____ (36, 42, or 50 ksi).

   design: Rolled steel section using allowable stress design.

   present: Final analysis sheet.

2-21 given: Same as in problem 2-20 except beam laterally supported at only $L/2$ and $3L/4$ from left support (not at $L/4$).

   design: Rolled steel section using allowable stress design.

   present: Final analysis sheet.

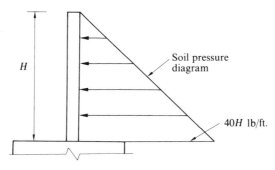

Fig. 2-P.7   Stem Wall

2-22 given: Stem wall of retaining wall shown in cross section in Fig. 2-P.7. $H =$ _____ (12, 14, or 16 ft).

   design: 1 ft strip of stem wall for bending. Space steel so that bar cutoff points are reasonable.

   present: Final analysis sheet.

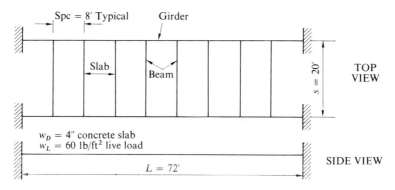

Fig. 2-P.8   Floor Framing Between Two Girders

2-23 given: Figure 2-P.8 where slab is simple reinforced concrete span between beams. Assume beams fixed at end where they are attached to girders. Although girder is subject to torsion, this is to be neglected until problems in Chap. 9. Also, assume complete lateral support.

    design: (a) Slabs ($f_y = 60$ ksi, $f'_c = 4$ ksi). (b) Beams and girders ($f_y = 60$ ksi, $f'_c = 4$ ksi). OR, beams and girders (A440 steel, $f_y = 50$ ksi).

    present: Final analysis sheets (a) for slab—show loads transmitted to beams; (b) for beams—show loads transmitted to girders; (c) for girders.

REFERENCES

1. Kuzmanović, B. O., "A Criticism of the Plastic Hinge Concept Used in the Plastic Theory of Structures," International Association for Bridge and Structural Engineers, Zurich (1959).

2. Whitney, Charles S., "Plastic Theory of Reinforced Concrete Design," *Transactions*, American Society of Civil Engineers, V. 107 (1942), pp. 251–326.

3. ACI-ASCE Committee 327, "Ultimate Strength," *Journal of the American Concrete Institute, Proceedings*, V. 52, No. 7 (January 1956), pp. 504–524.

4. Archibald, Raymond, *et al.*, "Report of ASCE-ACI Joint Committee on Ultimate Strength Design," *Proceedings*, American Society of Civil Engineers, Structural Division, V. 81, No. 809 (OCtober 1955), pp. 1–68. [*Note.* References 3 and 4 are the same report.]

5. Reese, Raymond C., *et al.*, "Explanatory Notes on Appendix (Ultimate Strength Design) to Building Code Requirements for Reinforced Concrete (ACI 318-56)," *Journal of the American Concrete Institute*, V. 29, No. 3 (September 1957), *Proceedings*, V. 54, Title No. 54-12, pp. 197–204.

6. Julian, Oliver G., "Synopsis of First Progress Report of Committee on Safety Factors," *Journal of the Structural Division, Proceedings*, American Society of Civil Engineers, V. 33, No. ST-4 (1957).

7. Pauw, Adrian, "Static Modulus of Elasticity of Concrete as Affected by Density," *Journal of the American Concrete Institute, Proceedings*, V. 57, No. 6 (December 1960), p. 685.

8. Brockenbrough, R. L. and B. G. Johnston, *Steel Design Manual*, Copyright U.S. Steel Corporation, Pittsburgh, Pennsylvania (1968).

9. Shank, M. E., *Control of Steel Construction to Avoid Brittle Failure*, Welding Research Council, New York (1957).

10. Merritt, Frederick S., "Bridge Collapse in Quebec Charged to Brittle Steel," *Engineering News Record*, V. 146, No. 6 (February 8, 1951), pp. 23–24.

11. Parker, Earl R., *Brittle Behavior of Engineering Structures*, Wiley, New York (1957).

12. *Steels for Elevated Temperature Service*, U.S. Steel Corporation, Pittsburgh, Pennsylvania (1965).

13. *USS Low Temperature and Cryogenic Steels—Materials Manual*, U.S. Steel Corporation, Pittsburgh, Pennsylvania (1966).

14. *Standard Speicfications for Highway Bridges*, 10th ed., The American Association of State Highway Officials, 341 National Press Building, Washington, D.C. 20004 (1969).

CHAPTER 3

CONCENTRICALLY LOADED COLUMNS

It is unlikely that concentrically loaded columns exist outside of research laboratories. It is virtually impossible to construct a truly straight column and actual end conditions generally cause eccentricity of loading, which produces a bending moment.

In column design effects of slight unintentional eccentricities can be ignored provided that the allowable stresses are adjusted accordingly. When bending moments become too important to neglect, the concentric capacity of the column still plays an important role in its design. More precise relationships, required for such beam-columns, are developed in Chap. 10.

3-1 IDEALIZATION OF CONCENTRICALLY LOADED COLUMNS—THE MATHEMATICAL MODEL

Intuitively, columns can be classified into three general categories. A *short column*, often called a *stub column*, fails if the stress reaches the compressive ultimate or yield strength of the material. A *long column* will fail by buckling at a stress dependent upon its stiffness and not the strength of the material. Between these extremes lies the *intermediate column*, the capacity of which depends on both strength and stiffness. Definition of columns as short, intermediate, or long requires detailed knowledge of the type of cross section, the structural material used, the unsupported column length, and conditions of end restraint.

3-1.1 Short Columns. Short columns are designed by keeping compressive stresses below a specified value. The familiar equation for the axial compressive stress is

$$\sigma = P/A, \tag{3-1.1}$$

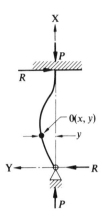

Fig. 3-1.1   Displaced Column

where $\sigma \equiv$ compressive stress, $P \equiv$ axial compressive load, and $A \equiv$ area of column cross section.

3-1.2 Long Columns.   The ultimate axial compressive strength of a long column is limited to the critical buckling load P_{CR}, first developed by Euler[1].

The formula for Euler critical buckling load is

$$P_{CR} = \pi^2 EI/(KL)^2, \tag{3-1.2}$$

where $E \equiv$ modulus of elasticity (ksi), $I \equiv$ moment of inertia (in.^{4}), $L \equiv$ unbraced column length (in.), $P_{CR} \equiv$ critical load (kips), and $K \equiv$ the ratio of effective column length to actual unbraced length, a function of column end restraint.

Typical of the development of K is Fig. 3-1.1 showing a laterally deflected column, with one end fixed and the other end hinged, subject to an axial load P. Note that the X axis is positive upward from the hinge while the Y axis is positive leftward from the hinge.† Horizontal reactions R are developed by the vertical load in conjunction with the displaced configuration.

Now, consider any point 0 located at coordinates (x, y). Bending moment is produced by the vertical force acting about the "arm" y caused by lateral displacement of point 0. Additional bending moment is produced by the horizontal reaction in conjunction with the "arm" x. Internally, the resisting moment is EIy'' (sec. A-3), so

$$\Sigma M_0 = -Py + Rx - EIy'' = 0. \tag{3-1.3}$$

---

† For discussion of sign conventions, see sec. E-2.

Rearranging, we have

$$y'' + (P/EI)y = (R/EI)x \qquad (3\text{-}1.4)$$

A solution for Eq. (3-1.4) is

$$y = A \sin(\sqrt{P/EI}\, x) + B \cos(\sqrt{P/EI}\, x) + Rx/P. \qquad (3\text{-}1.5)$$

To evaluate the unknown constants A and B and the reaction R, three boundary conditions are used.

At $x = 0$, the deflection $y = 0$. Therefore, substitution into Eq. (3-1.5) gives $0 = A(0) + B(1) + 0$, so $B = 0$.

At $x = L$, both deflection y and slope y' are zero. Thus, Eq. (3-1.5) with $y = 0$, $x = L$, and $B = 0$, becomes

$$0 = A \sin(\sqrt{P/EI}\, L) + (RL/P) \qquad (3\text{-}1.6)$$

from which

$$A = \frac{-RL}{P \sin(\sqrt{P/EI}\, L)}. \qquad (3\text{-}1.7)$$

Differentiation of Eq. (3-1.5) and substitution of $y' = 0$, $x = L$, and $B = 0$ yields

$$0 = \sqrt{P/EI}\, A \cos(\sqrt{P/EI}\, L) + (R/P) \qquad (3\text{-}1.8)$$

from which

$$A = \frac{-R}{P\sqrt{P/EI}\, \cos(\sqrt{P/EI}\, L)}. \qquad (3\text{-}1.9)$$

Equating the two expressions for A, Eqs. (3-1.7) and (3-1.9), we have

$$\sqrt{P/EI}\, L = \tan(\sqrt{P/EI}\, L). \qquad (3\text{-}1.10)$$

Because $\sqrt{P/EI}\, L$ must be positive, the tangent is in either the first or the third quadrant. In the first quadrant, the angle in radians goes from 0 to $\pi/2$ (1.57 radians) and the tangent varies from 0 to infinity, but at no place is the value of the tangent equal to the angle in radians. Therefore, the solution lies in the third quadrant, where

$$4.493 = \tan 4.493. \qquad (3\text{-}1.11)$$

By use of this equality, the right-hand side of the Eq. (3-1.10) can be replaced numerically by 4.493:

$$\sqrt{P/EI}\, L = 4.493. \tag{3-1.12}$$

Or, rearranging, we have

$$P = (4.493)^2(EI/L^2). \tag{3-1.13}$$

Although mathematically developed, this result can be interpreted physically. Because the solution satisfies all boundary conditions, the original equation is valid for these particular end conditions. Further, if P were less than the value specified by Eq. (3-1.13), then Eq. (3-1.11) would no longer be an equality. This would indicate that Eqs. (3-1.6) and (3-1.8) could be solved simultaneously only with R, and subsequently A, equal to 0. But, R could be 0 only if no lateral displacement occurred, that is, the column does not bend. Thus the value of P defined by Eq. (3-1.13) is the lowest value compatible with lateral displacement. If a given column were loaded to this capacity, it would become extremely pliable and hold any of an infinite number of displaced positions, all of the same general shape, without the need of a lateral load.†

This load, called the *critical buckling load* of the column, when used in Eq. (3-1.13) with π^2 factored out, yields the standard form

$$P_{CR} = \frac{\pi^2 EI}{(0.7\,L)^2}. \tag{3-1.14}$$

In the case of a column hinged at both ends, with the use of boundary conditions of $y = 0$ at $x = 0$ and $y'' = 0$ at $x = L$, Eq. (3-1.5) quickly reduces to

$$0 = -(P/EI)\, A \sin(\sqrt{P/EI}\, L). \tag{3-1.15}$$

If P and $A \neq 0$, Eq. (3-1.15) reduces to the nontrivial form,

$$\sin\sqrt{P/EI}\, L = 0. \tag{3-1.16}$$

This equation is interesting because it shows $\sqrt{P/EI}\, L = n\pi$ or that $P_{CR} = n^2\pi^2 EI/L^2$, which can be put into the general form of Eq. (3-1.2) by setting $n = 1/K$, where K is defined on the effective length factor. The number of buckles defined by n is shown in Fig. 3-1.2. With no intermediate support, the smallest value of P_{CR} is obtained with one buckle, $n = 1$. With the columns supported laterally at the mid-point, third points, or quarter points, the respective effective lengths of $L/2$, $L/3$, and $L/4$ correspond to values of P_{CR}, 4, 9, and 16 times as great as for a column with no lateral support. Such points of support represent *points of contraflexure* at which bending moments are zero.

---

†The relation between displaced position and applied load can be obtained by use of the more precise curvature equation discussed in sec. A-3. This formulation, termed the *elastica*, is developed by Timoshenko. [2]

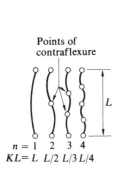

Points of contraflexure

$n = 1$ 2 3 4
$KL = L \; L/2 \; L/3 \; L/4$

Fig. 3-1.2 Buckled Mode Shapes

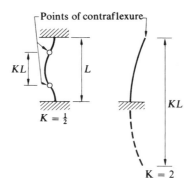

Points of contraflexure

KL L

$K = \frac{1}{2}$

KL

$K = 2$

Fig. 3-1.3 Effective Length Factor for Fixed and Cantilever Columns

With this knowledge, values may be extrapolated for both fixed and cantilever columns as shown in Fig. 3-1.3.

The Commentary on AISC Specifications, *Steel Construction*, page 5–138, recommends slightly higher values because full fixity of joints is seldom realized. Table 3-1.1, where

$$P_{CR} = \pi^2 EI/(KL)^2, \qquad [3\text{-}1.2]$$

summarizes the foregoing discussion.

Table 3-1.1 Effective Length Factors

End Conditions	Value of K	Recommended Value of K
Both ends hinged	1.0	1.0
Both ends fixed	0.5	0.65
One end fixed, one end free (cantilever)	2.0	2.1
One end fixed, one end hinged	0.7	0.8

When both sides of Eq. (3-1.2) are divided by the cross-sectional area A of the column and when I/A is replaced by r^2 (r defined as radius of gyration), the Euler equation can be expressed in terms of critical stress as

$$\sigma_{CR} = \pi^2 E/(KL/r)^2. \qquad (3\text{-}1.17)$$

This shows that both critical load and critical stress are inversely proportional to $(L/r)^2$.

Because a column can buckle about either axis, it is necessary to apply Eq. (3-1.2), or Eq. (3-1.17), twice. However, relationships between I_x and KL_x as compared to I_y and KL_y often allow mental elimination of one of these.

3-1.3 Intermediate Columns. Real columns can seldom be classified as either short or long, being in the transition range between compressive yielding and Euler buckling. Several forms for expressing column failure in this range have been developed, and although they do not differ greatly, it is interesting and instructive to trace their evolution to the current design form.

3-2 DEVELOPMENT OF CRC AND AISC COLUMN FORMULATIONS

A very complete development of the steel column design equations, along with an interesting discussion of their development, is presented in *Guide to Design Criteria for Metal Compression Members*† (and included references), edited by Bruce G. Johnson for the Column Research Council. [3]

In summary, Euler introduced the concept of critical load in 1744, although the procedure was not applied to design because proof tests often failed below P_{CR}. In 1935, Osgood [4] wrote a succinct history of subsequent events which is presented here with some editorial deletions.

... the double-modulus theory has frequently been called also the Considère-Engesser theory and Kármán's theory. In 1891 there was published a memoir included as an appendix to the proceedings of the Congrès International des Procédés de Constructions, held in Paris from the 9th to the 14th of September 1889, in which A. Considère pointed out that, as an ideal column stressed beyond the proportional limit begins to bend, the stress on the concave side increases according to the law of the compressive stress-strain diagram, and the stress on the convex side decreases according to Hooke's law, and that therefore the strength would be given by

$$P = \pi^2 \overline{E} I/L^2,$$

in which $\overline{E}$ is a modulus, the value of which lies between the modulus of elasticity and the tangent modulus. Considère realized that $\overline{E}$ was a function of P/A, the average stress in the column, but did not go further than to point out this fact.

Earlier in 1889 Fr. Engesser presented his tangent-modulus theory, and in 1895 Félix Jasinski pointed out that this theory was not correct and called attention to Considère's work. He stated that at that time it was impossible, however, to determine theoretically the form of the function $\overline{E}$. Thereupon Engesser acknowledged the error in his original theory and replied that the possibility of determining $\overline{E}$ theoretically was in no wise out of the question, and he determined it in the general form. Nothing further was done apparently until Kármán presented the theory again, added the actual evaluation of $\overline{E}$ for the rectangular cross section and the idealized H-section (consisting of infinitely thin flanges and negligible web), and gave the theory new life by making a series of careful tests designed to afford a check on the theory. Since then $\overline{E}$ has been evaluated for other cross sections by a number of writers.

---

†Often referred to as CRC Guide.

In May of 1947, Shanley[5] published an account which revalidated the tangent modulus (Engesser) equation. He did so by questioning the tacit assumption that perfect columns did not bend until after critical load, and then showed that it was possible to get a strain distribution without strain reversal. This was a major breakthrough because E_T is much easier to obtain than $\overline{E}$, the latter being partially dependent on member cross section. In the article he states: "It is interesting that aircraft engineers, in seeking greater accuracy in the inelastic range, have gone back to the formula that was first suggested by Engesser over 50 years ago."

Finally, in 1952 the CRC decided that working load formulas were to be based on the tangent modulus formula, with E_T established by a static compression test on a stub column rather than on tensile tests of coupons (see sec. 2-7.2). Such testing recognizes the importance of residual stresses in members produced in the rolling process.

3-2.1 Column Strength Curves. The curve resulting from basing calculations of critical stress on an Euler-type equation, with E_T used above the proportional limit, $(\pi^2 E_T)/(KL/r)^2$, is shown in the upper half of Fig. 3-2.1 as an Engesser-type curve. When L/r is large enough, the critical stress σ_{CR} is less than σ_p and the column fails by Euler buckling in the elastic range. As L/r becomes smaller, critical stress is greater than the proportional limit and the member fails inelastically. As discussed by Bleich[6], the shape of the curve is dependent upon only three parameters: yield stress σ_y, proportional limit σ_p, and elastic modulus E. Bleich suggested and the Column Research Council accepted, therefore, an alternate quadratic parabolic representation for the Engesser portion of the curve:†

$$\sigma_{CR} = \sigma_y - \frac{\sigma_y - \sigma_P}{(L'/r)^2}\left(\frac{L}{r}\right)^2, \qquad (3\text{-}2.1)$$

where σ_{CR} = critical stress in inelastic region, σ_y = yield stress, σ_p = proportional limit, r = radius of gyration, L = effective length, and L'/r = slenderness ratio for which $\sigma_{CR} = \sigma_P$.

Using the last definition and Eulers equation, we have

$$\sigma_{CR} = \sigma_P = \pi^2 E/(L'/r)^2, \qquad (3\text{-}2.2)$$

or, reordered,

$$(L'/r)^2 = \pi^2 E/\sigma_P. \qquad (3\text{-}2.3)$$

---

† Bleich's formula can be verified to yield identical curves with Engesser because (1) when $L/r = 0$, $\sigma_{CR} = \sigma_y$; (2) when $L/r = L'/r$, $\sigma_{CR} = \sigma_p$; and (3) both are quadratics.

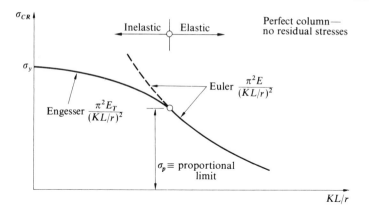

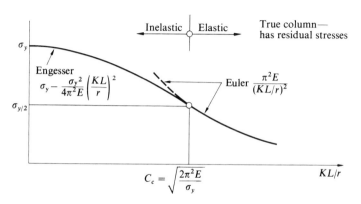

Fig. 3-2.1   Typical Column Curves

When this value of L'/r is substituted into Eq. (3-2.1),

$$\sigma_{CR} = \sigma_y - (\sigma_y - \sigma_P)\frac{\sigma_P}{\pi^2 E}\left(\frac{L}{r}\right)^2. \tag{3-2.4}$$

Equation (3-2.4) is used as the basis of the CRC equation except that L is replaced by KL.

If an axial load is distributed uniformly over the cross section of an ideal steel column having no residual stresses, the load deformation curve remains linear until an average stress of σ_p is reached. Subsequent nonlinearity is caused by yielding in portions of the column. By contrast, an actual steel column has residual stresses produced by the rolling process (Fig. 3-2.2). Hence, yielding will commence before σ_p is reached and the dividing line between linear and nonlinear is reduced to $\sigma_y - \sigma_{RS}$, where σ_{RS} is defined as the residual stress. It

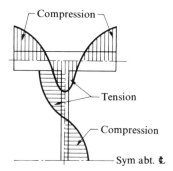

Fig. 3-2.2 Typical Residual Stresses

is therefore appropriate to replace σ_P by $\sigma_y - \sigma_{RS}$ in Eq. (3-2.4), resulting in

$$\sigma_{CR} = \sigma_y - \frac{\sigma_{RS}(\sigma_y - \sigma_{RS})}{\pi^2 E}\left(\frac{KL}{r}\right)^2. \tag{3-2.5}$$

To obtain the "CRC Column Strength Curve," σ_{RS} is taken equal to one-half of σ_y, yielding

$$\sigma_{CR} = \sigma_y - \frac{\sigma_y^2}{4\pi^2 E}\left(\frac{KL}{r}\right)^2. \tag{3-2.6}$$

This sets the dividing line between elastic and inelastic action at a ratio KL/r corresponding to a critical stress of $\sigma_y/2$ as computed by Euler's formula. Or

$$\frac{\sigma_y}{2} = \frac{\pi^2 E}{(KL/r)^2} \tag{3-2.7}$$

This ratio KL/r is called C_c, so

$$C_c = \sqrt{\frac{2\pi^2 E}{\sigma_y}}. \tag{3-2.8}$$

Thus, Eq. (3-2.6) represents Engesser's curve to C_c, where it becomes tangent to Euler's curve. These results are shown in the lower half of Fig. 3-2.1. Comparison of this curve with several test cases is shown in Fig. 3-2.3, as taken from the CRC Guide.

In Fig. 3-2.4 the CRC column strength curve is seen to be a conservative compromise where it is compared to theoretical solutions assuming various residual stress distributions for major and minor axis bending.

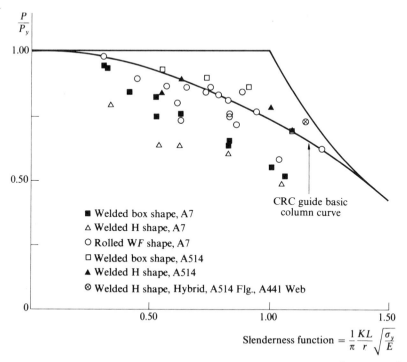

$$\text{Slenderness function} = \frac{1}{\pi}\frac{KL}{r}\sqrt{\frac{\sigma_y}{E}}$$

Fig. 3-2.3 Welded Columns and the CRC Guide Column Curve [3, page 27]

Of course, there are additional considerations affecting the strength of a column which include the following:

1. Effects related to material properties
     Nonlinearities in material properties
     Variation in yield strength at different points in a cross-section
     Creep

2. Effects related to accidental bending
     Accidental eccentricity and curvature
     Accidental lateral load
     Thermal effects

3. Effects related to cross-sectional shape
     Shear deformation
     Local buckling
     Torsional buckling

In general, these are accounted for by using a suitable factor of safety.

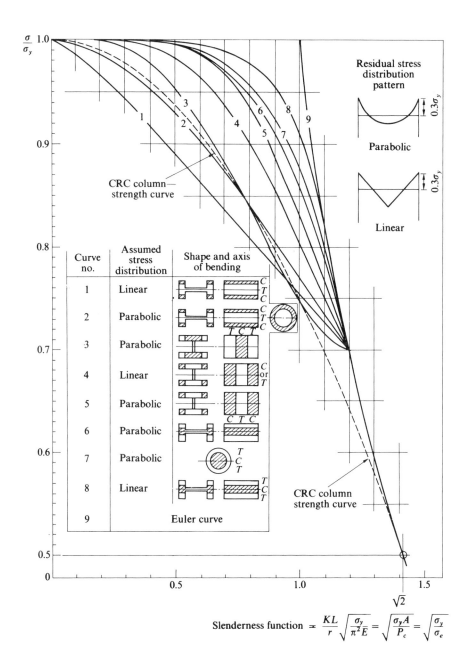

Fig. 3-2.4 Column Strength Curves for Various Patterns of Residual Stress in Various Cross Sections, Based on Buckling Load [3, page 23]

3-2.2 AISC Column Equation. The AISC specifications utilize two formulas to define the allowable loads on axially loaded compression members. AISC formula 1.5-1 is applied in the inelastic region when KL/r is less than C_c, where [compare to Eq. (3-2.8)].

$$C_c = \sqrt{2\pi^2 E/F_y}. \tag{3-2.8a}$$

With C_c substituted into the CRC Column Strength Curve Eq. (3-2.6) and with use of a safety factor, AISC formula 1.5-1 is obtained as:

$$F_a = \frac{\left(1 - \dfrac{(KL/r)^2}{2C_c^2}\right) F_y}{\dfrac{5}{3} + \dfrac{3(KL/r)}{8C_c} - \dfrac{(KL/r)^3}{8C_c^3}}, \tag{3-2.9}$$

where $F_a \equiv$ allowable column stress, the denominator $\equiv$ the factor of safety, $r \equiv$ radius of gyration, $\sqrt{I/A}$, and $KL \equiv$ the effective length of the column.

The AISC factor of safety varies from a low of $5/3(1.67)$ for $KL/r = 0$, to a high when $KL/r \geq C_c$ of $(5/3) + (3/8) - (1/8) = 23/12(1.92)$. The increase in the factor of safety as KL/r increases is related to the increasing slope of the P/A versus KL/r diagram up to the point at which KL/r equals C_c. A slight error in the estimation of KL/r is less serious for shorter columns than for longer ones. The factor of safety is constant at $23/12$ for columns with KL/r values larger than C_c. A plot of KL/r versus the actual factor of safety is shown in Fig. 3-2.5. When KL/r is greater than C_c, AISC specifies the Euler equation

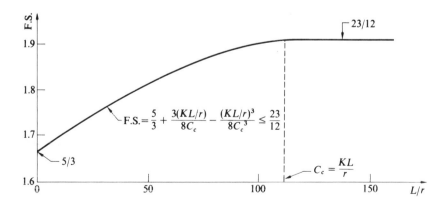

Fig. 3-2.5   Factor of Safety versus L/r Curve

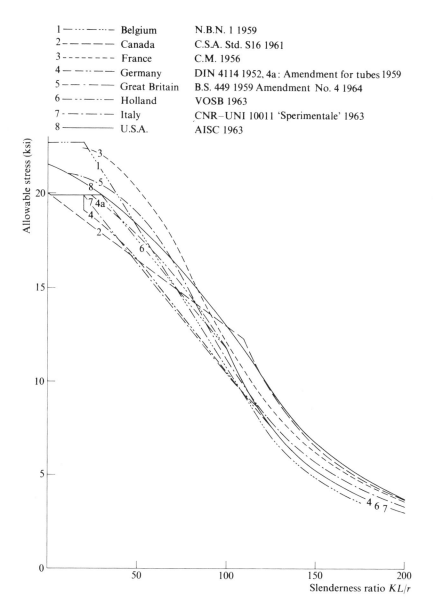

1 ——·——···—	Belgium	N.B.N. 1 1959
2 —— —— ——	Canada	C.S.A. Std. S16 1961
3 - - - - - - -	France	C.M. 1956
4 —— —·——	Germany	DIN 4114 1952, 4a : Amendment for tubes 1959
5 —— —·——	Great Britain	B.S. 449 1959 Amendment No. 4 1964
6 ——·——···—	Holland	VOSB 1963
7 -—·—·—	Italy	CNR–UNI 10011 'Sperimentale' 1963
8 ————————	U.S.A.	AISC 1963

Fig. 3-2.6 Allowable Stresses in Axially Loaded Structural Grade Steel Columns [3, page 36]

divided by the 23/12 factor of safety to yield code formula 1.5-2 as

$$F_a = \frac{\pi^2 E}{\text{F.S. } (KL/r)^2} = \frac{12\pi^2 E}{23(KL/r)^2} \qquad (3\text{-}2.10)$$

A comparison of design formulas is presented in Fig. 3-2.6 to indicate the range of variation among different groups of engineers.

Critical F_a is for the principal axis having the larger KL/r. Between principal axes, (1) K may differ because of differing end support conditions, (2) L can vary because of different points of lateral support, and (3) r_{xx} will seldom equal r_{yy}. Care must be exercised, then, to assure adequate stiffness about both axes.

3-3   ACTUAL, UNBRACED, AND EFFECTIVE LENGTH

Effective length is defined as KL, where for steel structures L is the laterally unbraced length of a member; for reinforced concrete structures L is the unbraced length, referred to in ACI 10.11.1 as ℓ_u and illustrated in Fig. 3-3.1. Sketches (a) and (b) show two views of the same column which has a different ℓ_u about each axis. For both steel and reinforced concrete structures, K is the effective length factor which can be obtained from monographs of Fig. 3-3.2.

To use these monographs, it is necessary to compute R_u (for the upper joint) and R_L (for the lower joint), where

$$R = \frac{\Sigma(E_c I_c)/L_c}{\Sigma(E_b I_b)/L_b} = \frac{\text{Sum of relative angular stiffness of columns at joint}}{\text{Sum of relative angular stiffness of beams at joint}}, \qquad (3\text{-}3.1)$$

and subscript c refers to columns, subscript b refers to beams, I is taken about the axis perpendicular to the plane of buckling, L is the unsupported length of member, Σ means either the column above and below the joint or the beams left and right of the joint, and R is figured for members in only one plane at a time.

Thus, for a column fixed at its lower joint, $R_L = 0$ (AISC recommends 1). For a hinged base, $R_L = \infty$ (AISC recommends 10). Unbraced frames sway laterally; braced frames do not.

The theoretical K values listed in Table 3-1.1 can be cross-checked by these nomographs:

Both ends hinged	Braced	$R_u = \infty$,	$R_L = \infty$	$K = 1$
Both ends fixed	Braced	$R_u = 0$,	$R_L = 0$	$K = 0.5$
Fix-Free	Unbraced	$R_u = 0$,	$R_L = \infty$	$K = 2.0$
Fix-Hinge	Braced	$R_u = 0$,	$R_L = \infty$	$K = 0.7$

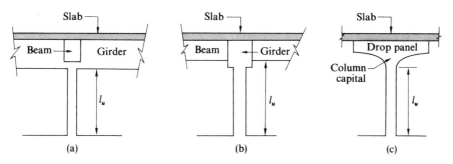

Fig. 3-3.1 Unbraced Length

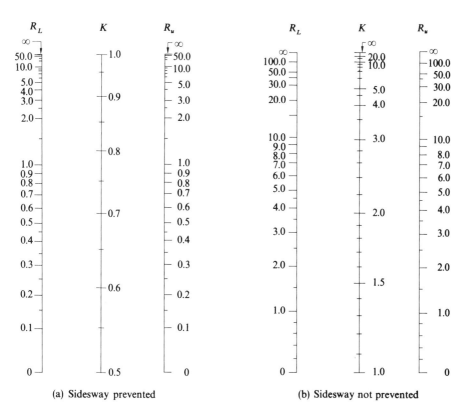

(a) Sidesway prevented   (b) Sidesway not prevented

Fig. 3-3.2 Alignment Charts for Effective Length of Column in Continuous Frames
[7, page 47]

The equations from which these nomographs were constructed are not derived in this text because required procedures are more akin to a course on frame analysis. An excellent, clearly written exposition can be found in the book by Salmon and Johnson. [8].

For preliminary design of reinforced concrete frames, Beck [9, page 10–8] suggests using $I_b/2$ and I_c, both figured using gross concrete dimensions. Use of $K = 1$ for braced frames (maximum) is conservative, but for unbraced frames $K = 1$ represents an unrealistically low minimum value.

3-4   STEEL COLUMNS

Both design and analysis of columns is greatly facilitated by tables provided in *Steel Construction*. Final analysis proof is obtained when actual stress, $f_a = P/A$, is less than allowable stress, F_a.

3-4.1 Allowable Stress Tables. To obtain a table of allowable stresses for columns of a specific steel, the value of C_c is first determined. For A36 steel,

$$C_c = \sqrt{2\pi^2(29,000)/36} = 126.1.$$

Then, F_a is evaluated by Eq. (3-2.9) for values of KL/r of 1, 2, 3, ..., 126 (that is, values less than C_c). For KL/r values greater than C_c (126.1 for A36), F_a is evaluated by Eq. (3-2.10) for 127, 128, ... to a maximum of 200 (AISC Specification 1.8.4). For bracing and secondary members, more or less fixed against rotation and translation at their braced points, and for which L/r exceeds 120 with K set at 1, values are increased by dividing by $1.6 - L/200r$. For A36 steel, these values are tabulated in Table 1-36 on page 5-84 of *Steel Construction*. Using this table, an A36 steel column with a value KL/r of 56.4 has an allowable stress of 17.77 ksi, obtained by interpolation between KL/r of 56 for which $F_a = 17.81$ and KL/r of 57 for which $F_a = 17.71$. (Care must be used in interpolation because F_a varies inversely as KL/r.)

3-4.2 Steel Column Analysis. Systematic analysis of a W14 x 370, A36 column is shown in Fig. 3-4.1. Total load is 1980 kips. The column is hinged so it can bend freely about the major axis and has no intermediate supports. With respect to the minor axis, however, ends of the column are fixed and intermediate support is provided at two points. About the weak axis, the upper 11 ft of the column may be considered fixed at one end and hinged at the other, with a theoretical K of 0.7, which is increased to 0.8 in accordance with Table 3-1.1 of this text and Table C1.8.1 of *Steel Construction*.

Under "Section Properties," the objective is to find the smallest allowable F_a. Thus KL is computed (L in feet) for all possible cases. Then KL/r (L multiplied by 12 to convert to inches) is computed for only the largest KL of

Load   1980^k Axial

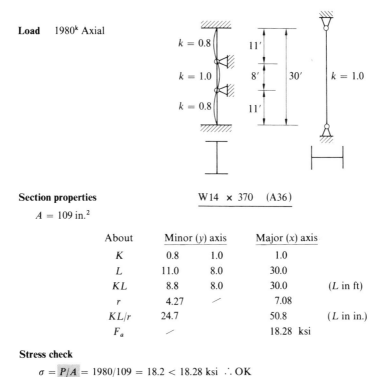

Section properties   W14 × 370   (A36)

$A = 109$ in.^{2}

About	Minor (y) axis		Major (x) axis	
K	0.8	1.0	1.0	
L	11.0	8.0	30.0	
KL	8.8	8.0	30.0	(L in ft)
r	4.27	/	7.08	
KL/r	24.7		50.8	(L in in.)
F_a	/		18.28 ksi	

Stress check

$\sigma = P/A = 1980/109 = 18.2 < 18.28$ ksi ∴ OK

Fig. 3-4.1   Steel Column Analysis

each axis. Finally, F_a is taken from Table 1-36 of *Steel Construction* for only the largest KL/r value. The latter step can be dangerous because it is easy to use a table for the wrong strength steel and obtain an incorrect value of F_a.

3-4.3 Steel Column Design.   For a given member, say the W14 × 370 section, considering only minor axis instability, KL could be incremented from 11 to 50. For each such KL, a corresponding KL/r and F_a could be obtained from which an allowable load could be computed. For instance, with $r_y = 4.27$, $A = 109$,

KL (L, ft)	KL/r_y (L, in.)	F_a (ksi)	P_{allow} (kips)
11	30.9	19.88	2167
12	33.7	19.68	2144
13	36.5	19.46	2121

Such values are tabulated for many sections starting on page 3-12 of *Steel Construction*. Important properties are also listed for each shape.

This tabulation is useful in designing an A36 column for an axial load of 1980 kips with column heights and intermediate supports as shown in Fig. 3-4.1. But tabular values are good for only minor axis instability, so critical minor axis KL is determined first.

Round KL of 8.8 to enter tables with KL of 9, and note that a W14 x 342 (page 3-13) is good for 2045 kips and a W14 x 320 for 1903 kips. The lightest adequate section is, therefore, a 14 WF 342.

Because $K_y L_y / r_y$ must be equal to or greater than $K_x L_x / r_x$, or else the major axis controls, we reorder and find that

$$K_y L_y (r_x / r_y) \geq K_x L_x. \qquad (3\text{-}4.1)$$

From the table (page 3-13 of *Steel Construction*) r_x / r_y is 1.65 for a W14 x 342. With $K_y L_y = 8.8$, a W14 x 342 is adequate only if $K_x L_x$ is equal to or less than $(1.65)(8.8) = 14.5$. But $K_x L_x = 30$ (see Fig. 3-4.1). Another trial must be made to satisfy major axis requirements.

Because values of r_x / r_y vary slowly from member to member, an equivalent minor axis KL can be obtained to replace the major axis KL by dividing $K_x L_x$ by r_x / r_y. Then,

$$\text{new } K_y L_y \text{ effective} = K_x L_x / (r_x / r_y), \qquad (3\text{-}4.2)$$

or

$$\text{new } K_y L_y \text{ effective} = 30/1.65 = 18.2.$$

Using this new KL of 18.2, enter the table again to find that a W14 x 370 is good for 1995 kips if $KL = 18$. In reality, this choice is for a fictitious minor axis effective length because the table is not developed for bending about the major axis. Final choice is a W14 x 370, already proved by analysis to be adequate (see Fig. 3-4.1).

To summarize the steps:

1. Compute critical KL minor and critical KL major.

2. Select trial column from table using KL minor.

3. Determine r_x / r_y for trial column and divide into KL major.

4. If the quotient is smaller than KL_{minor}, selection is adequate; if the quotient is larger than KL_{minor}, re-enter the table, use the quotient as the new equivalent KL minor, and determine the section actually required.

In practice, the loading shown in the column of Fig. 3-4.1 often is concurrent with bending moment. The analysis and design of such "beam-columns" is discussed in Chap. 10.

3-5   REINFORCED CONCRETE COLUMNS

As in design of reinforced concrete beams, columns are designed by either the strength method (USD) or working stress design (WSD). In reality, WSD equations for axially loaded columns are developed empirically from ultimate load tests and such members have equations which are, therefore, essentially identical. However, for more realistic cases that involve eccentricity of loading, only the USD equations provide a rational means for design (see Chap. 10).

3-5.1 Slenderness Considerations.   Slenderness, $K\ell_u/r$, defines the procedure to be used in designing reinforced concrete columns. Less than 10% [10] of the columns in a braced frame, and less than 60% in an unbraced frame, need to be designed for increased moments induced by axial force and eccentricity; therefore these columns are considered in Chap. 10. In this chapter, column slenderness is controlled by ACI 10.11.4, that is, for braced frames:

$$k\ell_u/r \leq 34 - 12\, M_1/M_2 \qquad (3\text{-}5.1)$$

where M_1 = the smaller end moment and M_2 = the large end moment using the beam sign convention (see sec. E-2.1). For single curvature, M_1/M_2 is positive. [*Note.* The symbols and sign convention are reversed for steel in Eq. (2-6.11).] Column slenderness, by ACI 10.11.4, for unbraced frames is

$$K\ell_u/r \leq 22. \qquad (3\text{-}5.2)$$

For calculating the radius of gyration $r = \sqrt{I/A}$, ACI 10.11.2 specifies $0.25h$ for circular members and $0.30h$ for rectangular members, the latter being slightly rounded.†

3-5.2 General Limitations and Requirements.   Although eliminated from ACI 10.8, certain limits presented for columns in sections 912 and 913 of the 1963 Code are paraphrased here so that the need for careful workmanship, and the increased significance of shrinkage, will be recognized if smaller sections are used. Present tense is used for ACI 10.8 and 10.9 requirements.

For rectangular tied columns, minimum dimension was 8 in. concurrent with minimum area of 96 in.², minimum bar size was #5, minimum number of bars is 4, and vertical reinforcement shall be between 0.01 and 0.08 times gross cross-sectional area.

For circular columns using spiral reinforcement, the minimum diameter was 10 in., minimum bar size was #5, minimum number of bars is 6, and

---

† $r = \sqrt{I/A} = \sqrt{(bh^3/12)/bh} = h\sqrt{1/12} = 0.289h = 0.3h.$

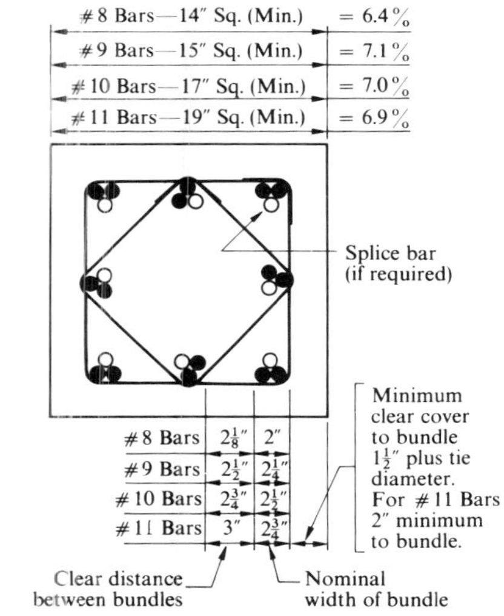

Bundled bars as column verticals*
Design and detail data

Bundle	Effective number of bars	Bar size	Total area (in.²)	Equiv. dia. (in.)	Effective perimeter of bundle (in.)			Minimum clear distance (in.)	
					At a splice bar		Without splice bar	Between bundles	Bundle to edge‡
					Splice bar	Remainder of bundle			
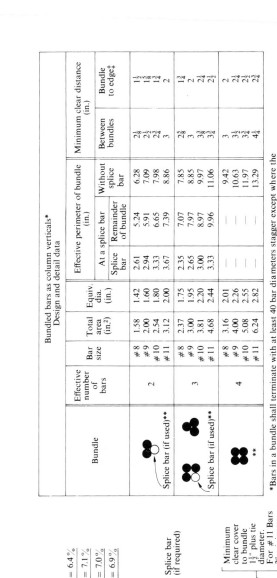 Splice bar (if used)**	2	#8	1.58	1.42	2.61	5.24	6.28	2⅛	1½
		#9	2.00	1.60	2.94	5.91	7.09	2½	1⅝
		#10	2.54	1.80	3.33	6.65	7.98	2¾	1¾
		#11	3.12	2.00	3.67	7.39	8.86	3	2
Splice bar (if used)**	3	#8	2.37	1.75	2.35	7.07	7.85	2⅝	1¾
		#9	3.00	1.95	2.65	7.97	8.85	3	2
		#10	3.81	2.20	3.00	8.97	9.97	3⅜	2¼
		#11	4.68	2.44	3.33	9.96	11.06	3¾	2½
**	4	#8	3.16	2.01	—	—	9.42	3	2
		#9	4.00	2.26	—	—	10.63	3½	2¼
		#10	5.08	2.55	—	—	11.97	3¾	2½
		#11	6.24	2.82	—	—	13.29	4¼	2¾

*Bars in a bundle shall terminate as shown by dashed lines when X distance is over 6 inches.
bundle terminates.

**Splice bars, welding, or positive connection must be provided for splices required' to carry
full tension or tension in excess of the capacity of the unspliced portion of the bundle.
Compression may be transmitted by end bearing of square-cut ends.

‡These minimum distances apply to bundles only. Where ties or spirals are present, the 1½ in.
minimum cover to them will control in some cases.

Notes: 1. These bars must be tied as shown by dashed lines when X distance is over 6 inches.
2. These bars need not be tied when X distance equals 6" or less.
3. Applicable to all tied columns.

A different pattern of ties may be substituted provided that details of the requirements are shown
on the contract drawings.

#8 Bars—14" Sq. (Min.) = 6.4%
#9 Bars—15" Sq. (Min.) = 7.1%
#10 Bars—17" Sq. (Min.) = 7.0%
#11 Bars—19" Sq. (Min.) = 6.9%

Splice bar (if required)

Minimum clear cover to bundle 1½" plus tie diameter.
For #11 Bars 2" minimum to bundle.

Nominal width of bundle

	Clear distance between bundles	Nominal width of bundle
#8 Bars	2½"	2"
#9 Bars	2½"	2¼"
#10 Bars	2¾"	2¼"
#11 Bars	3"	2¾"

Clear distance between bundles

Tied columns with 2-bar bundles

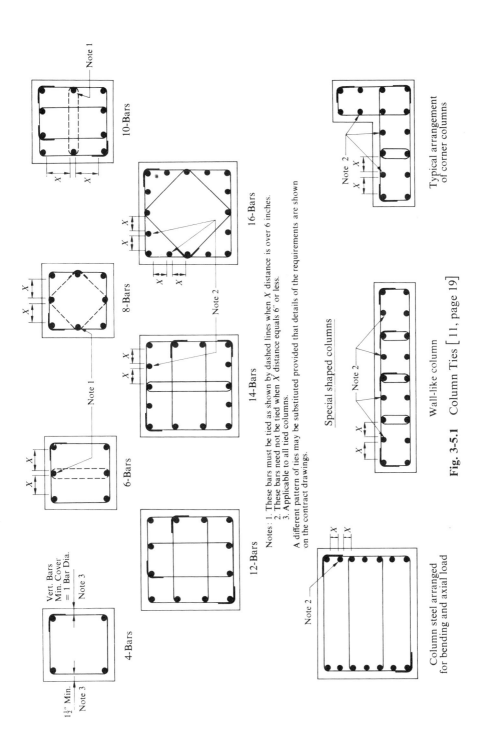

Notes : 1. These bars must be tied as shown by dashed lines when X distance is over 6 inches.
2. These bars need not be tied when X distance equals 6″ or less.
3. Applicable to all tied columns.

A different pattern of ties may be substituted provided that details of the requirements are shown on the contract drawings.

4-Bars

Vert. Bars
Min. Cover
= 1 Bar Dia.
Note 3

1½″ Min.
Note 3

6-Bars

Note 1

8-Bars

10-Bars

Note 1

12-Bars

14-Bars

Note 2

16-Bars

Note 2

Special shaped columns

Note 2

Wall-like column

Typical arrangement
of corner columns

Note 2

Column steel arranged
for bending and axial load

Note 2

Fig. 3-5.1 Column Ties [11, page 19]

vertical reinforcement shall be between 0.01 and 0.08 times the gross cross-sectional area.

Minimum 1% requirements are predicted upon maintaining reinforcement above yield stress even after shrinkage and creep have tended to transfer sustained load from concrete to reinforcement. The 8% upper bound is a practical limit based on placement requirements and overall economy.

Requirements for column ties (ACI 7.12.3) include:

1. Lateral ties must be (a) #3 for longitudinal bars < #10 in size, and (b)#4 for longitudinal bars > #10 or bundled bars.

2. Spacing of lateral ties must not exceed (a) 16 longitudinal bar diameters, (b) 48 tie bar diameters, or (c) the least dimension of the column.

Requirements for spiral reinforcement (ACI 7.12.2 and 10.9.2) specify that

$$p_s \leqq 0.45 \left(\frac{A_g}{A_c} - 1 \right) \frac{f_c'}{f_y}$$

where p_s = ratio of volume of spiral reinforcement to volume of core (diameter measured to outside of spirals), A_g = gross cross-sectional area (in.^{2}), A_c = area of core (in.^{2}) (same diameter as p_s), $f_y \leqq 60,000$ psi.

Figure 3-5.1 shows typical tie bar arrangements that satisfy the ACI Code.

3-5.3 The Basic Equations. The basis for computing concentric permissible loads on reinforced concrete columns is simply $\sigma = P/A$ or $P = \sigma A$. Because two materials are involved, typical equations tend to obscure this simplicity.

The fundamental equation, extensively verified by tests, is ACI-63 Eq. 19-7. (Because ACI-71 does not show equations, basic equations are keyed to ACI-63 for purposes of cross-referencing.) At ultimate strength,

$$P_o = \phi[0.85 f_c' (A_g - A_{st}) + A_{st}f_y], \tag{3-5.3}$$

where P_o = allowable ultimate concentric load (lb), ϕ = 0.75, the capacity reduction factor for spirally reinforced columns and ϕ = 0.70, the capacity reduction factor for tied columns, f_c' = ultimate compressive strength of concrete at 28 days (psi), A_g = gross area of column (in.^{2}), A_{st} = total area of steel reinforcing (in.^{2}), f_y = yield point of steel reinforcing (psi), and 0.85 is used with f_c' as in flexure (see sec. 2-5.2).

If the steel area is not deducted from the gross concrete area, that is, if the first A_{st} term is dropped, ϕ is discarded, and A_g is factored from Eq. (3-5.3), then

$$P = A_g \left(0.85 f_c' + \frac{A_{st}}{A_g}f_y \right). \tag{3-5.4}$$

Defining $p_g = A_{st}/A_g$ and adding a factor of safety, we see that Eq. (3-5.4) reduces to ACI-63 formula 14-1, at working stress:

$$P = A_g(0.25 f_c' + f_s p_g), \qquad (3\text{-}5.5)$$

where $f_s = 0.40 f_y \leqq 30{,}000$ psi and P is the allowable working load in pounds. Equation (3-5.5) is directly applicable to spirally reinforced columns, but only 85% of Eq. (3-5.5) is used to design tied columns.

Equations (3-5.3) and (3-5.5) theoretically are valid only when eccentricity is precisely zero. For WSD, because of inherent calculational difficulties, this assumption is often made. But for columns designed by USD, an eccentricity is always assumed as the maximum of 1 in. or $0.05h$ for spirally reinforced columns and as the maximum of 1 in. or $0.1h$ for tied columns. Ultimate strength design ACI-63 formula 19-10, for a rectangular tied column, is used in the next section:

$$P_u = \phi \left[\frac{A_s' f_y}{\dfrac{e}{d - d'} + 0.5} + \frac{bt f_c'}{(3\ te/d^2) + 1.18} \right], \qquad (3\text{-}5.6)$$

where $P_u \equiv$ allowable ultimate compressive load (lb); $A_s' \equiv$ area of *compressive* reinforcement (in.²); $b \equiv$ column width† (in.), parallel to bending axis; $t \equiv$ column depth (in.), parallel to plane of bending; $e \equiv$ eccentricity defined as greater than or equal to $0.1t$ or 1 in.; $d \equiv$ distance measured from extreme fiber in compression to centroid tensile steel (in.); and $d' =$ cover plus one-half bar diameter + tie diameter.

The simple $P = \sigma A$ relationship can be observed readily in Eq. (3-5.3), (3-5.4), and (3-5.5) if σA is replaced by concrete and steel components

$$P = \sigma_c A_c + \sigma_s A_s, \qquad (3\text{-}5.7)$$

where subscript c refers to concrete and subscript s to steel.

This relationship is valid also for Eq. (3-5.6), where the first term represents the steel-carrying capacity $P_s = \sigma_s A_s$ suitably modified, and the second term the concrete-carrying capacity $P_c = \sigma_c A_c$ [$bt = A_c$], also suitably modified.

This concept of concrete- and steel-carrying capacity, as exemplified by Eq. (3-5.7), simplifies proportioning of reinforced concrete columns when no design aids are used.

Chapter 10 treats the analysis and design of eccentrically loaded and slender columns with suitable derivations, including that for Eq. (3-5.6).

---

† Even when columns are designed for minimum eccentricity, it is often necessary to compute allowable P_u referred to each axis independently. In such case, dimensions b and t are interchanged between calculations.

3-5.4 Design Approach. Cost of an individual column is affected by the price of concrete per cubic yard, expense of steel reinforcing per pound, cost of formwork, and labor. But obtaining the most economical structure is not a matter of optimizing cost of individual columns, unless certain constraints are employed. Thus (1) material parameters are limited and specified (that is, f'_c and f_y), (2) shapes and dimensions are standardized and limited (circular columns normally are dimensioned to whole inches and rectangular columns to even inches), (3) steel percentage is established to limit column size and/or to allow for easier steel placement.

A particular design criteria might be $f'_c = 4000$ psi, $f_y = 60,000$ psi, restricted to rectangular columns with width b constant at 14 in., steel to between 2% and 3%, and all columns to be designed without being affected by slenderness ratios.

Steps in the design of an individual tied column in an unbraced frame in accordance with the criteria above are:

1. Determine design load
   $P_u = (1.4D + 1.7L) \div \phi$.
2. Determine R_L and R_u in terms of unknown t. For such a trial assume $I_c = $ gross I of column and $I_b = 1/2$ gross I of beams [8, page 10-8].
3. Assume trial t.
4. Compute actual R_L and R_u; obtain K from nomograph (Fig. 3-3.2).
5. Compute $k\ell_u/r$ and compare to 22, see Eq. (3-5.2).
   a) If $k\ell/r > 22$, increase t and return to step 4.
   b) If $k\ell_u/r \leqq 22$, proceed to next step.
6. Compute gross area.
7. Calculate amount of P_u which concrete can carry, assuming in second term of Eq. (3-5.6) that
   a) $e = 0.1t$ (but at least 1 in.) and
   b) $d = t - 2.5$ so
   c) $P_c = \dfrac{btf'_c}{(3te/d^2) + 1.18} = \dfrac{btf'_c}{0.3\left(\dfrac{t}{t - 2.5}\right)^2 + 1.18}$.
8. Compute amount steel must carry, $P_s = P_u - P_c$.
9. Calculate necessary steel by first term of Eq. (3-5.6):
$$P_s = \frac{A'_s f_y}{[e/(d - d')] + 0.5}$$
$$= \frac{A'_s f_y}{[(0.1t)/(t - 2.5 - 2.5)] + 0.5}$$

so

$$A_s = 2A_s' = \frac{2P_s}{f_y}\left(\frac{0.1t}{t-5} + 0.5\right).$$

10. Determine $p = A_s/A_g$.
    a) If less than 2%, reduce t and return to step 2.
    b) If greater than 3%, increase t and return to step 2.
    c) Otherwise, select ties.
11. Check by analysis.

The following example illustrates these steps using the previous design criteria and $\ell_u = 9$ ft, column fixed at base, $E_g I_g/L_g$ at top $= 36E$, $P_D = 340$ kips and $P_L = 210$ kips:

Design Load: $340 \times 1.4 = 476$ kips
$210 \times 1.7 = 357$
$\overline{833/0.7} = 1190$ kips.

Because column is fixed at base, $R_L = 0$. Beam stiffness is estimated at one-half its gross value; in this case, the actual value is known and used. I_c is computed for estimate as $I_{gross} = (14t^3)/12 = 1.167t^3$. Thus $\Sigma E_c I_c/L_c = E(1.167t^3)/(12)(9.5)$ and

$$R_u = \frac{E(1.167t^3)/(12)(9)}{36E} = 0.00030t^3.$$

The remaining steps are performed most efficiently in tabular form; see Table 3-5.1. For the first trial, t is assumed at 18 in., from which R_u is calculated at 1.75. Use the nomograph for unbraced frames of Fig. 3-4.2, with $R_u = 1.75$ and $R_L = 0$ and find that $k = 1.25$. The general equation of step 5 is

$$\frac{k\ell_u}{r} = \frac{k(12)(9)}{0.3t} = 360(k/t) = \frac{360(1.25)}{18} = 25.0 > 22.$$

Because the section is too slender, assume a second trial with $t = 20$ in. Again, the column appears to be too small, but calculations are continued based on the presumption that a more accurate I_c might show a $k\ell_u/r < 22$ and to get some idea of steel required. Gross area of step 6 is for use in computing p in step 10.

Step 7 is subdivided into three steps to simplify the calculations; 7(a) represents the denominator and 7(b) the numerator, their quotient being P_c. To simplify units, f_c' is represented in kips per square inch. With P_c determined as 712.5 kips, the load to be carried by the steel $P_s = P_u - P_c = 1190 - 712.5 = 477.5$ kips, from which the required area of steel is obtained as 10.08 in.^{2}. This trial is unsatisfactory because the criteria of $p < 0.3$ is exceeded.

Table 3-5.1 Column Design

$P_u = 1190^k$ $f'_c = 4$ ksi $f_y = 60$ ksi   ($b = 14$ in. constant)

Step	3	4	4	5	6	7(a)	7(b)	7(c)	8	9	10
Trial	Trial t (inches)	$R_u = 0.0003\,t^3$	k	$k\ell_u/r = 360\,k/t$	$A_g = 14\,t$	$0.3\left(\dfrac{t}{t-2.5}\right)^2 + 1.18$	btf'_c (kips)	P_c (kips)	$P_s = P_u - P_c$	$A_s = \dfrac{2P_s}{f_y}\left(\dfrac{0.1t}{t-5}+0.5\right)$	$P = A_s/A_g$
1	18	1.75	1.25	25.0							
2	20	2.40	1.32	23.8	280	1.572	1120	712.5	477.5	10.08	0.036
3	22	3.19	1.39	22.7	308	1.562	1232	788.7	401.3	8.42	0.027
4	24	4.15	1.47	22.0	336	1.554	1344	864.9	325.1	6.79	0.020

For the third trial, try $t = 22$ in., for which $k\ell_u/r = 22.7 > 22$ but $p = 0.027 < 0.03$. This probably is all right, so using equal reinforcement in each face, some possibilities are:

$$10 - \#9 = 10.00 \text{ in.}^2$$
$$8 - \#10 = 10.16 \text{ in.}^2$$
$$6 - \#11 = 9.36 \text{ in.}^2 \longleftarrow \text{USE.}$$

Then, tie size is #4 (main reinforcing is greater than #10) and spacing calculations are:

$$16 \times 11/8 = 22 \text{ in.}$$
$$48 \times 4/8 = 24 \text{ in.}$$
or
$$t = 22 \text{ in.,}$$

from which #4 ties are spaced at 22 in.

A fourth trial is shown as an alternate if the third trial is too slender when computed more accurately. In general, time spent in establishing a form for tabular solution is regained by more efficient and less error-prone calculations.

3-5.5 Analysis. An analysis suitable for permanent records is shown in Fig. 3-5.2. Effective length and end conditions are shown alongside the load caption. Section proportions are largely self-explanatory except that $E = 57,000\sqrt{f_c'} = 3,600,000$ psi, e of 2.2 is greater than 1 in. minimum so 2.2 is used, 8.31 on cross section is the distance from the centroid of the section to the centroid of the reinforcing bars, and middle bars do not need an extra tie because bars are spaced less than 6 in. apart (4.35 in.).

The slenderness check is performed using ACI formula 10-7,

$$EI = \frac{[(E_c I_g)/5] + E_s I_s}{1 + \beta_d} \tag{3-5.8}$$

rather than the more conservative ACI formula 10-8,

$$EI = \frac{[(E_c I_g)/2.5]}{1 + \beta_d}, \tag{3-5.9}$$

where $E_c \equiv$ modulus of elasticity of concrete, $I_g \equiv$ gross moment of inertia of concrete $[I_{c(\text{gross})}$ in Fig. 3-5.2$]$, $E_s \equiv$ modulus of elasticity of steel $= 29,000$ ksi, $I_s \equiv$ moment of inertia of steel about centroidal axis $= I_0 + A\bar{y}^2$, with I_0 neglected as insignificant $[(2)(4.68)(8.31)^2$ in Fig. 3-5.2$]$, and $\beta_d \equiv |$ratio: max DLM/max $TLM|$ (in this case directly proportional to loads). For simplicity of comparison and computation, E common to both girders and columns has been factored out. This implies that either (a) E is the same for

Loads $(l_u = 9.0'$, Bottom fixed, Top $\sum EgIg/Lg = 36E)$

DL $340^k \times 1.4 = 476$

LL $210^k \times 1.7 = \underline{357}$

$ 833^k$

Section Properties $f'_c = 4000$ psi $f_y = 60\,000$ psi $E = 3600$ ksi

$d' =$ cover $+ \frac{1}{2}$ bar diam $+$ stir diam

$d' = 1.5 + \frac{11}{16} + \frac{4}{8} = 2.69$ in.

$d = 22 - 2.69 = 19.31$ in.

$e = 0.1\,t = 2.2$ in.

$p_g = \dfrac{4.68 + 4.68}{(14)(22)} = 0.030$

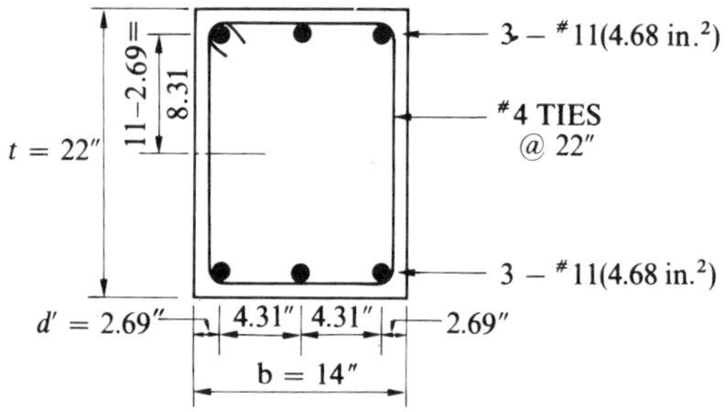

3 – #11(4.68 in.²)

#4 TIES @ 22"

3 – #11(4.68 in.²)

$t = 22"$

$11 - 2.69 =$

8.31

$d' = 2.69" \quad 4.31" \quad 4.31" \quad 2.69"$

$b = 14"$

Slenderness check

$I_{c(\text{gross})}/5 = 14(22)^3/(12)(5) \qquad = 2484$ in.⁴

$(E_s/E_c)I_s = \dfrac{29\,000}{3600}(2)(4.68)(8.31)^2 = \dfrac{5207}{7691\text{in.}^4} \qquad \left(\dfrac{bt^3}{12} = 12\,400 \text{ in.}^4\right)$

$1 + B_d = 1 + \dfrac{476}{833} = 1.571 \quad I_{\text{COL}} = \dfrac{7691}{1.571} = 4896$

$R_u = \dfrac{(E)4896/(12)(9)}{(E)36} = 1.26 \quad k = 1.19 \quad kl_u/r = \dfrac{1.19(12)(9)}{-0.3(22)} = 19.5 < 22$ OK

Check

$P_u \quad \dfrac{bt f'_c}{(3te/d^2) + 1.18} = \dfrac{(14)(22)(4)}{[3(22)(2.2)/19.31^2] + 1.18} = \dfrac{1232}{1.569} = 785^k$

$\dfrac{A'_s f_y}{\dfrac{e}{d-d'} + 0.5} = \dfrac{4.68(60)}{\dfrac{2.2}{19.31 - 2.69} + 0.5} = \dfrac{280.8}{0.632} = \dfrac{444}{1229} \times 0.7 = 860 > 833^k \therefore$ OK

Tie spacing

48 tie $\phi = 24"$

16 bar $\phi = 22"$ ◄── USE

$t = 22"$

Fig. 3-5.2   Column Analysis

girders and columns, or (b) E, although different, has been related to base E in $E_b I_b/L_b = 36E$. (*Note.* $I_b = $ moment of inertia of beam.)

Proof is secured when the allowable P_u exceeds the actual design load (including load factors). This is evaluated in a form suitable for slide-rule calculation. In offices, where it is more common to use electronic calculators with one or more memory locations, the check would appear as:

$$P_u = \left[\frac{\dfrac{14(22)4}{\dfrac{3(22)(2.2)}{(19.31)^2} + 1.18}} + \frac{(4.68)(60)}{\dfrac{2.2}{19.31 - 2.69} + 0.5} \right] 0.7 = 860^k > 833 \text{ OK.}$$

In practice, column design is faciliated by extensive design aids and/or computer programs. These are referred to in conjunction with the discussion of beam columns in Chap. 10. Timber column analysis and design is presented in Chap. 4.

PROBLEMS

3-1 given: Column_____ [(a), (b), or (c) of Fig. 3-P.1] with $H=$_____
(11, 14, or 16 ft), $a =$ _____ (0.2 or 0.3), and $b =$ _____ (0.3
or 0.4). Use _____ (A36 or A440 at $f_y = 50$ ksi).

design: Most economical W-shape steel column.

present: Complete analysis proof.

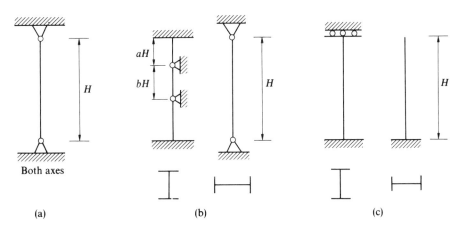

Fig. 3-P.1   Axial Loads on Steel Columns

3-2 given: $KL = 13$ ft; column type _____ (1, 2, 3, 4, 5, or 6) from the
list that follows.

Column type	Section	F_y	Area	r_y	r_x
1	Pipe 12 Std.	36	14.6	4.38	4.38
2	Pipe 8X-strong	36	12.8	2.88	2.88
3	Pipe 6XX-strong	36	15.6	2.06	2.06
4	TS 6 x 6 x $\frac{1}{2}$	36	10.1	2.19	2.19
5	TS 8 x 6 x $\frac{3}{8}$	36	9.33	2.34	2.92
6	TS 6 x 6 x $\frac{1}{2}$	46	10.1	2.19	2.19

find: F_x and P allowable.

3-3 given: $KL =$ _____ (11, 12.4, 15.2 ft), $P =$ _____ (170, 280, 340 kips), section is _____ (36 ksi pipe, 36 ksi tube, 46 ksi tube).

design: Most economical for type of section required. Use appropriate tables from *Steel Construction*.

present: Analysis proof.

3-4 given: $KL_{major} =$ _____ (10, 14, or 18 ft), $KL_{minor} =$ _____ (8, 10, or 12 ft), $F_y =$ _____ (36 or 50 ksi), Fig. 3-P.2, section _____ [(a), (b), or (c)] with _____ (C10 x 30, C12 x 30 or C15 x 33.9), $s =$ _____ (8 in., 9 in., or 10 in.) and $PL\frac{1}{2}x(s + 2b_f + 3)$ if used.

find: (a) A, I_x, I_y;
(b) r_x, r_y;
(c) F_a and P allowable.
(d) If lacing is prescribed, design according to AISC Specification 1-18.2.5, 6 and 7.

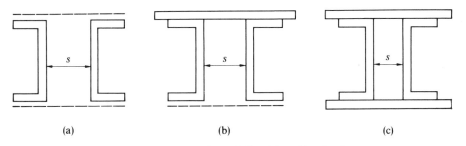

(a)   (b)   (c)

Fig. 3-P.2   Channels and Plate(s) and/or Lacing

3-5 given: $KL =$ _____ (12 or 16 ft); 2 L8 x 6 with short legs back-to-back but separated by $\frac{3}{8}$ in. (See Fig. 3-P.3.)
Thickness = _____ ($\frac{3}{4}$, $\frac{7}{8}$ or 1 in.); $F_y =$ _____ (36 or 50 ksi).

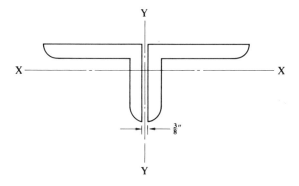

Fig. 3-P.3   Two Angles Back-To-Back with $\frac{3}{8}$ in. Clear Space

find: $F_{a(\text{minor})}$ and $F_{a(\text{major})}$, $P_{\text{allowable(minor)}}$ and $P_{\text{allowable(major)}}$.

3-6 given: $KL_x =$ _____ (0, 6, or 10 ft), $KL_y =$ _____ (0, 12, or 20 ft), $F_y =$ _____ (36 or 50 ksi), $P =$ _____ (60, 120, or 200 kips).

find: Suitable angles back-to-back using Double Angle Column Tables from *Steel Construction* (If no solution is found, satisfy design by increasing F_y; if still no solution, try another section.)

present: Analysis proof.

3-7 given: Frame of Fig. 3-P.4 with $L =$ _____ (40, 42, or 44 ft), $a =$ _____ (0.3, 0.4, or 0.6).

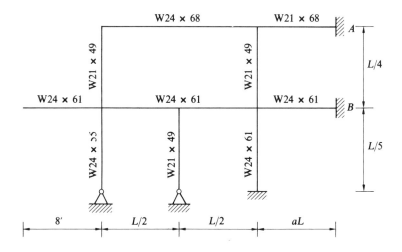

Fig. 3-P.4   Frame

find: *K* for each member using nomographs of Fig. 3-3.2 if all members are bending about their strong axes and sufficient bracing is provided perpendicular to the plane of the frame to preclude bucking about minor axes.

3-8 given: Same as in problem 3-7, except fixed ends at points *A* and *B* become free ends and *a* = _____ (0.1, 0.2, or 0.3).

find: Same as in problem 3-7.

3-9 given: *k* = _____ (1.1, 1.2, or 1.3), *L* = _____ (9, 10, or 11 ft), *P* = _____ (240, 300, or 360 kips), f_c' = _____ (3000, 4000, or 5000 psi), f_y = _____ (40, 50, or 60 ksi).

design: Square tied column having dimensions which are multiples of 2 in. Limit *p* to a minimum of _____ (2, $2\frac{1}{2}$, or 3%) and a maximum of _____ (3 or $3\frac{1}{2}$%).

present: Analysis proof.

3-10 given: Same as in problem 3-9.

design: Square column having longitudinal bars in a circular pattern. Limit *p* as in problem 3-9 and maintain dimensions which are multiples of 2 in.

present: Analysis proof.

3-11 given: Same as in problem 3-9.

design: Circular column with vertical bars in circular pattern. Limit *p* as in problem 3-9, but let diameter be any appropriate whole number (odd or even).

present: Analysis proof.

REFERENCES

1. Euler, L., *Sur la Force des Colonnes*, Mémoires de l'Académie de Berlin, V. 13, Berlin (1759).

2. Timoshenko, Stephen P. and James M. Gere, *Theory of Elastic Stability*, 2d ed., McGraw-Hill, New York (1961), Section 2.7.

3. Johnston, Bruce G., ed., *Guide to Design Criteria for Metal Compression Members*, 2d ed., Wiley (1966). (Third edition scheduled for 1973.)

4. Osgood, William R., "The Double-Modulus Theory of Column Action," *Civil Engineering*, V. 5, No. 3 (March 1935), pp. 173–175.

5. Shanley, F. R., "Inelastic Column Theory," *Journal of the Aeronautical Sciences*, V. 14, No. 5 (May 1947), pp. 261–268. ("Aerospace" has replaced "Aeronautical" in title of later publications.)

6. Bleich, Friedrich, *Buckling Strength of Metal Structures*, McGraw-Hill, New York (1952).

7. Johnston, Bruce G., ed., op. cit., p. 47.

8. Salmon, Charles G. and John E. Johnson, *Steel Structures—Design and Behavior*, Intext, Scranton, Pennsylvania (1971).

9. Beck. *Notes on ACI 318-71 Building Code Requirements with Design Applications*, Portland Cement Association, Skokie, Illinois (1972).

10. MacGregor, James G., James E. Breen and Edward O. Pfrang, "Design of Slender Concrete Columns," *Journal of the American Concrete Institute, Proceedings*, V. 67, No. 1 (January 1970).

11. *Manual of Standard Practice for Detailing Reinforced Concrete Structures*, ACI 315-65, American Concrete Institute (1965), p. 19.

CHAPTER 4

TOPICS PECULIAR TO TIMBER

National Design Specification for Stress-Grade Lumber and Its Fastenings (NDS), 1968 ed. [1], is the basic reference for this chapter. The several design aids used in this chapter are found in references 3 and 4.

Fundamental principles of analysis and design for wood are based on those for an ideally linearly elastic material. Thus bending stresses may be readily computed by the flexure formula, shearing stresses by $3V/2A$ (rectangular sections), and columns by a modified Euler formula or P/A. Timber also exhibits some peculiarities, however, which must be accounted for in the design procedure. This chapter, then, emphasizes these variations and illustrates their effect on calculations.

4-1 NOMENCLATURE AND ADJUSTMENTS

Lumber products are classified into three categories which depend upon dimensions and two additional divisions which depend upon fabrication:

1. Joist and Plank
   (J & P)

   $\geq$ 2 in. but < 5 in.

   $\geq$ 6 in.

   Suitable for bending about either axis.

2. Beam and Stringer
   (B & S)

   > 5 in. = t

   $\geq t + 2$ in.

   Suitable for bending about major axis only.

3. Post and Timber
   (P & T)

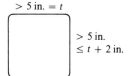

Suitable as columns
when bending is not
an important
consideration.

4. Laminated members are fabricated by doweling and gluing individual pieces of lumber into a single structural unit. In general, these members have higher allowable working stresses than either J & P or B & S of similar dimensions. Economy in manufacturing is obtained by minimizing waste and by using poorer grades of wood for web sections and better quality wood for flanges.

5. Plywood is a special type of laminated wood which has the configuration of a plate rather than a beam. Laminations alternate in direction, requiring that section properties be computed using only alternate layers rather than gross depth.

Structural cross sections termed *nominal*, are normally supplied in multiples of two inches. They are larger than the actual *dressed* dimensions, now called *surfaced sizes*. To determine surfaced sizes and section properties from nominal sizes, refer to reference 4.

In addition to gross classifications (J & P, B & S, P & T), lumber is graded according to its traits. Quality can be reduced by knots, wanes, splits, checks, shakes, and sloping grain. These typical imperfections are illustrated in Fig. 4-1.1. [2]

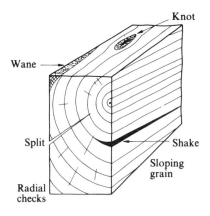

Fig. 4-1.1   Timber Faults

To find the allowable stresses, the nominal dimensions and the use of the wood must be known so classification can be made as to J & P, B & S, or P & T. Further, grades within classification must be selected from *select structural*, No. 1, No. 2, or No. 3. With this information, allowable stresses may be found for different representative types of wood in Tables 3, 4 and 5 of reference 3. In these tables, the modulus of elasticity and five allowable stresses are shown:

1. F_b allowable stresses for bending (previously called f),

2. F_t allowable tension parallel to the grain (previously t),

3. F_v allowable horizontal shear stress VQ/Ib or $3V/2A$ (previously called H),

4. $F_{c\perp}$ compression perpendicular to the grain applied to design of bearings and of joints (previously $c_\perp$),

5. F_c compression parallel to the grain applied to the design of columns and to joints (previously c),

6. E the modulus of elasticity.

In design, the species and commercial grade of wood to use depends largely on economics. Thus the use of larch in Georgia would be impractical because it is grown only in the far west. In addition to geography, local usage affects the best choice. If one is in an unfamiliar region, help in making the proper selection can be obtained by calling the area representatives of the National Forest Products Association. Once the choice of material is made and the allowable stresses determined, two possible adjustments in these stresses must be considered: moisture and duration of load.

Under continuously dry conditions, such as in most covered structures, the stresses may be used as tabulated. However, under *wet* conditions, moisture in wood at or above the *fiber saturation point* (assumed at 19%), reductions similar to the following are required:

F_c —reduce allowable tabulated stress by 10%;
$F_{c\perp}$—reduce allowable tabulated stress by one-third;
E —reduce tabulated value by 1/11.

Conditions of alternate wetting and drying should be avoided, if possible.

Tabulated values (not the modulus of elasticity) are altered also for members based on the duration of fully stressed condition:

Shortest duration in the combination of loads being considered	Factor by which allowable stresses are to be multiplied (or loads are to be divided if tabulated values are used without adjustment)
Permanent	0.90
Normal	1.00
Two months	1.15
Seven days	1.25
Wind or Seismic	1.33
Impact	2.00

For long-term loading, a permanent set, about equal to the original elastic deflection, will be obtained.

4-2 DESIGN OF BEAMS

Beams are readily designed by the flexure formula, with shear checked by $3V/2A$, if consideration is taken of adjustments to allowable stresses. However, two deviations may be important: (1) a form factor C_f used with bending calculations and (2) a modification for computing the actual shear force.

Appendix D of NDS specifies form factors used in the modified flexure formula,

$$F_b = \frac{Mc}{IC_f},$$

(4-2.1)

where $M \equiv$ allowable bending moment, $C_f \equiv$ form factor, $I \equiv$ moment of inertia, and $c \equiv$ distance from extreme fiber to neutral axis.

Equation (4-2.1) can be reordered (with I/c replaced by section modulus S) as:

$$F_b = \frac{M}{SC_f},$$

(4-2.2)

where for rectangular members whose depth $d \geq 14$ in. or so,

$$C_f \equiv 0.81 \left(\frac{d^2 + 143}{d^2 + 88} \right),$$

(4-2.3)

for circular section (same strength as square beam of same area)

$$C_f = 1.18,$$

(4-2.4)

for bending about diagonal of square section (same strength as if used with axes vertical and horizontal)

$$C_f \equiv 1.414,$$

(4-2.5)

and for I and box beams,

$$C_f = 0.81 \left[1 + \left(\frac{d^2 + 143}{d^2 + 88} - 1 \right) \left(p^2 (6 - 8p + 3p^2)(1 - q) + q \right) \right]. \quad (4\text{-}2.6)$$

Here, $p \equiv$ ratio depth of compression flanges to full depth of beam and $q \equiv$ ratio thickness of web (or webs) to full width of beam.

Equations are presented in NDS which allow for reduction in total shear. The reaction due to a uniform load is multiplied by $(1 - 2d/L)$ and each reaction due to a concentrated load is multiplied by $10(x/d)^2/9[2 + (x/d)^2]$, where x is the distance from the point of support to that particular concentrated load. When shear does not control member size, it is unnecessary to use these equations.

A typical analysis proof of a timber beam design is given in Table 4-2.1. In addition to the normal (of normal duration, that is, not less than two months or greater than several years) load of 0.84 k/ft, the beam is subject to four other loadings of various durations. Possible subtotals are obtained by accumulating all loads which *could* act concurrently during the minimum time duration, and the resulting sums are divided by factors that correspond to this minimum duration. The largest value (1.26 k/ft) is critical and is the *design load.* (It is assumed that seismic and impact loads would not occur simultaneously.)

Calculations for shear and bending moment are standard. Under Section properties, however, the shear reduction factor is computed as 5/6. This reduction makes unnecessary the use of a larger section as dictated by unmodified shear calculations.

Section properties and allowable stresses are read from appropriate tables of reference 4. Except for the form factor C_f used in the flexure formula and the shear reduction factor of 5/6 used in the maximum shear stress equation, the final check is straightforward.

4-3   BEARING

Bearing plate area is established by limiting P/A to an allowable $F_{c\perp}$ value. If a three-in. overhang is used beyond the bearing, and if the plate is less than 6 in. long, then

$$F_{c\perp_{\text{adj}}} = F_{c\perp_{\text{table}}} \left(\frac{\ell_b + 0.375}{\ell_b} \right), \quad (4\text{-}3.1)$$

where $\ell_b \equiv$ length of bearing plate. For round washers, bearing length is taken as the diameter of the washer.

A typical bearing plate area calculation is shown in Fig. 4-3.1 for the beam of Table 4-2.1. Computations are presented for both dry and saturated conditions to illustrate the marked effect of moisture content.

Table 4-2.1 Analysis Proof of a Timber Beam

Loads (20 ft Simple Span, Dry Condition)

| | | Normal | Combined Load Durations | | | |
			2 months	7 days	Seismic	Impact
	Normal	0.84	0.84	0.84	0.84	0.84
Individual	2 month		0.435	0.435	0.435	0.435
Load	7 day			0.29	0.29	0.29
Durations	Seişmic				0.105	
	Impact					0.84
	Total	0.84	1.275	1.565	1.67	2.405
	Coef	1.00	1.15	1.25	1.33	2.00
	Design	0.84	1.11	1.25	1.26	1.20

USE: 1.26 k/ft

V & BM

$$V = 1.26 \times 20/2 = 12.6^k$$
$$M = 1.26 \times 20^2/8 = 63.0^{\prime k}$$

Section Properties

(Douglas Fir South, Select Structural, B & S)

$$F = \boxed{0.81\left(\frac{d^2 + 143}{d^2 + 88}\right)} = 0.81\left(\frac{400 + 143}{400 + 88}\right) = 0.90 \qquad \underline{\mathbf{10'' \times 20''}}$$

$S = 602$ in.^{3} $F_b \equiv 1550$ psi
$A = 185$ in.^{2} $F_v \equiv 85$ psi
$I = 5870$ in.^{4} $E \equiv 1200$ ksi

Shear reduction factor:

$$\boxed{\left(1 - \frac{2d}{L}\right)} = 1 - \frac{2(20)}{12(20)} = \frac{5}{6} \qquad \Delta_{ALL} = \boxed{\frac{L}{360}} = \frac{12 \times 20}{360} = 0.67 \text{ in}$$

Check

$$\sigma_B = \boxed{\frac{M}{FS}} = \frac{12(63.0)}{0.90(602)} = 1.40 < 1.55 \text{ ksi} \therefore (OK)$$

$$\sigma_v = \boxed{\frac{3V}{2A}\left(\frac{5}{6}\right)} = \frac{3(12600)5}{2(185)6} = 85 \text{ psi} = 85 \therefore (OK)$$

$$\Delta = \boxed{\frac{5wL^4}{384EI}} = \frac{5(1.26)(20)^4 1728}{384(1200)(5870)} = 0.64 < 0.67 \text{ in.} \therefore (OK)$$

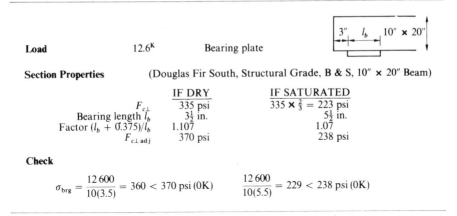

| Load | 12.6^K | Bearing plate | |

Section Properties   (Douglas Fir South, Structural Grade, B & S, 10″ × 20″ Beam)

	IF DRY	IF SATURATED
$F_{c\perp}$	335 psi	$335 \times \frac{2}{3} = 223$ psi
Bearing length l_b	$3\frac{1}{2}$ in.	$5\frac{1}{2}$ in.
Factor $(l_b + 0.375)/l_b$	1.107	1.07
$F_{c\perp \, adj}$	370 psi	238 psi

Check

$$\sigma_{brg} = \frac{12\,600}{10(3.5)} = 360 < 370 \text{ psi (OK)} \qquad \frac{12\,600}{10(5.5)} = 229 < 238 \text{ psi (OK)}$$

Fig. 4-3.1   Bearing Plate Area Calculation

4-4   COLUMNS

Strength and stiffness requirements for wood columns are specified in sec. 401 of NDS. Sufficient strength is assured of P/A is less than F_c. Sufficient stiffness is obtained if P/A is less than

$$F'_c = \frac{\pi^2 E}{2.727(L/r)^2} = \frac{3.619E}{(L/r)^2} \tag{4-4.1}$$

Of course, moisture and load duration adjustments must also be applied.

For rectangular columns, with $r^2 = I/A = (bd^3/12)/bd = d^2/12$, Eq. (4-4.1) reduces to

$$F'_c = \frac{0.3E}{(L/d)^2}. \tag{4-4.2}$$

To eliminate trial and error in the design procedure for square and rectangular wood columns, it is appropriate to introduce new terms and rearrange the equations. Therefore, define $L_b \equiv$ effective length of column for buckling about minor axis, using b as the depth for flexure; $L_d \equiv$ effective length of column for buckling about major axis, using d as the depth for flexure; $C_L \equiv L_b/L_d$ which is equivalent to b/d.† Equation (4-4.2) can then be written in terms of buckling about either axis as:

$$F'_c = \frac{P}{bd} = \frac{0.3E}{(L_b/b)^2} \qquad \text{or} \qquad \frac{P}{bd} = \frac{0.3E}{(L_d/d)^2}. \tag{4-4.3}$$

---

† From Eqs. (4-4.3) that follow, $P/0.3E = bd(b/L_b)^2 = bd(d/L_d)^2$, so with b replaced by $C_L d$, $(C_L d)/L_b = d/L_d$ or $C_L = L_b/L_d$.

Replacing b by $C_L d$ in both equations and solving for d yields [note that $L_b^2/C_L^3 = L_b^2/(L_b^2/L_d^2)C_L = L_d^2/C_L$]

$$d = \sqrt[4]{\frac{PL_d^2}{0.3EC_L}} \qquad (4\text{-}4.4)$$

The direct procedure is:

1. Compute L_b, L_d, and $C_L = L_b/L_d$ (note that for a square column with the same effective length about each axis, $C_L = 1$).
   Determine $d = \sqrt[4]{PL_d^2/0.3EC_L}$.
3. Determine $b = d(C_L)$.
4. Select size [4].
5. Check $P/bd < F_c$? If so, the design is complete. If not, obtain appropriate area, keeping the ratio of b to d close to C_L.

Design Example. Assume that for dry conditions a 20 ft column is to be designed where the axial load is 145 kips and intermediate lateral support is provided at mid-height in one plane only. Use Douglas fir, select structural, P & T, with $F_c = 1150$ psi and $E = 1600$ ksi. Then:

1. $L_d = 20$ ft x $12 = 240$ in.; $L_b = \frac{1}{2}$ x 20 ft x 12 in. $= 120$ in. (lateral support mid-height).
2. $C_L = L_b/L_d = 0.5$; $d = \sqrt[4]{PL_d^2/0.3EC_L} = \sqrt[4]{145(240)^2/0.3(1600)(0.5)} = 13.7$ in.
3. $b = dC_L = 13.7(0.5) = 6.9$ in.
4. Try [4] 8 x 14 with $A = 101.25$ in².
5. Check $P/A = 145/101.25 = 1.432$ ksi > 1.150 (NG); $P/F_c = 145/1.15 = 126$ in.²; select 10 in. x 14 in. from reference 4.

An analysis of this choice is given in Table 4-4.1. Notice, that L/r is computed about each axis, but only the larger value is used to compute F_c' of 1675 psi. Because F_c is less than F_c' the lower value of 1150 psi is the controlling allowable stress.

For a circular column, the design formula to be used in lieu of Eq. (4-4.4) is

$$d = \sqrt[4]{\frac{5.63\,PL^2}{E}}. \qquad (4\text{-}4.5)$$

Table 4-4.1 Timber Column Design

Load (Intermediate Support for $10''$ $d \Rightarrow L = 10'$; No Support for $14''$ $d \Rightarrow L = 20'$.)
    145^k Axial.

Section Properties

(Douglas Fir Select Structural, P & T)

$$I_x = \frac{9\frac{1}{2} \times 13\frac{1}{2}^{3}}{12} = 1948 \text{ in.}^4 \qquad\qquad \underline{10'' \times 14''} \; A = 128.25 \text{ in.}^2$$
$$E = 1760 \text{ ksi}$$

$$I_y = \frac{13\frac{1}{2} \times 9\frac{1}{2}}{12} = 965 \text{ in.}^4$$

$$r_x = \sqrt{1948/128.3} = 3.89 \qquad\qquad r_y = \sqrt{965/128.3} = 2.75$$

$$L_x/r_x = \frac{12 \times 20}{3.89} = 61.7 \text{ critical} \qquad L_y/r_y = \frac{12 \times 10}{2.75} = 43.7$$

$$F'_c = \boxed{\frac{3.619\,E}{(L/r)^2}} = \frac{3.619(1760)}{61.7^2} = 1.675 \text{ ksi}$$

$$F_c = 1.150 \text{ ksi} \longleftarrow \text{critical}$$

Check

$$\sigma = \boxed{P/A} = 145/128.3 = 1.13 \text{ ksi} < 1.15 \;\therefore\; \text{OK}$$

4-5   COMPRESSION AT ANGLE TO GRAIN

Figure 4-5.1 illustrates a connection between two members causing compression at an angle to the grain. In such a case, failure may occur because of (1) compression, parallel to grain, in member A; (2) compression, at an angle to the grain, in member B; (3) shear along distance a, and (4) shear along distance b (unlikely). All are amenable to standard techniques except compression at an angle to the grain for which Hankinson's formula is used:

$$F_n = \frac{F_c F_{c\perp}}{F_c \sin^2 \theta + F_{c\perp} \cos^2 \theta}, \qquad (4\text{-}5.1)$$

where F_n = allowable compressive stress acting perpendicular to inclined surface, and θ = the angle between the direction of load and the direction of the grain.

Assume the members in Fig. 4-5.1 are of western cedar, No. 2, J & P, and that the inclined member A is a 2 x 6 (area = 8.25 in.^{2} from reference 4). Then with $F_{c\perp} = 295$ psi and $F_c = 825$ psi [3, Table 5], the allowable stress in horizontal member B, with $\theta = 30°$, is

$$F_n = \frac{(295)(825)}{825(\sin 30)^2 + 295(\cos 30)^2} = 569 \text{ psi.}$$

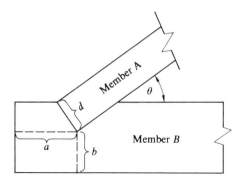

Fig. 4-5.1   Compression at Angle to Grain

This limits the maximum force in inclined member A, as controlled by compression acting at a 30° angle to the grain in member B, to $(569)(8.25) =$ 4690 lb.

Because F_c is greater than 569 psi, inclined member A would not be overstressed in compression parallel to the grain. However, because the allowable shear stress is only 75 psi, dimensions a and b would be critical.

Connections are further treated in Chap. 5.

PROBLEMS

4-1 given: Timber beam of Fig. 4-P.1 with $L =$ _____ (26, 30, or 34 ft), $a =$ _____ (6, 8, or 10 in.), $b =$ _____ (16, 18, or 20 in.). Material _____ (1, 2, 3, or 4 of Fig. 4-P.1). Deflection is limited to $L/360$. Normal climatic and loading conditions.

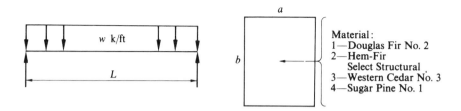

Fig. 4-P.1   Timber Beam

find: Uniform load w that the beam can carry, including itself.

4-2 given: Same as in problem 4-1.

find: Same as in problem 4-1 except the beam is continuously submerged in water.

4-3 given: Beam _____ [(a), (b), (c), (d), (e), or (f) of Fig. 4-P.2] , where
$L =$ _____ (18, 20, or 22 ft), $a =$ _____ (0.1, 0.15, or 0.2),
$b =$ _____ (0.25, 0.3, or 0.35), $c =$ _____ (0.25, 0.3, or 0.4),
$w =$ _____ (0.6, 0.7, or 0.8 k/ft); 40%—normal, 30%—two
months, balance—seven days. $P =$ _____ (6, 7, or 8 kips);
80%—normal, 20%—impact.

select: A timber species.

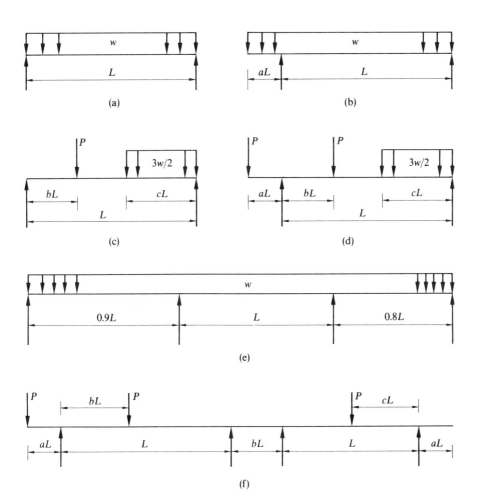

(a)

(b)

(c)

(d)

(e)

(f)

Fig. 4-P.2   Beam Configurations

> *design*: Most economical beam of constant cross section. Limit deflection to $L/360$. Normal climate.

> *present*: Analysis proof.

> **4-4**: Identical to problem 4-3 except beams are continuously submerged in water.

4-5 given: A column with an effective length of _____ (10, 11, or 12 ft) and a design load of _____ (90, 100, or 110 kips).

> *select*: A timber species.

> *design*: Economical column.

> *present*: Analysis proof.

4-6 given: The configuration of Fig. 4-P.3 with $g =$ _____ (6, 8, or 10 in.) and $\theta =$ _____ (40, 45, or 50 degrees).

> *select*: Timber species.

> *find*: (a) Maximum allowable load per inch of width, and (b) Minimum h corresponding to that maximum.

4-7 given: The configuration of Fig. 4-P.4 where beam width is _____ (6, 8, or 10 in.). Reaction is _____ (12, 14, or 16 kips).

> *find*: Reasonable bearing plate length ℓ_b.

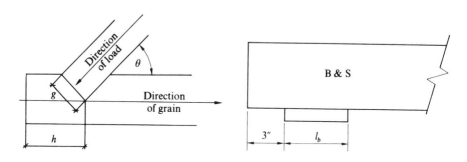

Fig. 4-P.3   Bearing at an Angle   **Fig. 4-P.4**   Bearing Plate

REFERENCES

1. *National Design Specification for Stress-Grade Lumber and its Fastenings*, National Forest Products Association, Washington, D.C. (1968).

2. Adopted from definitions in *Standard Grading Rules for Western Lumber*,

Western Wood Products Association, Yeon Buildings, Portland, Oregon 97204 (1970).

3. *Product Use Manual*, Catalog A 6.1/We, Western Wood Products Association, Yeon Building, Portland, Oregon 97204.

4. *Form TG-2*, Western Wood Products Association, Yeon Building, Portland, Oregon, 97204.

CHAPTER 5

CONNECTORS, CONNECTIONS, AND DETAILS

Fascination for design of main structural members often leads to disdain for details and connections. Reflection by an observant student quickly can change this viewpoint because it is evident that no structure can be better than its connections. In practice this conclusion is verified, it not being unusual for a preponderance of office time to be spent in seeking creative solutions to the design of complex connections. Space in this chapter is too limited to dwell upon such unique situations, but mastery of the principles developed herein will provide an ample basis for such solutions.

PART A
CONNECTORS

Before the study of connections is undertaken, it is important for one to become familiar with typical types of commercially available connectors, all of which can be classified broadly as either *discrete* or *continuous*. This distinction is very important because design techniques are categorized similarly.

A *discrete connector* is one which has its *holding action* concentrated at a single *point*. Nails, bolts, and rivets are examples of discrete connectors. Glue and welding are examples of *continuous connectors* which have their holding action distributed over some finite area.

5-1 DISCRETE CONNECTORS

Discrete connectors, as discussed in this chapter, apply directly to steel, timber, and reinforced concrete structures. However, extension to other materials is parallel.

5-1.1 Discrete Connectors for Steel Members. Three types of discrete steel connectors predominate: erection bolts, rivets, and high-strength bolts. Before the advent of a high-strength bolts and welds, rivets were the only generally accepted connectors, with erection bolts required in conjunction with fabrication. Today, rivets are being used less and less, with erection bolts used primarily as temporary fasteners prior to completion of welding.

Erection bolts, also called "fitting-up" bolts, are made from low carbon steel and have a semi-cone point to facilitate penetration. Bolt heads are square and the threads are coarse and free-running. A typical erection bolt is pictured in Fig. 5-1.1.

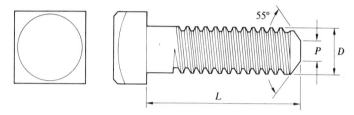

Fig. 5-1.1   Erection Bolt [1, page 12]

Rivets have a long history of success; however, they are used less because for many applications they no longer make the most economical connection. Steel rivets come in standard diameters from $\frac{1}{2}$ in. to $1\frac{1}{4}$ in., varying by $\frac{1}{8}$-in. increments. Several different head types are available as may be seen in Fig. 5-1.2.

Field riveting has always been a colorful sight. At a base location, one man heats the rivets till they glow red hot. Then he extracts a rivet from the heat source by a pair of tongs and "pitches" it through the air to his "catcher." The catcher receives the red-hot rivet with a metal, cone-shaped device and places it into a waiting hole. On that same side of the connection another man, called the bucker, "bucks" the head of the rivet against thrust from a rivet gun applied by a fourth man on the far side of the connection.

As the rivet is hammered on one side and bucked on the other, it deforms (flows) so that it completely fills the hole at the same time a second head is being formed. As it cools, it contracts and develops residual tensile stresses which are normally not an important design parameter. Because a good rivet fills its hole completely and is in tension, it is readily checked by a hammer tap. Any loose-sounding rivets are pulled and replaced. Ease in checking rivets and their concurrent implied reliability allowed rivets to hold a comparatively favorable position over welding for many years.

A later innovation in erection of structural steel is the *high-strength bolt.* Two types are in common use: high-strength structural bolts and structural rib bolts.

Button (round) head

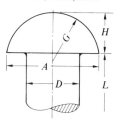

Dimensions
$A = 1.75D$ $H = 0.75D$
$G = 0.885D$

High button (acorn) head

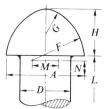

Dimensions
$A = 1.50D + \frac{1}{32}$ $G = 0.75D - \frac{9}{32}$
$H = 0.75D + \frac{1}{8}$ $F = 0.75D + \frac{9}{32}$
$M = 0.50$ in. $N = 0.094$ in.

Countersunk head

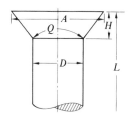

Dimensions
$A = 1.81D$ $H = 0.50D$
$Q \doteq 78$ deg.

Cone head

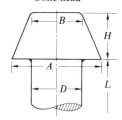

Dimensions
$A = 1.75D$ $B = 0.938D$
$H = 0.875D$

Round-top countersunk head

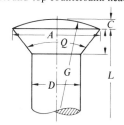

Dimensions
$A = 1.81D$ $H = 0.50D$
$C = 0.19D$ $G = 2.25D$
$Q \doteq 78$ deg.

Pan head

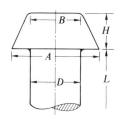

Dimensions
$A = 1.60D$ $B = D$
$H = 0.70D$

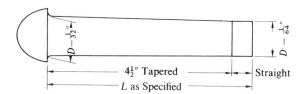

Fig. 5-1.2   Rivets $[1,$ page $11]$

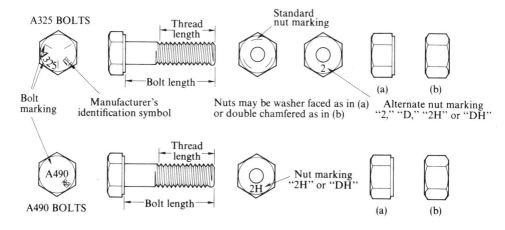

Fig. 5-1.3   High-Strength Bolts [1, page 4]

High-strength structural bolts (Fig. 5-1.3) obtain their holding power by inducing high friction in the joined plates. That is, plates are held very tightly together by tension applied from the nuts tightening on the bolt. For safety, bolts are individually marked. In inspection, an early difficulty was encountered in keeping bolts from being overtightened. Presently several reliable tightening techniques, including the use of a calibrated wrench which slips at a specified value, are employed.

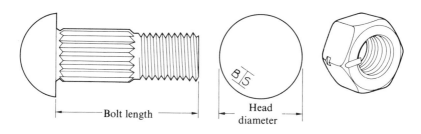

Fig. 5-1.4   Rib Bolt [1, page 8]

The structural rib bolt (Fig. 5-1.4) is very similar to the high-strength structural bolt. However, it has tapered ribs which deform as it is driven into standard drilled or punched holes so that it makes a body-tight fit, holding the bolt firmly in position. Such bolts are termed *bearing type*, and the threads must therefore be excluded from bearing against connected plates.

5-1.2 Discrete Connectors for Timber Members. For lighter construction, timber is connected by nails, spikes, wood screws, and lag screws. For heavier construction, it is common practice to use bolts or timber connectors.

Two common types of *timber connectors* are pictured in Fig. 5-1.5. A toothed-ring connector has sharpened teeth which cut into the wood as the bolts are tightened. The smooth ring is used by preparing a groove in each face of two members to be joined. The groove must be deep enough to seat the ring completely. Shear is transferred from one member to another through the steel ring, not by the bolt alone. Rings provide a much larger bearing area and consequently are much more efficient connections.

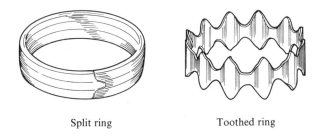

Split ring   Toothed ring

Fig. 5-1.5 Timber Connectors (Courtesy of Timber Engineering Company, Washington, D.C.)

5-1.3 Discrete Connections for Concrete. Hooking and bending reinforcing bars (ACI 12.8) provide discrete connections for tension. Anchor bolts are used for similar purposes.

The distinction between discrete and continuous is not clearcut, however, because of bonding between steel and concrete. In ACI-63, 918 (h), single discrete values were given for hooks, but ACI-71 uses such values in conjunction with embedment length, as discussed in sec. 5-12.

5-1.4 Strength of Discrete Connectors. Most discrete connectors transmit loads by bearing and shear (sometimes tension). Permissible capacities are listed in *Steel Construction* for rivets and high-strength bolts. For timber, the *National Design Specifications* provide the following tables: 4a, 4b, 4c for timber connectors, 6, 7, 8, 9, 10 for timber connectors with lag screws or bolts, 12 for bolts, 13, 14, 15 for lag screws, 16, 17 for nails and spikes, and 18 and 19 for wood screws.

Shear capacity of a solid discrete connector is its allowable shear stress multiplied by its own cross-sectional area.

Bearing capacity of a discrete connector is the minimum product of bearing area times (1) allowable bearing stress of members being joined or (2) allowable bearing stress of discrete connector. Bearing area is computed as connector diameter times distance connector penetrates into joined materials.

A distinction (no longer made by AISC) is sometimes made between *single-shear* bearing and *double-shear* bearing (Fig. 5-1.6). If the plates in tension are lapped over each other and stitched together by discrete connectors, allowable bearing is classified as single shear and allowable stress is reduced. Such a reduction is made because bending produced by inherent eccentricity causes bearing stress to "peak" near the junction of the two plates.

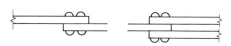

Fig. 5-1.6   Single and Double Shear

If three plates in tension are lapped (two outside plates extend in one direction with the inside plate extending in the opposite direction), bearing is more nearly uniform over the middle plate and a higher allowable bearing stress may be assumed for it than for the outside plates. (This distinction is no longer made in most steel design.) The connectors themselves are still in double shear, however, and therefore are twice as effective as those spanning only one shear plane.

Typically, allowable loads for discrete connectors are made in terms of *kips per connector*: k/R, where k means kips and R means connector. (R is used in honor of historically important rivets.) In the following sections, this unit is used extensively.

5-1.5 Net Section, Steel.   In the recent past, holes in steel were often obtained by punching. More precise holes were either drilled or subpunched (punched smaller than desired) and reamed to precise sizes. These punched holes were typically made $\frac{1}{16}$ in. greater in diameter than their connectors. However, an additional $\frac{1}{16}$ in. of material was considered to be damaged by the punching process. Thus, a hole was typically considered to be $\frac{1}{8}$ in. greater in diameter than its connector. As an example, if three $\frac{7}{8}$in. connectors (side by side) were used in a 10-in. wide plate, the effective plate width left to carry tension was $10 - 3(\frac{7}{8} + \frac{1}{8}) = 7$ in.

This is altered somewhat by AISC provisions 1.23.4 and Supplement No. 2 which replaces 1.14.5. Under these criteria all holes, in plates greater than the maximum of (a) $\frac{1}{8}$ in. or (b) the nominal diameter of the connection, must be subpunched and reamed or drilled. Thus in computing net section, $\frac{1}{16}$ in. is added to *hole* diameter. Depending upon local practice, this may or may not be $\frac{1}{8}$ in. greater than the nominal diameter of the connector.

Many times connectors are staggered rather than used directly in line. In such applications total width is taken as the minimum of either (1) 85% of gross plate width or (2) gross width of plate less summation of hole diameters in any chain + $s^2/4g$, where $s \equiv$ pitch (longitudinal spacing) and $g \equiv$ gage (transverse spacing), both in inches, and calculated for the same two consecutive holes. A curve to aid in these calculations is available on page 4-87 of *Steel Construction*. An example problem is shown in Fig. 5-1.7. Five holes exist in the plate. Possible paths are charted in tabular form composed of (1) gross width, (2) summation of hole diameters, and (3) $s^2/4g$ for each diagonal distance between holes. On path A–D–C, between holes A and D, pitch is 2 in., gage is 2 in. and $s^2/4g = 2^2/4(2) = 0.5$. Between holes D and C, pitch is also 2 in. but gage is 6 in. so $s^2/4g = 2^2/4(6) = 0.17$. For this path, therefore, effective net width is gross $w - \sum$ holes $+ \sum s^2/4g = 12.00 - 3(1.0) + [0.5 + 0.17] = 9.67$ in.

Final critical net width is the smallest value determined for any path, or 85% of gross width if the latter is smaller.

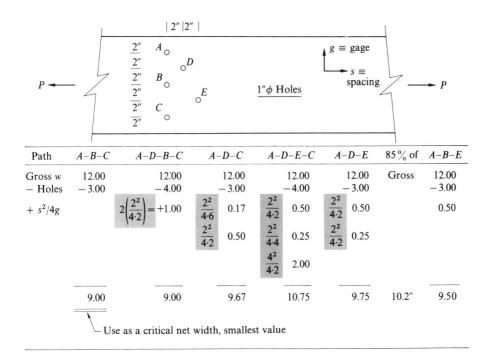

Path	A–B–C	A–D–B–C	A–D–C		A–D–E–C		A–D–E		85% of	A–B–E
Gross w	12.00	12.00	12.00		12.00		12.00		Gross	12.00
$-$ Holes	-3.00	-4.00	-3.00		-4.00		-3.00			-3.00
$+ s^2/4g$		$2\left(\dfrac{2^2}{4 \cdot 2}\right)=+1.00$	$\dfrac{2^2}{4 \cdot 6}$	0.17	$\dfrac{2^2}{4 \cdot 2}$	0.50	$\dfrac{2^2}{4 \cdot 2}$	0.50		0.50
			$\dfrac{2^2}{4 \cdot 2}$	0.50	$\dfrac{2^2}{4 \cdot 4}$	0.25	$\dfrac{2^2}{4 \cdot 2}$	0.25		
					$\dfrac{4^2}{4 \cdot 2}$	2.00				
	9.00	9.00	9.67		10.75		9.75		10.2″	9.50

Use as a critical net width, smallest value

Fig. 5-1.7   Net Section

5-2   CONTINUOUS CONNECTORS

Continuous connectors include welding (many metals), glue (timber and some metals), epoxy (special "glue" for use with concrete), and bond between concrete and reinforcing bars.

5-2.1 Welding of Steel. Many methods of welding are employed in current technology; they include diffusion, explosive, electroslag, oxyacetylene welding, and a host of others. For this discussion, however, welding refers to electric-arc welding because it is commonly used for most structural work. Electric-arc welds are of three primary types: (1) *Full-strength* (full penetration groove welds), (2) *fillet*, and (3) *auxiliary*. Full-strength welds (assuming full penetration) require no design calculations because they obtain the same strength as the parent material. Fillet welds are commonly used structural welds which transmit loads by shear through the effective throat. Auxiliary welds occasionally require design calculations but are used more often for tacking sections together.

Larger welds cannot be made completely at one time, so weld metal is built up by depositing it in several passes. Typical full-strength welds are shown in Fig. 5-2.1. The relative position of joined plates distinguishes (a)-(d) as tee joints and (e)-(k) as butt joints. The first part of each designation results from the shaping of the plates; square means no plate shaping, bevel means a straight taper, and J a curved taper. For each type, the several weld "passes" are indicated by small circles on the leftmost sketch which depicts how the plates actually are shaped prior to welding. Sketches to the right show how welds appear and should be called out on plans.

Note in Fig. 5-2.1 the following characteristics:

1. A weld arrow always has a sharp break to distinguish it from other lines on the drawing.

2. The arrowhead points toward the piece which is shaped.

3. The symbols *always* flag right from the vertical stem, whether left or right of the arrowhead.

4. A symbol above the line means *far side*; contrastingly, a symbol below the line indicates *near side*.

5. Full-strength welding is not allowed if there is no root opening (space between pieces being joined). See AISC Specification 1.14.7.

6. A solid black circle at the sharp break nearest an arrowhead signifies *weld in the field*.

7. An open circle at the sharp break nearest an arrowhead signifies *weld all around*.

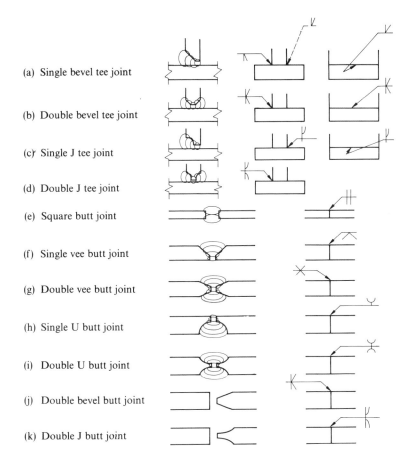

(a) Single bevel tee joint

(b) Double bevel tee joint

(c) Single J tee joint

(d) Double J tee joint

(e) Square butt joint

(f) Single vee butt joint

(g) Double vee butt joint

(h) Single U butt joint

(i) Double U butt joint

(j) Double bevel butt joint

(k) Double J butt joint

Fig. 5-2.1   Full-Strength Welds

8. The latter two symbols can be combined so that

signifies *weld all around, in the field.*

These characteristics also apply to fillet and auxiliary weld symbols. Figure 5-2.2 indicates various fillet welds. The strength of fillet welds depends upon the effective throat area, measured perpendicular to the weld face, as shown in Fig. 5-2.3. Thus, effective area is the product of effective throat area times effective length.

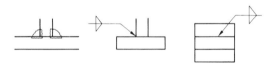

Fig. 5-2.2   Fillet Welds

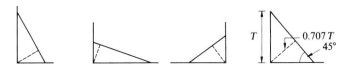

Fig. 5-2.3-   Effective Throat

Involved calculations are seldom required because of the following rule: *An E70XX electrode is good for 928 pounds per linear inch for each one-sixteenth inch of fillet weld size (assuming the ordinary isosceles right triangular weld).* It is easy to verify the rule because the allowable stress for the E70XX electrode is 21 ksi. Effective throat is 0.707 times weld size, so

$$0.707 \times \tfrac{1}{16} \times 21{,}000 = 928 \text{ lb/in. per } \tfrac{1}{16} \text{ in. of fillet weld size.} \quad (5\text{-}2.1)$$
$$(800 \text{ for E60XX.})$$

This is true only if the base metal matches the weld material; for more detail, see AISC Table 1.5.3.

In joining two plates, a requirement is that the maximum size of fillet welds must be limited to plate thickness for plates less than one-quarter inch thick. For plates one-quarter inch thick or more, fillet welds must be one-sixteenth inch less in size than the corresponding plate thickness.

Two common auxiliary welds are *plug* (in round holes) and *slot*. A typical plug weld and its symbol are shown in Fig. 5-2.4. The bottom of the plug has a diameter of one inch. The plug welds are placed on four-inch centers. The plug itself slopes at 60°/2 and weld metal fills only the bottom three-eighths inch of the hole. This, of course, may be read directly from the symbol.

To indicate more clearly the meaning of near side and far side, Fig. 5-2.5 shows symbolic notation and a schematic with the welds·hash-marked.

Figure 5-2.6 indicates an intermittent fillet weld. The notation $\tfrac{1}{4}$ indicates that one-quarter inch fillet welds are to be used. The numerical code 2–8 signifies each weld is to be two inches long and that welds are to be placed on eight-inch centers. The fillet weld symbols are staggered to indicate that the two-inch weld increments are to be staggered in spacing between the near and the far side.

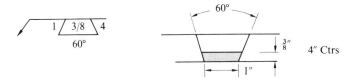

Fig. 5-2.4   Plug Weld

Fig. 5-2.5   Near–Far Side Welding

Fig. 5-2.6   Staggered Intermittent Weld

Figure 5-2.7 shows a dimensioned butt weld. The notation $\frac{1}{8}$ signifies the depth of weld is to be one-eighth inch. The plates are to be held in a position one-sixteenth inch apart during the welding process as indicated by the $\frac{1}{16}$ between two vertical lines. (A more complete listing of symbols is found in *Steel Construction*.)

Fig. 5-2.7   Butt Weld

5-2.2 Glue and Epoxy. The difficulties encountered in using glue or epoxy are due to cost, design data, and reliability. When adequate design data are known, design is relatively uncomplicated.

5-2.3 Bonding of Reinforcing Bars in Tension. In ACI-71, bond is no longer referred to directly. (The bonding of reinforcing bars in tension is developed in sec. 5-12, where embedment lengths are specified.) This section is included, however, for four reasons: (1) historically, bond is important; (2) under several codes, bond is still a design entity; (3) bond calculations provide the logic for embedment length specifications; and (4) when the phenomenon of bond between concrete and reinforcing bars is more completely understood, bond calculations may be extended to include splitting characteristics and once again become code specifications.

Bond is the term applied to holding action between concrete and embedded reinforcing bars. Sufficient bond, therefore, is a requirement for reinforced concrete members so that they perform adequately. In general, bond has two constituents—*inherent adhesion* ("gluing" action) and *mechanical interface* (produced by filling voids between reinforcing bar lugs with concrete).

The bond stress equations in ACI-63 were derived by observing (see Fig. 5-2.8) that any change in resultant tensile force is transferred by bond. Mathematically, this can be stated as

$$\Delta T = \mu \, \Sigma_0 \, dx, \tag{5-2.2}$$

where $\Delta T \equiv$ change in resultant tensile force over dx; $dx \equiv$ incremental length; $\mu \equiv$ bond stress; and $\Sigma_0 \equiv$ circumference of reinforcing bars. Concurrently,

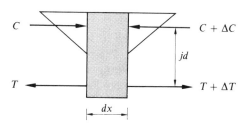

Fig. 5-2.8   Increment of RC Beam

change in the bending moment over this same incremental length is evaluated by the area under the shear diagram (sec. A-3):

$$\Delta M = V \, dx, \tag{5-2.3}$$

where $V \equiv$ average shear over dx, and $\Delta M \equiv$ change in the bending moment over dx.

This change in bending moment can also be expressed as a couple,

$$\Delta M = \Delta T \, jd, \tag{5-2.4}$$

where $jd \equiv$ the moment arm between resultant tensile and compressive forces. Rearranged,

$$\Delta T = \Delta M/jd. \qquad (5\text{-}2.5)$$

If ΔM is replaced by Eq. (5-2.3), then

$$\Delta T = V\,dx/jd. \qquad (5\text{-}2.6)$$

Equating both expressions, equations (5-2.2) and (5-2.6), for ΔT:

$$\mu \Sigma_0\,dx = V\,dx/jd \qquad (5\text{-}2.7)$$

and

$$\mu = V/\Sigma_0 jd \qquad (5\text{-}2.8)$$

which is ACI-63 Eq. 13-1 (suggesting ACI-63 Eq. 18-1).

By Eq. (5-2.8), bond becomes increasingly critical as shear forces increase, but bond stress decreases when total bar perimeters increase. Thus when bond is critical, smaller diameter bars are desirable because they furnish more circumference than larger bars that provide the same area.

Extensive testing established allowable bond stresses as a function of concrete strength and reinforcing bar diameter:

$$\frac{nc\,\sqrt{f'_c}}{D} \leqq UV \qquad (5\text{-}2.9)$$

where $nc \equiv$ appropriate numerical coefficient, taken from ACI-63, 1301 or 1801; $f'_c \equiv$ ultimate strength of concrete; $D \equiv$ nominal diameter of bar, that is, bar size; and $UV \equiv$ maximum upper value as established by ACI-63, 1301 or 1801.

In connection with large bars having deformations corresponding to ASTM A408, and all bars in compression, the bar size term was not used. Standard numerical coefficients are reduced for top bars because it is difficult to maintain effective contact between concrete and the underside of such bars.

Equating actual and allowable stresses yields

$$\frac{V}{\Sigma_0 jd} \leqq \frac{nc\,\sqrt{f'_c}}{D} \leqq UV. \qquad (5\text{-}2.10)$$

For design purposes this is expressed in more convenient form as

$$\frac{V}{(\Sigma_0/D)jd} \leqq nc\,\sqrt{f'_c}, \qquad (5\text{-}2.11)$$

where the term on the right becomes a constant and Σ_0/D can be read directly from design aids. No examples are included because ACI-71 uses the methods of sec. 5-12 instead.

PART B
CONNECTIONS

5-3 DESIGN ASSUMPTIONS

It is now necessary to consider how commercially available connectors, classified as either discrete or continuous, can be used in connections. Basic assumptions are (1) Members and/or plates to be joined are rigid; (2) connectors are linearly elastic; and (3) no friction is induced between the connected members.

None of these assumptions is correct, but the assumption that members to be joined are rigid induces relatively little error in calculated strengths.

An assumption of linearity of connectors is, however, both *physically* undesirable and impossible. Consider, for example, the lapped splice of Fig. 5-3.1. As a load is gradually applied and plates are put in tension, plate elongation is PL/AE. If only one connector were used, plate elongation would have no effect on the connection. However, when more than one connector is used in a longitudinal file, each connector carries only a portion of the load (or none) at first. Yet, distance between connectors varies in proportion to load carried over that same distance ($P'L'/AE$). It is apparent, therefore, that connectors themselves displace significantly and reach their yield points at different times. In fact, all connectors can act as a group *only* after most have actually reached their yield strengths.

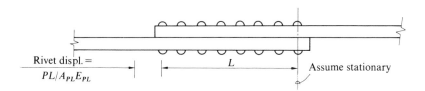

Rivet displ. = $PL/A_{PL}E_{PL}$

L

Assume stationary

Fig. 5-3.1   Nonlinear Nature of Connections

The assumption of no friction between joined members is physically incorrect for high-strength bolts because that is precisely why they work. However, friction calculations are not required in design. Each connector has a specified prescribed strength to account for the available friction force under a variety of conditions.

Interestingly, even though the three basic assumptions given above are not *physically* correct, they do lead to reasonable designs as verified by a multitude of tests and years of experience. Further, they allow for a reasonable design process. Although a more rational theory may be accepted in the future, at present the design of reasonable, safe, and economical connections is dependent upon those three basic assumptions.

5-4 ANALYSIS OF DISCRETE ECCENTRIC CONNECTIONS WITHOUT TENSION

Based on the assumptions set forth in sec. 5-3, forces applied to individual connectors imposed by shear and torsional moment applied to a total connection are combined vectorially. Connector force induced by shear is simply total shear divided by the number of connectors. Connector force induced by torsion is computed by the simple torsion formula $T\rho/J$, where T is defined as torsional moment, ρ is the distance from the centroid of the connection to the centroid of an individual connector, and J is the equivalent polar moment of inertia expressed in units of R-in.^{2} rather than in.^{4}. This definition of J allows results to be stated in values of force per connector (k/R) rather than in values of stress per square inch per connector, k/in.^{2}/conn. Expressed vectorially, then,

$$\overrightarrow{F_R} = \frac{\overrightarrow{V}}{N} \pm \frac{\overrightarrow{(Ve)\rho}}{J}, \tag{5-4.1}$$

where an arrow over a value indicates a vector; $F_R \equiv$ force per connector (k/R) (R is used in honor of the historically important rivet and k for kip); $V \equiv$ shear force (kips); $N \equiv$ number of connectors; $(Ve) \equiv$ torsional moment induced by shear acting with eccentricity e (in.-kips); $\rho \equiv$ distance from the centroid of the connection to the centroid of the connector; and $J \equiv$ equivalent polar moment of inertia, in R-in.^{2}, computed as $I_{xN} + I_{yN}$ with equivalent moments of inertia about the x and y axes and using *number* of connectors in place of *area* of connectors.

5-4.1 General Case. To illustrate the general case, a discrete eccentric connection is shown to scale in Fig. 5-4.1. To apply Eq. (5-4.1) it is necessary to determine the centroid and J. This is performed by the usual A, y, Ay, Ay^2 table (see sec. A-2), except that area is replaced by number; the calculations are given in table 5-4.1. (Note that equivalent moment of inertia of each connector about its own axis is consciously neglected as being insignificant.)

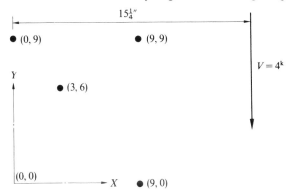

Fig. 5-4.1   Eccentric Connection

Table. 5-4.1 Tabular determination of centroid and equivalent moments of Inertia for connection of Fig. 5-4.1

N	y	Ny	Ny^2	N	x	Nx	Nx^2
1	0	0	0	1	0	0	0
1	6	6	36	1	3	3	9
2	9	18	162	2	9	18	162
4	6	24	198	4	5.25	21	171
	$\bar{y}$		-144		$\bar{x}$	21	-110
			$I_{Nx} = 54$ R-in.^{2}				$I_{Ny} = 61$ R-in.^{2}

From Table 5-4.1 it may be seen that the centroid is located at coordinates $\bar{x} = 5.25$, $\bar{y} = 6$. Eccentricity e is calculated as $15.25 - 5.25 = 10$ in., and $J = I_{nx} + I_{ny} = 54 + 61 = 115$ R-in.^{2}. To check the force carried by the lower right connector, it is necessary to compute $\rho = \sqrt{3.75^2 + 6^2} = 7.07$ in. (where $3.75 = 9 - 5.25$) and the angle denoted α in Fig. 5-4.2.

Direct shear force loads the lower right connector by V/N or $4/4 = 1$ k/R. Torsional moment loads the same connector by $(Ve)\rho/J$ or $(4)(10)(7.07)/115 = 2.46$ k/R. However, these two forces must be added vectorially. (These dimensions and values are plotted to scale in Fig. 5-4.2.) Note that shear force produces a force paralleling itself on connectors, whereas torsional moment produces forces perpendicular to a radial line between the connection centroid and the connector centroid. Thus, final force acting on the connector nearest the connection centroid is only about one-quarter kip downward; the torsional moment force has almost balanced the shear force.

By the laws of cosines, connector force in the lower right connector is

$$F_R = \sqrt{1^2 + 2.46^2 - (2)(1)(2.46) \cos(90 + \alpha)} = 3.1 \text{ kips},$$

where $-\cos(90 + \alpha) = \sin \alpha = 3.75/7.07 = 0.530$. In general form, this is equivalent to

$$F_R = \sqrt{F_V^2 + F_T^2 + 2F_V F_T \sin \alpha}, \tag{5-4.2}$$

where $F_V \equiv$ force on connector caused by shear force V and $F_T \equiv$ force on connector caused by torsional moment (Ve).

5-4.2 Simplified Formulation. Figure 5-4.3 illustrates geometric equivalence of the preceding basic formulation and an alternate simplified formulation. The actual plot is to scale for the connector analyzed in sec. 5-4.1.

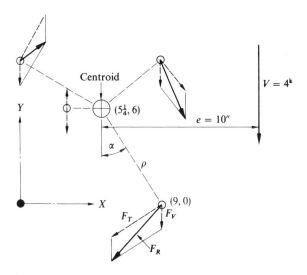

Fig. 5-4.2   Forces Acting on Connectors

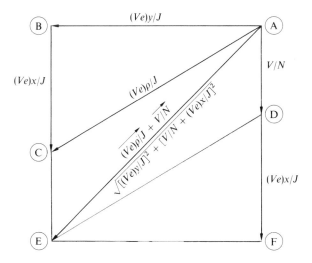

Fig. 5-4.3   Simplified Vector Solution

The value $(Ve)y/J$ is plotted horizontally from point A leftward to point B. From point B, $(Ve)x/J$ is plotted vertically downward to point C. However, because ρ is defined as $\sqrt{x^2 + y^2}$, the vector from A to C must represent $(Ve)\rho/J$. Now, from point A, the value of V/N is plotted vertically downward to point D. Adding vectors AC and AD yields vector AE which is $\overrightarrow{(Ve)\rho/J} + \overrightarrow{V/N}$, the final result.

Using an alternate approach, $(Ve)x/J$ can be plotted vertically from D to F. But then, the sum of vectors AB + (AD + DF) is equal to vector AE, again the final result. Because vector AB is perpendicular to vector (AD + DF), a simple alternate formulation for F_R by the Pythagorean theorem is

$$F_R = \sqrt{[(Ve)y/J]^2 + [V/N + (Ve)x/J]^2}. \qquad (5\text{-}4.3)$$

By inspection, the equation may be more completely generalized as

$$F_R = \sqrt{(V_x/N + Ty/J)^2 + (V_y/N + Tx/J)^2}, \qquad (5\text{-}4.4)$$

where $F_R \equiv$ force per connector (k/R), $V_x \equiv$ shear force component in x direction (kips), $N \equiv$ total number of connectors, $T \equiv$ total torsional moment Ve (in.-k), $y \equiv y$ component of distance from the centroid of connection to the centroid of connector (in.), $J \equiv$ total equivalent polar moment of inertia (R-in.²), $V_y \equiv$ shear force component in y direction (kips), and $x \equiv x$ component of distance from the centroid of connection to the centroid of connector (in.).

Equation (5-4.4), in general, is superior to both Eq. (5-4.1) and Eq. (5-4.2) because it does not require calculation of radial distance or angles. It also lends itself to ready visualization and to quick simplification for specialized cases.

5-4.3 Specializations. Two equations can be specialized from Eq. (5-4.4) when all connectors are in a single line.

When the *load V is perpendicular to a single vertical row* of connectors (Fig. 5-4.4), both V_y and x are zero. Then

$$F_R = V_x/N + Ty/J. \qquad (5\text{-}4.5)$$

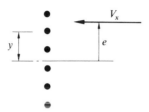

Fig. 5-4.4   Load Perpendicular to Single Row of Connectors

When the *load V is parallel to a single vertical row* of connectors (Fig. 5-4.5), both V_x and x are zero. Then

$$F_R = \sqrt{(V_y/N)^2 + (Ty/J)^2}. \qquad (5\text{-}4.6)$$

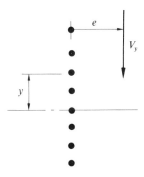

Fig. 5-4.5   Load Parallel to Single Row of Connectors

For more common connections which involve connectors symmetrically placed in rows, center of gravity is determined by inspection. I_{yn} is computed simply as Ny^2, where y is measured from the centroid. If connectors have uniform pitch p, then for a single vertical row

$$I_{xn} = np^2 (n^2 - 1)/12. \qquad (5\text{-}4.7)$$

All other values then can be determined easily as shown by the following example and in Fig. 5-6.2.

To prove Eq. (5-4.7), first consider an odd number of equally spaced (uniformly pitched) connectors with their centers of gravity about the center connector. Distance to each successive connector is p, $2p$, $3p$, the last distance being $(n - 1)p/2$, where n is total number of connectors in a vertical row. Thus

$$I_{xn} = 2 \left\{ p^2 \left[1^2 + 2^2 + 3^2 + \cdots + \left(\frac{n-1}{2} \right)^2 \right] \right\}. \qquad (5\text{-}4.8)$$

But, because $\sum_1^m m^2 = m(m + 1)(2m + 1)/6$, Eq. (5-4.8) can be rewritten, with $(n - 1)/2$ replacing m, as

$$I_{xn} = \frac{2p^2 \left[\frac{n-1}{2} \right] \left[\frac{n-1}{2} + 1 \right] \left[2 \left(\frac{n-1}{2} \right) + 1 \right]}{6} \qquad (5\text{-}4.9)$$

and this is reducible to Eq. (5-4.7).

Center of gravity for an even number of uniformly pitched connectors is midway between the two innermost connectors. Thus, distances to each successive connector are $p/2$, $3p/2$, $5p/2$, the last distance being $(n - 1)p/2$. This allows the expression:

$$I_{xn} = \frac{2p^2}{4}\left[1^2 + 3^2 + 5^2 + \ldots + [2(n/2) - 1]^2\right]. \tag{5-4.10}$$

But $\sum_1^m (2m - 1)^2 = m(2m + 1)(2m - 1)/3$, which, when m is replaced by $n/2$, leads to

$$I_{xn} = \frac{2p^2}{12}\left(\frac{n}{2}\right)\left[2\left(\frac{n}{2}\right) + 1\right]\left[2\left(\frac{n}{2}\right) - 1\right],$$

which again reduces to

$$I_{xn} = np^2(n^2 - 1)/12. \tag{5-4.7}$$

5-5   ANALYSIS OF CONTINUOUS ECCENTRIC CONNECTIONS WITHOUT TENSION

A typical continuous connection with holding action along a line, can be analyzed by using identical assumptions to those made for discrete connections: (1) connected members are rigid, (2) connectors are linearly elastic, and (3) friction between connected members is neglected.

As a typical general case, maximum allowable V is sought for the fillet weld of Fig. 5-5.1. Centroidal axis and polar moment of inertia are computed, first with respect to the y axis and then the x axis, being sure to include moments

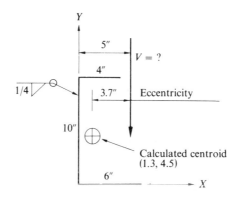

Fig. 5-5.1   Eccentric Welded Connection

of inertia of each line about its own centroid. These tabulated values are shown in Table 5-5.1.

Table 5-5.1 Tabular determination of centroid and moment of inertia for connection of Fig. 5-5.1.

Weld Length (wℓ)	x	(wℓ)x	(wℓ)$x^2 + I_0$
4	2	8	16
			5.3 (4³/12)
10	0	0	0
6	3	18	54
—	—		18 (6³/12)
20	1.3	26	93.3
	$\overline{x}$		− 33.8
			$I_y = 59.5$

Weld Length (wℓ)	y	(wℓ)y	(wℓ)$y^2 + I_0$
4	10	40	400
10	5	50	250
			83.3 (10³/12)
6	0	0	0
20	4.5	90	733.3
	$\overline{y}$		− 405.0
			$I_x = 328.3$

Thus, the centroid of the weld group is located at coordinates (1.3, 4.5) and polar moment of inertia is 59.5 + 328.3 or 387.8 in.³ (*not* in.⁴). Assume the one-quarter fillet weld is made with an E60XX electrode; then allowable stress per inch is 4 (the number of sixteenths in the weld) times 800 or 3.2 k/in. [Eq. (5-2.1)]. The shear force has an eccentricity of 3.7 in. (Fig. 5-5.1), and maximum weld stress will occur at one of the four corners. Because direct shear and torsion combine at the lower right, stress is critical. From inspection of Eq. (5-4.4), a general expression can be written:

$$F_L = \sqrt{(V_x/L + Ty/J)^2 + (V_y/L + Tx/J)^2}, \tag{5-5.1}$$

where $F_L \equiv$ force in continuous connection per inch (k/in.), $V_x \equiv$ shear component in x (horizontal) direction (kips), $L \equiv$ total length of weld (in.), $T \equiv$ total torsional moment $Ve = $ (in.-k), $y \equiv y$ component (vertical) of distance from centroid of connection to the point where F_L is computed (in.), $J \equiv$ linear polar moment of inertia (in.³), $V_y \equiv$ shear component in y (vertical)

direction (kips), and $x \equiv x$ component (horizontal) of distance from centroid of connection to the point where F_L is computed (in.).
Substituting numerical values in Eq. (5-5.1) to get

$$3.2 = \sqrt{[0/20 + (V)3.7(4.5)/387.8]^2 + [V/20 + (V)3.7(4.7)/387.8]^2},$$

shows that maximum allowable shear is 30.7 kips.

A more usual continuous connection is one with legs of equal length (Fig. 5-5.2). The X axis is taken through weld pattern centroid and the Y axis through the vertical weld. Legs are related to length of the vertical weld by a constant k. Thus total weld length $L \equiv \ell + 2k\ell = \ell(1 + 2k)$. By use of these symbols, a simplified equation for J can be obtained, as shown in Table 5-5.2, first with respect to the Y axis, then the X axis.

Table 5-5.2 Generalize solution for centroid and moment of inertia for channel shape weld with equal legs

Weld Length $(w\ell)$	x	$(w\ell)x$	$(w\ell)x^2 + I_0$
$k\ell$	$k\ell/2$	$k^2\ell^2/2$	$k^3\ell^3/4$
			$k^3\ell^3/12$
ℓ	0	0	0
$k\ell$	$k\ell/2$	$k^2\ell^2/2$	$k^3\ell^3/4$
			$k^3\ell^3/12$
$\ell(1 + 2k) \equiv L$	$k^2\ell^2/L$	$k^2\ell^2$	$2k^3\ell^3/3$
	$\overline{x}$		$-k^4\ell^4/L$
			I_y
$k\ell$	$\ell/2$	$k\ell^2/2$	$k\ell^3/4$
ℓ	0	0	0
			$\ell^3/12$
$k\ell$	$-\ell/2$	$-k\ell^2/2$	$k\ell^3/4$
L	$\overline{y} = 0$	0	$I_x = k\ell^3/2 + \ell^3/12$

Thus, the polar moment of inertia for the connection of Fig. 5-5.2 is

$$J = 2k^3\ell^3/3 - k^4\ell^4/L + k\ell^3/2 + \ell^3/12.$$

But

$$L^3/12 = \frac{[\ell(1 + 2k)]^3}{12} = \ell^3/12 + k\ell^3/2 + k^2\ell^3 + 2k^3\ell^3/3$$

so that

$$J = (L^3/12 - k^2\ell^3) - k^4\ell^4/L = L^3/12 - (k^2\ell^3 L + k^4\ell^4)/L.$$

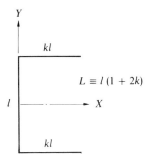

Fig. 5-5.2   Continuous Connection with Equal Legs

or

$$J = L^3/12 - \{k^2\ell^3[\ell(1 + 2k)] + k^4\ell^4\}/L = L^3/12 - k^2\ell^4(1 + 2k + k^2)/L$$

so, finally,

$$J = L^3/12 - k^2\ell^4(1 + k)^2/L. \tag{5-5.2}$$

To complete this simplified approach, note that

$$\overline{x} = (k\ell)^2/L. \tag{5-5.3}$$

If a continuous connector is used with respect to area instead of linearly, a solution is similar except that J is the true polar moment of inertia in in.[4]. Then calculations for the stresses would be in k/in.[2]. In the following section, several examples of design and analysis proof are shown.

5-6   DESIGN OF CONNECTIONS—CONNECTORS NOT IN TENSION

In general, the design of connections is by trial and error, although many tables and charts are available to assist during initial sizing. Once a connection has been selected, an analytical proof is desirable.

5-6.1 Framed Beam Connections. To obtain maximum economy in steel construction, it is advisable to use standard connections as much as possible. A *simple beam* connection for a W24 x 84 spanning 11 ft-3 in. and carrying a total uniform load of 24.2 k/ft serves to illustrate calculation procedures. An additional requirement is that ASTM A325 HS bolts (bearing type) are to be used in the field and E70XX fillet welds in the shop. Connecting angles may be fastened either to the beam itself or to its supporting column. Because a "simple beam" support is desired, this connection carries only shear and is therefore attached to just the web. Thus flanges are not restrained from rotation and the joint carries little (assumed no) moment. Total design shear is 136 kips.

By Table I of *Steel Construction* (p. 4-17), observe that ten rows of connectors need a 36-in. deep beam. The maximum number of rows for a 24-in. beam is seven (p. 4-18), and the total shear capacity for $\frac{3}{4}$-in. diameter ASTM A325-X HS bolts is 136 kips. If the connectors are used through the 0.47-in. thick web, the allowable bearing is $(0.47)(255) = 120$ kips, where 255 is obtained from Table I-B7 for $F_y = 36$ and fastener diameter is $\frac{3}{4}$ in. If this were adequate, which it is not, a check would be made for allowable shear on the web net section, where hole diameter $+\frac{1}{16}$ in. is deducted for each connector in a vertical row. Although larger HS bolts would be satisfactory, they tacitly have been excluded from consideration. Angle thickness for A36 steel is read from Table I-A7 as $\frac{5}{16}$ in. Observe in Fig. 5-6.1 that suitable angles are $L4 \times 3 \times \frac{5}{16} \times 1$ ft-$8\frac{1}{2}$ in.

Another possibility is to weld the angles to beam webs. Turn to Table III on p. 4-31 of *Steel Construction*, use the column entitled Weld A; the allowable load is 149 kips for $\frac{1}{4}$-in. welds used in conjunction with seven rows of fasteners. Weld length is 1 ft-$8\frac{1}{2}$ in. Minimum web thickness is 0.43 in., which is less than the 0.47 in. furnished, so the connection is satisfactory.

If the web were only 0.37-in. thick, then allowable shear would be $149(0.37/0.43)$. Such reductions assure that webs do not become overstressed. This minimum value is found by equating the shear capacities of two welds (both sides) versus web, as

$$2(0.928)(4)(20\tfrac{1}{2}) = 14.5 \; t \; (24),$$

where $0.928 \equiv$ allowable for E70XX electrode per $\frac{1}{16}$ in. per inch, $4 \equiv$ number of one-sixteenths in the weld used, 14.5 ksi $\equiv$ allowable web shear, $20\frac{1}{2}$ in. $\equiv$ weld length, 24 in. $\equiv$ nominal depth of web, and t in. $\equiv$ minimum web thickness, yielding $t_{min} \doteq 0.43$ as shown in the table.

Figure 5-6.1 shows the analytical check of this connection, which is straightforward.

5-6.2 An Eccentric HS Bolted Connection. Several tables are provided in *Steel Construction* to aid in the design of brackets. The general design procedure is:

1. Determine $C = P/r_v$, where $P \equiv$ load (kips) and $r_v \equiv$ permissible load on one connector (kips).

2. Use eccentricity to locate the row, then use C to locate the column which specifies the required number n of rows of connectors.

3. Compute a new effective length (eccentricity), which is based on n, using the equation provided.

4. Return to step 2 and repeat the procedure until solution converges.

As an example, a bracket is designed to support 155 kips with an actual eccentricity of 13 in. A bolt pattern is chosen, using 13.23 kips per $\frac{7}{8}$-in.

Load 136^k Reaction of W24 × 84 spanning 11'-3"

Section Properties (A36 W24 × 84; E70XX electrode; ASTM A325-X$\frac{3}{4}$"ϕ HS bolts, bearing type)

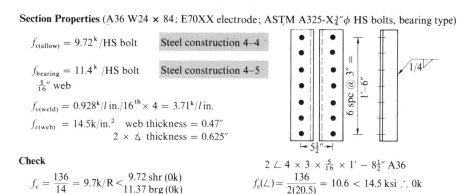

$f_{v(allow)} = 9.72^k$/HS bolt   Steel construction 4-4

$f_{bearing} = 11.4^k$ /HS bolt   Steel construction 4-5
$\frac{5}{16}$" web

$f_{v(weld)} = 0.928^k/l$ in./16th × 4 = 3.71$^k/l$ in.
$f_{v(web)} = 14.5$k/in.^{2} web thickness = 0.47"
                      2 × 4 thickness = 0.625"

Check

$f_v = \dfrac{136}{14} = 9.7$k/R $<\begin{array}{l}9.72 \text{ shr (0k)}\\11.37 \text{ brg (0k)}\end{array}$

$f_{v(weld)} = \dfrac{136}{2(20.5)} = 3.32$k/$l$ in. < 3.7 (0k)

2 ∟ 4 × 3 × $\frac{5}{16}$ × 1' − 8$\frac{1}{2}$" A36

$f_t(\angle) = \dfrac{136}{2(20.5)} = 10.6 < 14.5$ ksi ∴ 0k

$f_{v(web)} = \dfrac{136}{(0.47)(24.09)} = 12.0$ ksi < 14.5 ∴ 0K

Fig. 5-6.1   Analysis Check of a Framed Beam Connection

diameter A325-X HS bolt (with threads excluded from shear planes; see *Steel Construction*, page 4-6).

Turning to Table XIII of *Steel Construction* on page 4-65, the preliminary choice for C is computed as $P/r_v = 155/13.23 = 11.7$. With eccentricity ℓ_{act} of 13 in., the number of fasteners in a vertical row would be eight (interpolating between ℓ_{eff} of 11 and 12 with C values of 12.4 and 11.6, respectively, for $n = 8$). With $n = 8$, effective eccentricity ℓ_{eff} is computed as $\ell_{act} - (1 + n)/2 = 13 - (9/2) = 8\frac{1}{2}$ in. Using this effective length of 8$\frac{1}{2}$ in., the table shows that for $n = 7$, $C = 12.1 > 11.7$, therefore try seven rows.

As a converging step, with $n = 7$, $\ell_{eff} = 13 - [(1 + 7)/2] = 9$ in. With $\ell_{eff} = 9$ in. and $n = 7$, $C = 11.7$. The design is complete and uses a total of 4 × 7 = 28 HS bolts.

Use of effective length (eccentricity) is based upon a large amount of experimental data which seem intuitively plausible.

Analytical proof for this design is shown in Fig. 5-6.2. The polar moment of inertia is easily computed by using Eq. (5-4.7) for I_{xn}. Critical HS bolt force is found by Eq. (5-4.4).

5-6.3 An Eccentric Fillet Weld Connection. Tables for various weld patterns used for eccentric fillet connections are published in *Steel Construction*. In general, one dimension is used as a base length, called ℓ. Other weld lengths are then defined as $k\ell$ and eccentricity as $a\ell$, P is the load in kips, D is the number of one-sixteenths of fillet weld used, and C_1 is a conversion factor for different strength welds which is defined as 1 for E70XX electrodes. Tables give C values based on a and k, where $C = P/C_1D\ell$.

Load 155^k @ $13''$ eccentricity $l_{eff} = 13 - \dfrac{1+7}{2} = 9''$

T $9 \times 155 = 1395''^k$

Section Properties (A325-X$\frac{7}{8}''\phi$ HS bolts—bearing type)

$f_{v(allow)}$(single shear) $= 13.23$k/R

$N = 28$ HS bolts

$I_x = \boxed{np^2(n^2 - 1)/12} = 4[7(3)^2(7^2 - 1)/12] = 1008$ R in.^{2}

$I_y = 2[7(2\frac{3}{4})^2 + 7(5\frac{3}{4})^2] \qquad\qquad = \underline{569}$

$J = 1577$ R in.^{2}

Check

$V_x/N + T_y/J = \qquad 0 + 1395(9)/1577 \qquad = 7.96$ sqd $\to$ 63.4

$V_y/N + T_x/J = 155/28 + 1395(5.75)/1577 = 10.62$ sqd $\to$ 112.8

$\qquad\qquad\qquad\qquad\qquad\qquad\qquad F_R^2\ 176.2 \therefore F_R = 13.27$ k/R $\doteq 13.23 \therefore$ say 0k

Fig. 5-6.2   Eccentric High-Strength Bolted Connection

As an example, a fillet weld E70XX electrode pattern (shown in Fig. 5-6.3) is used (similar to that shown in Table XVII on page 4-71 of *Steel Construction*), with height limited to 12 in. and width to 10 in. Load P is 26.2 kips, located 12 in. from the vertical weld (not weld centroid), and a $\frac{1}{4}$ in. fillet weld is considered desirable.

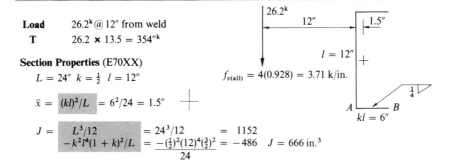

Load 26.2^k @ $12''$ from weld

T $26.2 \times 13.5 = 354''^k$

Section Properties (E70XX)

$L = 24'' \quad k = \frac{1}{2} \quad l = 12''$

$f_{v(all)} = 4(0.928) = 3.71$ k/in.

$\bar{x} = \boxed{(kl)^2/L} = 6^2/24 = 1.5''$

$J = \boxed{\begin{array}{l} L^3/12 \\ -k^2l^4(1+k)^2/L \end{array}} = \dfrac{\begin{array}{l} 24^3/12 \\ -(\frac{1}{2})^2(12)^4(\frac{3}{2})^2 \end{array}}{24} = \begin{array}{l} 1152 \\ = -486 \end{array} \quad J = 666$ in.^{3}

Stress

$F_L: V_x/L + T_y/J = \qquad 0 + (354)(6)/666 \quad = 3.19$ sqd $\to$ 10.17

$\qquad V_y/L + T_x/J = 26.2/24 + (354)(1.5)/666 = 1.89$ sqd $\to \underline{3.57}$

$\qquad\qquad\qquad\qquad\qquad\qquad F_L^2 = 13.74\ F_L = 3.71 = 3.71$ ksi 0k

Fig. 5-6.3   Eccentric Fillet Weld Connection

To illustrate rapid conversion, a first attempt is $\ell = 6$ in., $k = \frac{1}{2}$ (6-in. vertical weld with 3-in. legs). By Eq. (5-5.3), the weld centroid is computed as $(k\ell)^2/L = [\frac{1}{2}(6)]^2/(6 + 3 + 3) = \frac{3}{4}$ in. Addition of this to the given eccentricity makes $a\ell = 12.75$ in. so that $a = a\ell/\ell = 12.75/6 \doteq 2.2$. Minimum permissible C is determined as $P/C_1 D\ell = 26.2/(1 \times 4)(6) = 1.09$. Entering the table with $a = 2.2$ and $C = 1.09$ shows that $k \doteq 2$, but k was originally set at $\frac{1}{2}$, so the solution is invalid.

Because a larger weld is needed, try $\ell = 12$ in., $k = \frac{1}{2}$, so $k\ell = 6$ in. Then, $\bar{x} = [\frac{1}{2}(12)]^2/(12 + 6 + 6) = 1.5$ in., thus $a\ell = 12 + 1.5 = 13.5$ in., $a = 13.5/12 = 1.12$, $C_{min} = 26.2/(1)(4)(12) = 0.55$. Entering the table with $a = 1.12$ and $C = 0.55$ yields $k \doteq 0.5$ as was previously assumed. This appears to be a reasonable answer, requiring a total of 24 in. of weld.

Analytical proof is presented in Fig. 5-6.3. J is computed by use of Eq. (5-5.2), $\bar{x}$ by Eq. (5-5.3), and the critical weld force per linear inch by Eq. (5-5.1).

5-6.4 A Glued Eccentric Connection. Figure 5-6.4 shows a tapered wooden eccentric connection subject to a load V. The glue has an allowable shear stress of 100 psi and glue is applied on both sides of the leftmost 6 in. of the connection. What is the maximum value of V?

Computations, first about the X axis, then about the Y axis, yield the results shown in Table 5-6.1.

Table 5-6.1 Centroid and moment of inertia for glued eccentric connections of Fig. 5-6.4.

Segment	A	y	Ay	$Ay^2 + I_0$
A_1	54	-4.5	-243	1092
				364 $[6(9)^3/12]$
A_2	9	-10.0	-90	900
				5 $[6(3)^3/36]$
	63	-5.3	-333	2361
		$\bar{y}$		-1761
				$I_y = 600$ in.^{4}

Segment	A	x	Ax	$Ax^2 + I_0$
A_1	54	3	162	486
				162 $[9(6)^3/12]$
A_2	9	2	18	36
				18 $[3(6)^3/36]$
	63	2.86	180	702
		$\bar{x}$		-515
				$I_x = 187$ in.^{4}

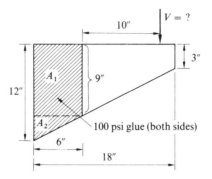

Fig. 5-6.4   Glued Eccentric Connection

From the computation about the Y axis, actual eccentricity is $10 + 6 - 2.86 = 13.14$ in. so that torsional moment is $13.14V$. With $\bar{y} = 5.3$, maximum stress occurs in the upper right-hand corner. Using Eq. (5-4.4), with N replaced by area and J by actual polar moment of inertia $I_p = 600 + 187 = 787$ in.^{4},

$$100 = \sqrt{(0/63 + 13.14\,V\,5.3/787)^2 + (V/63 + 13.14\,V\,3.14/787)^2}$$

or $V = 890$ lb (actually, mathematics show 894.4). Because both sides are glued, this connection can carry a 1.78-kip load at this eccentricity.

5-6.5 An Eccentric Nailed Connection.   Which is better, four nails in a square pattern with an additional one at the centroid, or five nails equally spaced around a circle with diameter equal to the diagonal of the previous square? These two configurations are pictured in Fig. 5-6.5, where 10d (10-penny) nails in commercial red oak are shown. This is a type I wood (NDS Table 13) and the allowable lateral load per nail is 116 lb (NDS Table 17) if penetration is ten diameters. The diameter of a 10d nail is 0.148 in., (NDS 800-A-2), therefore, $1\frac{1}{2}$-in. penetration is just barely adequate.

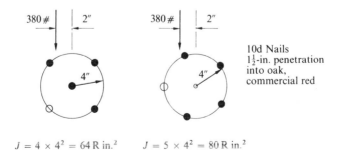

Fig. 5-6.5   Eccentric Nailed Connections

For the more conventional pattern, the critical nail (shown unshaded) has a force on it $[(\text{Eq. 5-4.4})]$ of

$$F_R = \sqrt{[(380)(2)(2.828)/64]^2 + [380/5 + (380)(2)(2.828)/64]^2}$$
$$= 115 \text{ lb/nail} < 116. \text{ Therefore it is OK.}$$

For the circular pattern, the critical nail (also shown unshaded) will be considered to be in the most unfavorable position so that the torsional moment force and shear force are directly additive. Then

$$F_R = 380/5 + 380(2)(4)/80 = 114 \text{ lb/nail} < 116 \text{ (OK)}.$$

Although these results are nearly identical, comparison between the equations shows that when eccentricity increases, the pattern chosen becomes increasingly important.

5-6.6 An Anomaly. Can *adding* a rivet make an eccentric riveted connection weaker? Consider a $\frac{7}{8}$-in. A502-1 rivet, placed as shown in Fig. 5-6.6, with an allowable single shear of 9.02 kips (*Steel Construction*, page 4-6). If the fifth rivet (shown encircled) were *not* used, this connection would have no eccentricity and the force would be $36/4 = 9$ k/rivet.

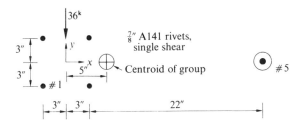

Fig. 5-6.6   An Anomaly

Using the axes indicated in the figure, polar moment inertia of the four-rivet group is $4(3\sqrt{2})^2 = 72 \text{ R in.}^2$. The completed computation for addition of the fifth rivet is given in Table 5-6.2. Under these circumstances, the 36-kip force has an eccentricity $\bar{x}$ of 5 in. producing a torsional moment of 180 in.-k. Critical rivets are now either those in the left row, or possibly the lone rivet no. 5. By Eq. (5-4.4),

$$F_{no.1} = \sqrt{[180(3)/572]^2 + [36/5 + 180(8)/572]^2} = 9.77 \text{ k/R},$$

and concurrently,

$$F_{no.5} = 36/5 - 180(20)/572 = 7.2 - 6.3 = 0.9 \text{ k/R}$$

Question. Is this five-rivet connection *really* able to carry less load than its four-rivet counterpart? Your answer, and substantiating argument, will indicate the depth of your knowledge as related to the design of connections.

Table 5-6.2 Adding fifth rivet to four-rivet group (see Fig. 5-6.6)

N	ρ	$N\rho$	$N\rho^2$
4	0	0	0
			72
$\dfrac{1}{5}$	$\dfrac{25}{5}$	$\dfrac{25}{25}$	$\dfrac{625}{697}$
	$\overline{x}$		-125
			$J = \overline{572} \text{ R in.}^2$

5-7 CONNECTIONS WITH CONNECTORS IN TENSION

When connectors are subject to combined shear and tension, use of Mohr's circle suggests itself. Rather than adding this complexity, however, AISC 1.6.3 requires that tensile and shear stresses be checked for adequacy against separate allowables. The following analysis example is for the bracket of Fig. 5-7.1. The portion which is in compression below the bottom rivet has been purposely and conservatively neglected because it might not bear on the support.

Before calculations can begin, a position for the neutral axis must be assumed. If this assumption of position is not verified by subsequent calcula-

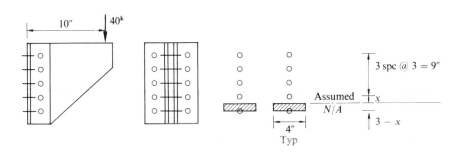

Fig. 5-7.1 Bracket with Connectors in Tension

tions, another position must be assumed for the neutral axis and calculations must be done again. If x has a computed value between 0 and 3 in., then the initial assumption of Fig. 5-7.1 is verified.

Take first moments about the assumed neutral axis (area of one rivet is 0.6013):

$$2(4)(3 - x)^2/2 = 2(0.6013)\big[x + (x + 3) + (x + 6) + (x + 9)\big]$$

so that $x = 0.54$ in., and the assumed position of the neutral axis is verified.

Compute the noment of inertia about this neutral axis.

plate: $2(4)(2.46)^3/3$ $= 39.7$ in.^{4}
rivets: $2(0.6013)(0.54)^2$
          $2(0.6013)(3.54)^2$
          $2(0.6013)(6.54)^2$
          $2(0.6013)(9.54)^2$ or $2(0.6013)(146.6) = 176.3$
moment of inertia: $I = 216.0$ in.^{4}

This connection is adequate if it meets two tests. First, rivet shear stress must be less than 15 ksi in accordance with AISC Specification 1.6.3 and Table 1.5.2.1. Thus:

$$\sigma_v = 40/(10 \times 0.6013) = 6.65 \text{ ksi} < 15 \text{ (OK)}.$$

Second, the concurrent allowable tensile stress permitted is calculated using this shear stress of 6.65 in the equation

$$F_t = 28.0 - 1.6 f_v \le 20.0$$

or

$$F_t = 28 - 1.6(6.65) = 17.3 \text{ ksi}.$$

From the flexure formula, tensile stress in the extreme rivet is computed and compared with F_t as

$$\sigma_T = \frac{10(40)(9.54)}{216.0} = 17.7 \text{ ksi} > 17.3 \text{ by about } 2\%$$

and this connection is nearly adequate.

5-8   RESTRAINED MEMBER CONNECTIONS

For moment connections, bending moment is transferred by flange plates and shear through a web connection. Flange area and weld requirements are determined by equating external bending moment to an internal couple formed by the plate, thus obtaining the magnitude of tensile stress block T:

$$T = 12M'^k/(\text{distances center-to-center of plate centroids}). \qquad (5\text{-}8.1)$$

Shear connections should also be checked for adequacy due to combined shear and appropriate portion of bending moment.

When such members frame into I or H shapes, AISC Specification 1.15.5 requires stiffeners to be used on column webs in accordance with the following formulas:

opposite compression flange of beam if

$$t < \frac{C_1 A_f}{t_b + 5k} \quad \text{(AISC formula 1.15-1)} \quad (5\text{-}8.2)$$

or if

$$t \leq d_c \sqrt{F_y}/180 \quad \text{(AISC formula 1.15-2)}; \quad (5\text{-}8.3)$$

opposite tension flange of beam if

$$t_f < 0.4 \sqrt{C_1 A_f} \quad \text{(AISC formula 1-15.3)}; \quad (5\text{-}8.4)$$

where $t \equiv$ thickness of column web to be stiffened, $C_1 \equiv F_y$(beam flange)/F_y(column web), $A_f \equiv$ area of flange plate transmitting load, $t_b \equiv$ thickness of flange plate transmitting load, $k \equiv$ distance from outer face of flange to web toe of fillet of rolled shape which can be read from "Dimensions for Detailing" in *Steel Construction*, $d_c \equiv$ column depth $- 2k$, $t_f \equiv$ thickness of column flange. Areas of stiffeners is specified by AISC formula 1.15-4 as $A_{st} \geq [C_1 A_f - t(t_b + 5k)] C_2$, where $C_2 \equiv F_y$(column)/F_y(stiffener plate).

Connections considered in these last three sections are all relatively simple. Yet the principles developed and illustrated make it possible to design splices, fixed joints, partially restrained joints, etc. With the use of these basic ideas and the exercise of reasonable care, creative design work for all sorts of connections is possible. Productivity is a matter of time and experience.

PART C
DETAILS

Every code specifies a number of rules which must be followed to obtain adequate structures. Because reinforced concrete is normally a field manufactured product, the ACI code is particularly comprehensive in defining good practice.

5-9   OVERALL SAFETY

At the time when USD was introduced to the code as an alternate method, considerable debate ensued over a bifurcated approach to safety factors.

Finally, it was decided to use (1) the capacity reduction factor ϕ to account for variations in materials and construction tolerances, and (2) load factors to account for variations from prescribed loadings. The author understood, at that time, that load factors would also reflect uncertainties in theory and related calculations. By this means, it was hoped that ϕ would eventually be a factor reflecting construction quality, and that it would enhance the position of contractors doing outstanding work. To date, however, ϕ is varied only by the type of stress, as summarized in sec. 2-5.1 and ACI 9.2.

Among load factors specified in ACI 9.3, the most common is $U = 1.4D + 1.7L$, which has been illustrated previously. For frames, additional considerations are:

$$
\left.
\begin{array}{l}
D \equiv \text{dead loads,} \\
E \equiv \text{earthquake,} \\
F \equiv \text{liquid pressure,} \\
H \equiv \text{lateral earth pressure,} \\
L \equiv \text{live loads,} \\
T \equiv \text{cumulative effects of temperature,} \\
\quad\ \text{shrinkage and differential settlement,} \\
W \equiv \text{wind,}
\end{array}
\right\}
\begin{array}{l}
\text{or internal} \\
\text{moments and} \\
\text{forces developed} \\
\text{by these loads.}
\end{array}
\quad (5\text{-}9.1)
$$

These are used in the following combinations, and every frame must have sufficient strength and stiffness to satisfy all of them:

$$U \geq 1.4D + 1.7L, \tag{5-9.2}$$
$$U \geq 0.75(1.4D + 1.7L + 1.7W), \tag{5-9.3}$$
$$U \geq 0.9D + 1.3W, \tag{5-9.4}$$
$$U \geq 0.75(1.4D + 1.7L + 1.87E), \tag{5-9.5}$$
$$U \geq 0.9D + 1.43E, \tag{5-9.6}$$
$$U \geq 1.4D + 1.7L + 1.7H, \tag{5-9.7}$$

$\overset{\nwarrow}{0.9} \quad \overset{\nwarrow}{0}$ if opposing H,

$$U \geq 1.4D + 1.7L + 1.4F, \tag{5-9.8}$$
$$U \geq 0.75\left[1.4(D + T) + 1.7L\right]. \tag{5-9.9}$$

5-9.1 Span Lengths. ACI 8.5.2 specifies design span lengths as follows:

1. Beams not integral with supports; *clear span + d but less than or equal to center-to-center of supports;*
2. Members of continuous frames *c-to-c of supports,* but *use moments at faces of supports;*
3. Solid or ribbed slabs ≦ 10 ft clear span and integral with supports; *use clear span and neglect beam width.*

5-9.2 Lateral Support. ACI 10.4 requires lateral supports for beams at distance ≦ $50b$, where $b \equiv$ the least width of compression flange or face.

5-10   MINIMUM REINFORCING REQUIREMENTS

These requirements insure ductile behavior of reinforced concrete frames. Although used in several examples, they are summarized here for completeness.

5-10.1 Beams. When reinforcement ratios are very small, beams have computed strengths less than as if they were designed of plain concrete. An excessive load on such a beam can cause quite sudden failure. ACI 10-5 requires that a conservative maximum value be derived in accordance with the rule stipulating that reinforced beams should have strengths at least equal to their strengths as plain concrete.

In very large beams, this requirement is relaxed if one and one-third times the reinforcement obtained by calculations is used at *every* section. Then

$$p_{\min} = \frac{200}{f_y} \quad \text{or} \quad \frac{4}{3}A_{\mathrm{SR}} \qquad (5\text{-}10.1)$$

at every section where $A_{\mathrm{SR}} \equiv$ area of steel required by calculations.

For T-beams, required steel is based on web, rather than flange, width.

5-10.2 Slabs. Restrictions applied to beams are somewhat relaxed when used for floor and roof slabs. Basic control is by the value z which is discussed in sec. 5-11.3 and illustrated in Fig. 5-11.5. But for slabs, usual limitations are imposed by the additional requirement that reinforcement cannot be less than that required for shrinkage and temperature, in accordance with ACI 7.13 and summarized below.

Minimum Shrinkage and Temperature Reinforcement	Grade of Steel
$0.0014\,A_g$	any
$0.0020\,A_g$	40 or 50
$0.0018\,A_g$	60
$0.0018 \times 60{,}000\,A_g/f_y$	more than 60

5-10.3 Webs Greater Than Three Feet. ACI 10.6.5 specifies that steel be placed in faces of webs which exceed 3 ft. This requirement insures that a few wide cracks do not replace the more desirable numerous and well distributed fine (hairline) ones. This extra steel is to be at least 10% of main steel required, spaced no further apart (in both faces) than web width or 12 in., as summarized in Fig. 5-10.1. This steel may *not* be used in carrying loads unless strain compatability is used to determine the stresses in reinforcing.

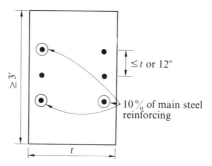

Fig. 5-10.1   Deep Web Face Steel

5-10.4 Shear Reinforcement. ACI 11.1 states a few exceptions to the shear reinforcement requirement. (In sec. 6.2 web reinforcement and design are discussed.) No minimum is required:

1. for slabs and footings,
2. for concrete joist floors,
3. for beams if overall depth is $h \leq 10$ inches or $h \leq$ two and one-half times the flange thickness or $h \leq$ one-half of web width,
4. when actual shear is less than or equal to allowable concrete shear, that is, $v_u \leq v_c/2$.

When minimum stirrups are necessary, formulas are dependent upon torsional stresses v_{tu}. If

$$v_{tu} < 1.5 \sqrt{f'_c} \quad \text{then} \quad A_v = 50 \frac{b_w s}{f_y}, \quad (5\text{-}10.2)$$

and

$$v_{tu} > 1.5 \sqrt{f'_c} \quad \text{then} \quad A_v + 2A_t = 50 \frac{b_w s}{f_y}, \quad (5\text{-}10.3)$$

where $v_{tu} \equiv$ torsional stress, $A_v \equiv$ area of shear reinforcement within spacing s, $A_T \equiv$ longitudinal reinforcement to resist torsion (see Chap. 9), $b_w \equiv$ web width, and $s \equiv$ spacing of shear reinforcement along the member.

Maximum spacing of such reinforcement is the smaller of $d/2$ or 24 inches [ACI 11.1.4(b)].

Additional provisions are required when compressive reinforcement is used or when stress reversals can occur.

Ties may be #3 for vertical steel #10 and smaller. But, by ACI 7.12.3, #4 ties must be used with #11, #14, #18, *and* bundled bars of any size. The maximum spacing of ties is based on the minimum of:

1. 16 longitudinal bar diameters,
2. 48 tie bar diameters, or
3. least dimension of the beam.

5-11   PLACEMENT

When a contractor orders reinforcing steel for his job, the supplier prepares shop drawings which include provision for mechanical devices to properly space reinforcing bars. Even so, placement is not precise.

5-11.1 Beams Slabs, and Walls. By ACI 3.3.2, maximum aggregate size is tied to minimum clear space between bars. The criteria are summarized in Fig. 5-11.1, and are that maximum aggregate size must be less than or equal to: (a) one-fifth of the distance between forms, (b) one-third of the depth of slabs and walls, and (c) three-quarters of the minimum clear spacing between reinforcing bars.

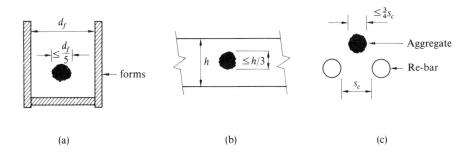

Fig. 5-11.1   Maximum Sized Aggregate

Clear distance between parallel reinforcing bars is specified by ACI 7.4.1 and illustrated in Fig. 5-11.2. Distance between parallel bars in the same layer must be spaced the greater of (a) 1 in., (b) maximum bar diameter, bar ϕ, or (c) one and one-third times maximum aggregate size. Clear spacing between layers must be greater than or equal to (a) 1 in. or (b) the largest bar diameter. In walls and slabs, however, reinforcing bars cannot be placed farther apart than the minimum of (a) 18 in. or (b) three times the slab thickness (ACI 7.4.3).

Cover is required around reinforcing bars, by ACI 7.14, to insure that they will not be affected adversely by environment. These criteria are also shown in Fig. 5-11.2. Members subject to the greatest exposure have the thickest cover and larger bars require extra cover. When stirrups are used

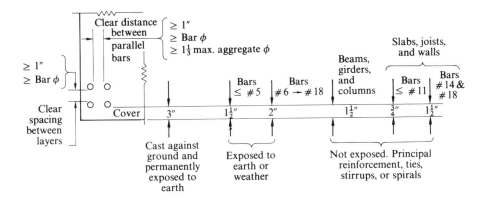

Fig. 5-11.2   Cover and Bar Spacing

in a member, they must not extend into minimum outside cover distances. With or without stirrups, bar sizes to be used in conjunction with Fig. 5-11.2 are determined by longitudinal bars.

Cover is specified as at least 3 in. for members cast against exposed earth. If otherwise exposed, cover is reduced to 2 in. for most bars ($1\frac{1}{2}$ in. for #5 and smaller). For unexposed reinforcing, $1\frac{1}{2}$ in. is typical for beams and $\frac{3}{4}$ in. for slabs and walls.

Less cover is required for precast concrete (ACI 7.14.1.2), which is manufactured under plant control conditions, and for prestressed concrete (ACI 7.14.1.3). Further, by ACI 7.4.6, clear spacing for prestressed steel is less than or equal to one and one-third of the maximum aggregate size and less than or equal to 4 ϕ for individual size, or less than or equal to 3 ϕ for strands (woven wire).

All of these clear distance limitations also apply between a contact lap splice and adjacent bars of any kind (ACI 7.4.5).

5-11.2 Tolerances. Figure 5-11.3 illustrates tolerances as prescribed by ACI 7.3.2.1. Both distance d *and* cover can vary by $\pm\frac{1}{4}$ in. when $d \leq 8$ in., so long as the fixed cover is not less than two-thirds that specified (as illustrated in Fig. 5-11.2). As d increases, so tolerances increase to a maximum of $\pm\frac{1}{2}$ in.

Longitudinal bars must be bent or cut off within a tolerance of ± 2 in., according to ACI 7.3.2.2. This is reduced to $\pm\frac{1}{2}$ in. near discontinuous ends, but not at member ends that are continuously connected to each other.

ACI 7.1.4 restricts field bending to that shown on plans or permitted by the engineer.

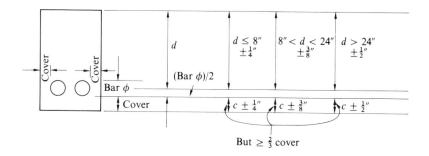

Fig. 5-11.3   Placing Tolerances, d and cover c

5-11.3 Crack Control. When f_y exceeds 40 ksi, crack widths can become excessive unless special care is taken. ACI 10.6.3 and 10.6.4 control this problem by limiting z to 175 k/in. for interior exposure and 145 k/in. for exterior. Here

$$z = f_s \sqrt[3]{d_c A}, \tag{5-11.1}$$

and $f_s \equiv 0.60 f_y$ or $M_{SL}/A_s jd$, where $M_{SL} \equiv$ service load M, without load factors; $A_s \equiv$ area of steel; $jd \equiv$ moment arm; $d_c \equiv$ cover + stirrup diameter (if stirrups exist) + one-half the maximum diameter of the longitudinal bar; and $A \equiv$ effective tension area per number of bars.

Effective tension area is two times the *distance between centroid of tensile steel and extreme concrete fiber, also in tension* times *web width*.

When bars are of different sizes, number of bars is defined as *total area of steel* divided by *area of largest bar used*.

A sample calculation is shown in Fig. 5-11.4 for a 10 in. x 21 in. beam having 3-#7s as its bottom layer of tensile steel and 2-#8s in top tensile layer. Initial computations involve the dimensions: d_c from the extreme fiber to the centroid of near bar is equal to cover ($1\frac{1}{2}$) + stirrup diameter ($\frac{3}{8}$) + one-half the longitudinal bar diameter ($\frac{7}{16}$) = 2.31 in. The second step is to obtain the center-to-center distance of bars in two layers as 1.94 in. Distance to the tensile steel centroid is found from an A, y, Ay chart (sec. A.2) to be 3.22 in., making $d = 21 - 3.22 = 17.78$ in. The equivalent number of bars 4.28 is found as the quotient of tensile steel area 3.38 and maximum bar area 0.79. Effective area A is 15.05, computed as twice 3.22 times beam width b divided by equivalent bar area. Finally, z is computed by Eq. (5-11.1) as 118.

When the same number of the same sizes bars are used in each layer, computations are much simpler. If the beam shown in Fig. 5-11.4 had its tensile steel replaced by two layers of 3-#7 bars, $d_c = 2.31$ in. Distance to centroid of steel is equal to cover ($1\frac{1}{2}$) + stirrup diameter ($\frac{3}{8}$) + longitudinal bar diameter

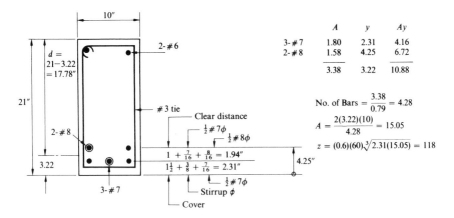

	A	y	Ay
3-#7	1.80	2.31	4.16
2-#8	1.58	4.25	6.72
	3.38	3.22	10.88

No. of Bars $= \dfrac{3.38}{0.79} = 4.28$

$A = \dfrac{2(3.22)(10)}{4.28} = 15.05$

$z = (0.6)(60)\sqrt[3]{2.31(15.05)} = 118$

Fig. 5-11.4 Crack Control Computations

$(\frac{7}{8})$ + one-half of the clear distance between layers $(\frac{1}{2})$, for $3\frac{1}{4}$ in. Thus $A =$ $2(3.25)(10)/6 = 10.83$ and $z = (0.6)(60)\sqrt[3]{(2.31)(10.83)} = 105$.

A more accurate calculation of f_s is probably justified only if z exceeds prescribed limits when using $f_s \equiv 0.6 f_y$.

To facilitate computations, these criteria are summarized in Fig. 5-11.5.

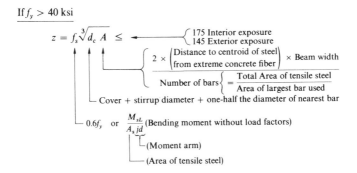

Fig. 5-11.5 Crack Control Criteria

5-11.4 Columns. For columns, clear distance limitations are increased to a minimum of the larger of (a) one and one-half the longitudinal bar diameters or (b) $1\frac{1}{2}$ in. (ACI 7.4.4).

For tied columns ties may be #3 for vertical steel #10 and smaller. But, by ACI 7.12.3, #4s must be used with #11, #14, #18, *and* bundled bars of any size. Such ties have a maximum spacing based on the minimum of (a) 16 longitudinal

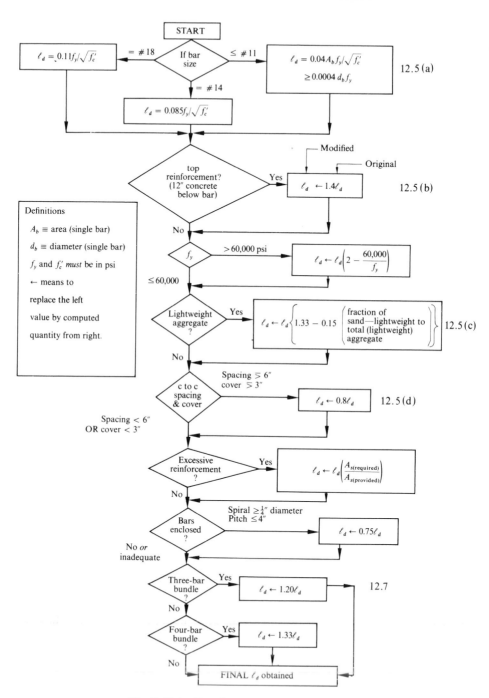

Fig. 5-12.1   Development Length in Tension

bar diameters, (b) 48 tie bar diameters, or (c) the least dimension of the column.

ACI 7.12.2 defines provisions for spirals and ACI 7.10 provisions for column steel passing from one floor to the next.

5-11.5 Bundled Bars. ACI 7.4.2 allows bars to be grouped together only in bundles of no more than four. Whenever such bundles are used, they must be enclosed by stirrup or ties. In computations which involve bar diameters, equivalent bundle diameter is computed as $\sqrt{\phi_1^2 + \phi_2^2 \cdots}$. Typical bundle configurations are two bars side by side, three bars nestled together in a crude triangular shape, or four bars nestled together in a crude square shape.

5-12   DEVELOPMENT LENGTHS AND SPLICES

Requirements for development length are contained in Chap. 12 and for splices in Chap. 7 of the ACI Code. The capacity reduction factor ϕ is excluded from both of these chapters. Rather than computing bond stress, as developed in sec. 5-2.3, minimum development length is based on average attainable bond stress over the development length ℓ_d. Development lengths are predicated primarily upon the tendency of highly stressed reinforcing bars to split their thin restraining sections of concrete.

Flexural reinforcement, in tension, can be terminated only if any one of the three criteria of Table 5-12.1 are met.

5-12.1 Tension Development Length ℓ_d. Figure 5-12.1 is a flow diagram indicating the procedure for computing ℓ_d. Diamond-shaped boxes represent questions and rectangular boxes operations to be performed. In every calculation, ℓ_d appears on both sides of the $\leftarrow$ symbol. This means that the current value of ℓ_d to the right of the arrow is used to compute a new value for ℓ_d left of the $\leftarrow$.

As an example, consider sketch (a) of Fig. 5-12.2 in conjunction with the diagram of Fig. 5-12.1. The first question after START in the flow chart refers to bar size. For #8 bars,

$$\ell_d = 0.04 \, A_b f_y / \sqrt{f_c'}$$
$$= 0.04 \, (0.79)(80,000)/\sqrt{5000} = 35.9,†$$

($> 0.0004 \, d_b f_y = 0.0004 (1)(80,000) = 32$ in.), where steel and concrete stresses must be expressed in psi. These are top bars and it is difficult to get good bond

---

† The following computations are expedited by using mini-calculators; however, computations are shown *only* to slide rule accuracy to emphasize their true lack of preciseness.

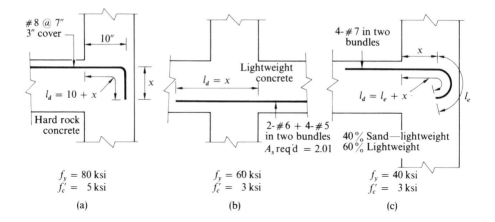

Fig. 5-12.2 Bar Configurations

on their undersides, therefore $\ell_d \leftarrow 1.4\ell_c = 1.4(35.9) = 50.2$ in. The third question considers the yield strength of steel, which is greater than 60,000 psi, so $\ell_d \leftarrow \ell_d (2\text{-}60,000/f_y)$, $\ell_d = 50.2(2\text{-}60,000/80,000) = 62.6$ in. Concrete is not lightweight, so the fourth question is answered affirmatively. Basic development lengths correspond to ACI-63 permissible bond stresses, increased by 20%. If bars are spaced 6 in. or more apart and cover is 3 in., then this increase is not applied because the bars have enough concrete surrounding them to act essentially independent of one another. Thus $\ell_d \leftarrow 0.8 \ell_d = 0.8(62.6) = 50.1$ in.

No reference is made to the actual area of steel required and bars are not enclosed by spirals or in bundles. Therefore, final ℓ_d is 50.1 rounded upward to the nearest inch, or $\ell_d = 51$ in. Typically, these computations can be performed collectively so that

$$\ell_d = (0.04 A_b f_y/\sqrt{f_c'})(1.4)(2 - 60,000/f_y)(0.8)$$

or

$$\ell_d = [0.04(0.79)(80,000)/\sqrt{5,000}](1.4)(2 - 60,000/80,000)(0.8)$$
$$= 50.1, \text{ or } 51 \text{ in., and } x = 51 - 10, \text{ or } 41 \text{ in. } [(\text{Fig. } 5\text{-}12.2(a)].$$

Two computations are required for (b) of Fig. 5-12.2 because two bar sizes are involved. Combine the appropriate coefficients from Fig. 5-12.1:

$$\ell_d = 0.04 A_b f_y/\sqrt{f_c'} \ (> 0.0004 \ d_b f_y),$$
$$\ell_d \leftarrow \ell_d [1.33 - 0.15(0)],$$
$$\ell_d \leftarrow \ell_d (A_s(\text{required})/A_s(\text{furnished})),$$
$$\ell_d \leftarrow 1.20 \ \ell_d.$$

So,

$$\ell_d = 0.04 A_b(60,000)/\sqrt{3,000})(1.33)(2.01/2.12)(1.20),$$
$$\ell_d = 66.5 A_b$$

yielding 21 in. (20.6) for #5 bars and 30 in. (29.2) for #6 bars.

Because $0.04 A_b f_y/\sqrt{f_c'}$ is greater than $0.0004 d_b f_y$ for #6 bars, $\ell_d = 30$ in. is correct. For #5 bars, however, $0.04 A_b f_y/\sqrt{f_c'}$ yields 13.6 in. whereas $0.0004 d_b f_y$ results in 15 in. Thus, for the #5 bars, ℓ_d must be increased by the ratio of 15/13.6 so that $\ell_d = 20.6(15/13.6) = 22.5$, rounded to 23 in.

Sequential computations for ℓ_d, in accordance with Fig. 5-12.1 for sketch (c) of Fig. 5-12.2, appear as:

$$\ell_d = 0.04 A_b f_y/\sqrt{f_c'} = 0.04(0.60)(40,000)/\sqrt{3,000} = 17.6,$$
$$\ell_d \leftarrow 1.4\ell_d = 1.4(17.6) = 24.6,$$
$$\ell_d \leftarrow \ell_d[1.33 - 0.15(\text{fraction})] = 24.6[1.33 - 0.15(4/10)] = 31.2,$$
$$\ell_d \leftarrow 1.33\ell_d = 1.33(31.2) = 41.5 \text{ in.}$$

Effective length of a *standard hook* is computed in a similar manner except that f_y is replaced by $f_h = \xi\sqrt{f_c'}$ and the result is called ℓ_e rather than ℓ_d. From Table 5-12.2 (ACI Table 12.8.1), ξ, defined by ACI as "constant for standard hook," may be read. Thus with $\xi = 360$ (#7 bars and $f_y = 40$ ksi), Fig. 5-12.2(c),

$$f_h = 360\sqrt{3,000} = 19,700.$$

Then

$$\ell_e = 0.04 A_b f_h/\sqrt{f_c'} = 0.04(0.60)(19,700)/\sqrt{3,000} = 8.6,$$
$$\ell_e \leftarrow 1.4\,\ell_e = 1.4(8.6) = 12.1,$$
$$\ell_e \leftarrow \ell_e[1.33 - 0.15(\text{fraction})] = 12.1[1.33 - 0.15(4/10)] = 15.4,$$
$$\ell_e \leftarrow 1.33\,\ell_e = 1.33(15.8) = 20.4 \text{ inches.}$$

Finally, length x must be at least $\ell_d - \ell_e$ or $41.5 - 20.4 = 21.1$ inches for ℓ_d to be adequate.

These computations can be combined and expedited as

$$x = [0.04 A_b(f_y - f_h)/\sqrt{f_c'}](1.4)[1.33 - 0.15(\text{fraction})](1.33),$$

or

$$x = [0.04(0.60)(40,000 - 19,700)/\sqrt{3,000}](1.4) \times [1.33 - 0.15(4/10)](1.33) =$$
$$= 21.1 \text{ inches,}$$

so

$$x = 22 \text{ inches.}$$

5-12.2 Positive Moment Reinforcement. At least

one-third of the positive reinforcement for simple spans,

one-quarter of the positive reinforcement for continuous spans

must extend

at least 6 in. into the support, or

ℓ_d if the member is a part of the primary lateral load resisting frame. In addition,

at simple supports (Fig. 5-12.3) and

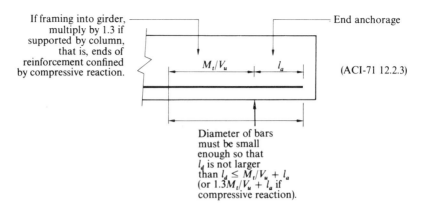

Fig. **5-12.3**   Maximum Bar Size at Simple Support

at points of inflection (Fig. 5-12.4) positive moment tension reinforcement must be limited to a diameter such that

$$\ell_d \leqq M_t/V_u + \ell_a, \qquad (12\text{-}1.1)$$

where $M_t \equiv$ theoretical strength $\left[A_s f_y(d - a/2)\right]$ for actual steel without ϕ, $V_u \equiv$ maximum ultimate shear existing in length ℓ_d, and $\ell_a \equiv$ additional embedment length at the support or point of inflection (in inches), or

$$\ell_d \leqq 1.3 \, M_t/V_u + \ell_a \qquad (12\text{-}1.2)$$

if reaction confines ends of reinforcement.

For example, at a simple support, if $M_t = 150$ in.-kips, $V_u = 10$ kips and $\ell_a = 10$ in., then $\ell_d \leqq 150/10 + 10 \leqq 25$ in. Assumiyg no modifications, $\ell_d = 0.04 \, A_b f_y/\sqrt{f_c'}$ or $0.04 \, A_b$ (say $(60,000)/\sqrt{5,000}) = 34.0 \, A_b \leqq 25$, so $A_b \leqq 0.735$, signifying that bars must be #7 or smaller.

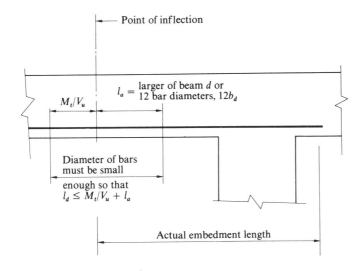

Fig. 5-12.4   Maximum Bar Size at Point of Inflection

Positive bars in tension can be terminated (Fig. 5-12.5) only if:

1. a requirement of Table 5-12.1 is met,

Table 5-12.1   Termination of Flexural Reinforcement in Tension

Bars may be terminated if EITHER

    (1) ACI-71, 12.1.6.1:
$$V_{\text{actual}} \leqq \tfrac{2}{3} V_{\text{allowable}} \text{ (including web reinforcement)}$$

OR (2) ACI-71, 12.1.6.2:

        Additional stirrups are provided
          (a) $3d/4$ beyond terminated bar;
          (b) $A_v f_y / b_w s \geqq 60$ psi;
          (c) $s \leqq d/8\,\beta_b$, where $\beta_b \equiv$ ratio $\dfrac{\text{area of bars cut off}}{\text{total bar area at section}}$.

OR (3) ACI-71, 12.1.6.3:

          (a) Bars $\leqq$ #11 in size;
          (b) $A_{s\text{(continuing bars)}} \geqq 2A_{s\text{(required)}}$
          (c) $V_{\text{actual}} \leqq \tfrac{3}{4} V_{\text{allowable}}$.

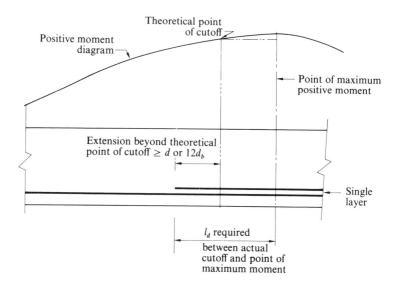

Fig. 5-12.5   Actual versus Theoretical Bar Cutoff Points

2. ℓ_d is furnished between the actual cutoff point and the maximum moment, and

3. the bar is extended beyond the theoretical point of cutoff the larger of d or 12 bar diameters.

5-12.3 Negative Moment Reinforcement.   Because maximum negative moment occurs at the face of a support, ℓ_d must be supplied, as illustrated in Fig. 5-12.6. In addition (ACI-71, 12.3.3), at least one-third of total negative reinforcement must extend beyond the inflection point by the maximum of:

1. d, the effective member depth,

2. 12 d_b, 12 bar diameters, or

3. one-sixteenth of clear span, $\ell_n/16$.

To illustrate, bar lengths of Fig. 5-12.6 are computed; for #7 top bars,

$$\ell_d = \left[0.04(0.60)60,000/\sqrt{4,000}\right]1.4 = 32.0 \text{ in.}$$

and $\ell_d = 42.2$ in. for #8s. Because $p = 3.38/(12.5 \times 19) = 0.0142$, it can be seen that $K_u = 673$ (D-6) and $F = 0.376$ making $M_u = 254$ ft-kips.

For two short #7 bars, then (using actual versus allowable moment reduction) left of the support $\ell_d = 32.0(200/254) = 25.3$, say, 26 in., and right of the support $\ell_d = 32.0(230/254) = 27.9$, say, 28 in.

One #7 bar in combination with two #8 bars provide 2.18 in.² for $p =$

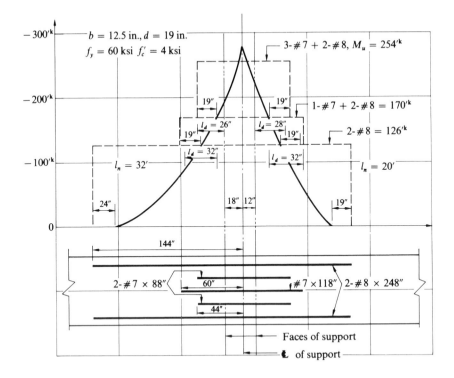

Fig. 5-12.6   Bar Cutoff

0.0092; good for $M_u = 170$ ft-kips. Thus the two #7 bars must also extend past this theoretical point of cutoff the maximum of $d = 19$ in. or $12d_b = 12(\frac{7}{8}) = 10.5$ in. These values are plotted in Fig. 5-12.6, from which the minimum length of the two short #7 bars is seen to be as follows.

ℓ_d (left)	= 26 in.
Left face to center line	= 18
Center line to theoretical cutoff	= 25
Extension beyond theoretical cutoff	= 19
Total bar length	= 88 in.

The 2-#8 bars acting alone are good for $M_u = 126$ ft-kips. From the theoretical point of cutoff ℓ_d is measured at 100% allowable moment, so has a value of 32 in. Extension beyond the theoretical cutoff controls to the left and ℓ_d on the right for the single #7 bar, with its length expressed as follows.

Extension beyond theoretical cutoff	= 19 in.
Theoretical cutoff to center line	= 41
Center line to *previous* theoretical cutoff	= 26
ℓ_d (right)	= 32
Total bar length	118 in.

Finally, at least one-third of steel must extend to a maximum of

$$d = 19 \text{ in.,}$$
$$12d_b = 12(1) = 12 \text{ in., or}$$
$$\ell_n/16 = 32 \times 12/16 = 24 \text{ in. left,}$$
$$\text{and } 20 \times 12/16 = 15 \text{ in. right}$$
$$2 - \#8 = 1.58 \text{ in.}^2;$$
$$2 - \#8 + 3 - \#7 = 3.38 \text{ in.}^2;$$
$$1.58/3.38 = 0.467 > \tfrac{1}{3}; \text{ therefore, OK.}$$

A reasonable alternative is to extend $2 - \#7s$ (1.20), or 0.356 of steel, past inflection points. In any case, the two longer bars should be on the outside.

Lengths have been shown in inches for ease in following computations. In practice, they should be shown in feet and inches and some rounding off (upward) might be appropriate.

Table 5-12.2 ζ Values [2, page 44]

Bar size	$f_y =$ 60 ksi		$f_y =$ 40 ksi
	Top bars	Other bars	All bars
#3 to #5	540	540	360
#6	450	540	360
#7 to #9	360	540	360
#10	360	480	360
#11	360	420	360
#14	330	330	330
#18	220	220	220

5-12.4 Compressive Development Lengths. In compression,

$$\ell_d \geq 0.02\, f_y d_b / \sqrt{f_c'}$$
$$\ell_d \geq 0.0003\, f_y d_b$$
$$\geq 8 \text{ in.}$$

In addition, development length can be reduced by the ratio $A_{s(\text{required})}/$

$A_{s(\text{provided})}$ and by 25% if spiral reinforcement of $\frac{1}{4}$ in. or greater diameter, with 4 in. or less pitch, is used. If three-bar bundles are used, ℓ_d is increasep by 20% and by 33% if four-bar bundles are used.

Hooks are *not* effective for compressive stresses.

5-12.5 Development Length of Stirrups. Provisions of sec. 12.13.1 of ACI-71 are illustrated in Fig. 5-12.7. Obviously, beams may have to be deeper if standard 180° hooks are not used over the top longitudinal bars.

For simple placement, it is often advantageous to use a pair of U-stirrups

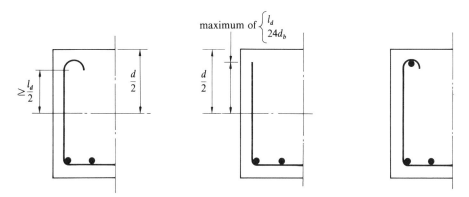

Fig. 5-12.7   Stirrup Anchorage [3, page 61]

rather than a tie. If $A_b f_y$ for one leg is less than 9000 lb, and if the member is 18 in. deep or more, legs extending the full available depth are considered adequate. In any case, a splice of $1.7\ell_d$ is sufficient.

5-12.6 Splices in Tension. Tension tie members must have staggered lap splices of $2.0\ell_d$ in conjunction with a spiral enclosure and 180° hooks.

In beams and girders, it is desirable to keep lap splices out of highly stressed regions and to stagger them. Appropriate lap splice lengths are illustrated in the flow diagram of Fig. 5-12.8.

5-12.7 Splices in Compression. Minimum lap length by ACI-71, 7.7.1.1 is

$$\geq \ell_d,$$
$$\geq 0.0005\, f_y d_b \text{ (if } f_y \leq 60{,}000 \text{ psi)},$$
$$\geq (0.0009\, f_y - 24)d_b \text{ (if } f_y > 60{,}000 \text{ psi)},$$
$$\geq 12 \text{ in.}$$

For concrete below 3000 psi, minimum lap length is to be increased by one-third.

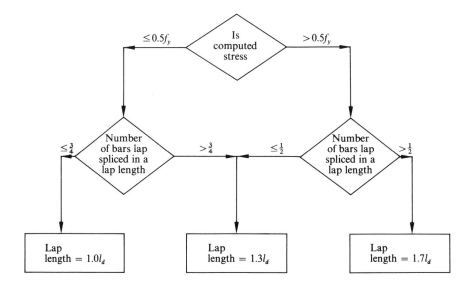

Fig. 5-12.8   Tension Lap Lengths

Reductions to 83% and 75% are allowed for tied and spiral columns, respectively, but final lengths cannot be less than 12 in. To be applicable to tied columns, ties throughout the lap length must have *effective area* ≥ 0.0015 hs; where h ≡ overall member thickness (inches) and s ≡ tie spacing (inches).

End bearing is allowed to transmit compressive stress by ACI-71, 7.7.2 if (a) the ends are within $1\frac{1}{2}°$ of square cut; (b) bars fit within 3° of full bearing; (c) suitable devices hold ends; and (d) closed ties, stirrups, or spirals are used.

PROBLEMS

5-1 given: Section _____ [(a) or (b) of Fig. 5-P.1] with plate thickness
_____ ($\frac{3}{4}, \frac{7}{8}$, or 1 in.), holes_____ ($\frac{3}{4}, \frac{7}{8}$, or 1 in. in diameter),
$a =$ _____ (2, $2\frac{1}{2}$, or 3 in.), $b =$ _____ (1, 1.5, or 2) times
$a, c =$ _____ (1.5, 2, or 3) times a, f_y_____ (36 or 50 ksi).

find: Allowable P based on net section.

5-2 given: Eccentric connection pattern _____ [(a), (b), (c), or (d) of
Fig. 5-P.2] with $V =$ _____ (40, 83.2, 116.7, or 121 kips), $\theta =$
_____ (0, 21.6, 43.7, 51.9, or 90°).

find: (a) Polar moment of inertia and centroid of connector pattern.
(b) Determine eccentricity of load V.

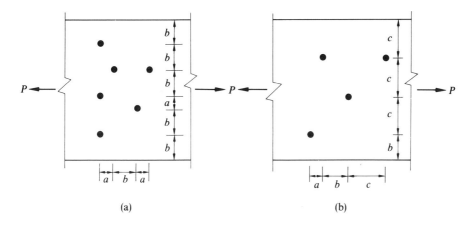

Fig. 5-P.1   Net Section

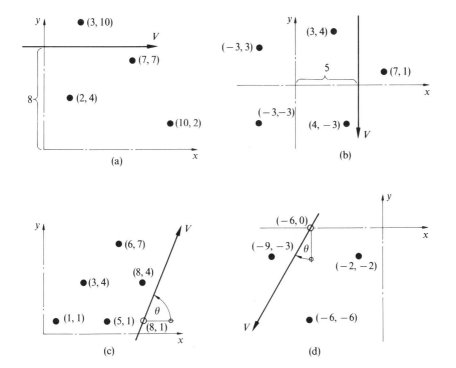

Fig. 5-P.2   Eccentric Connector Patterns

(c) Determine maximum force on critical connector (1) by general approach, and (2) by simplified formulation.

5-3 given: Pattern _____ [(a), (b), (c), or (d) of Fig. 5-P.3] with $V =$ _____ (15.4, 25, 57.6, or 113 kips), $e =$ _____ (2, 3.95, 4.6, or 7 in.), $s =$ _____ (3, 4, 5, or 5.5 in.).

   find: Maximum force on critical connector (1) by general approach, and (2) by simplified formulation.

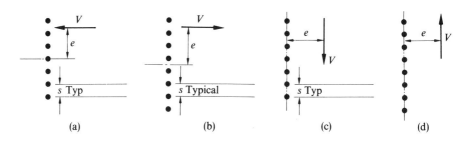

Fig. 5-P.3   Single Row of Connectors

5-4 given: Continuous connector pattern _____ [(a), (b), (c), (d), (e), or (f) of Fig. 5-P.4] with $\ell =$ _____ (8, 10, 12, or 14 in.); $k =$ _____ (0.4, 0.7, 0.9, or 1.2); $f =$ _____ (0.1, 0.23, 0.41 or 0.5) times ℓ; $V =$ _____ (18, 21.7, 46.9, or 83 kips).

   find: (a) Polar I and centroid of connector pattern; (b) eccentricity of load; (c) critical stress. (1) Use general approach and then (2) simplified formulation.

5-5 given: Lightest weight_____ (W36, W33, W24, W18, or W14). Fastener _____ (A307, A325-N, A325-X, or A490-X). E70XX electrodes. Load _____ (60, 120, 180, 240, 300, or 360 kips).

   select: A standard framed beam connection using (a) fasteners only; (b) weld only; (c) fasteners to column, weld to beam; (d) weld to column, fasteners to beam.

   present: Analysis proof.

5-6 given: Table XIII, p. 4-65 of *Steel Construction*. Fasteners _____ ($\frac{7}{8}$ -in. diameter A502-1 rivets, 1-in. diameter A307 unfinished bolts, $\frac{3}{4}$ -in. diameter A490F HS bolts, $1\frac{1}{2}$ -in. diameter A325-X HS bolts, bearing type, threads excluded, or 16d nails). *Total* number provided_____ (16, 28, 32, or 44). Eccentricity _____ (3, 4.6, 7.9, or 10 in.).

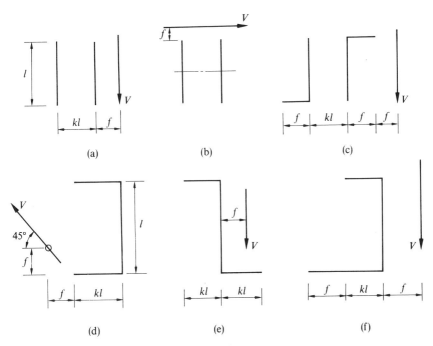

Fig. 5-P.4 Continuous Connector Patterns

find: Allowable P.

present: Analysis proof.

5-7 given: $\frac{7}{8}$ -in. diameter A325-X HS bolts, bearing type, threads excluded. Eccentricity $e =$ _____ (3, 5.2, 6.8, or 9 in.). P _____ (40, 96.3, 118, or 170 kips).

find: n from Table XIII of *Steel Construction*, p. 4-65.

present: Analysis proof.

5-8 given: Same as in problem 5-7.

find: Eccentric fastener group having *minimum* number of HS bolts with not over twelve (and same number) in any vertical row.

present: Analysis proof.

5-9 given: E70XX electrodes; eccentricity _____ (4, 6.1, 8.3, or 10 in.); P _____ (21.8, 43.7, 81.6, or 121.3 kips).

select: Weld size.

find: (a) ℓ and k by *Steel Construction* Table XVII, p. 4-71.
           (b) Determine weld configurations using other tables provided in *Steel Construction*, as assigned.

present: Analysis proof.

5-10 given: Eccentricity of _____ (4, 6, 8, or 10 in.) and E70XX electrodes.

   select: *P* and pattern.

present: Plot of total weld length versus number of sixteenths of an inch of fillet weld used for this load and pattern.

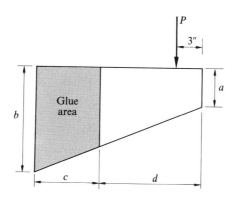

Fig. 5-P.5   Glued Bracket

5-11 given: Glued bracket of Fig. 5-P.5 with *c* = _____ (6, 7, or 8 in.), *d* = _____ (6, 10, or 14 in.), *P* = _____ (1, 1.5, or 2 kips). Glue has _____ (200, 250, or 300) psi strength.

   find: *a* and *b* where $a \leqq 2b/3$.

present: Analysis proof.

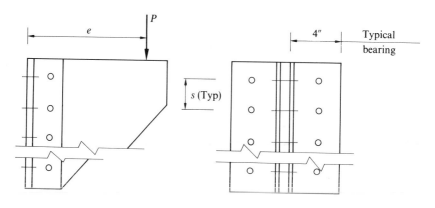

Fig. 5-P.6   Bracket—Connectors in Tension

5-12 given: Bracket of Fig. 5-P.6 with _____ (4, 5, 6, or 7) rows of
_____ ($\frac{3}{4}$, $\frac{7}{8}$, or 1 in.) A325-F HS bolts. $s = $ _____ (3,
3.5, or 4 in.), $e = $ _____ (10, 11.63, 12.71, or 14 in.).

find: Maximum permissible P (A36 steel).

present: Analysis proof.

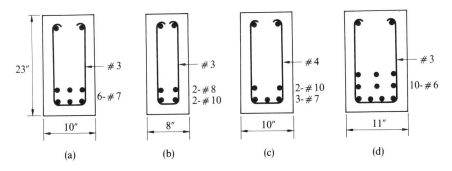

Fig. 5-P.7 RC Beam Cross Sections

5-13 given: Beam _____ [(a), (b), (c), or (d) of Fig. 5-P.7].

find: d', d, d_c, A, and z.

5-14 given: Figure 5-P.8 with $f_y = $ _____ (40, 50, 60, 70, or 80 ksi), $f'_c = $
_____ (3, 4, or 5 ksi). Bars _____ (#6, #7, or #9). c-to-c
spacing _____ (3, 5, 7, or 9 in.) with corresponding cover.
$t = $ _____ (10, 14, 18, or 22 in.)

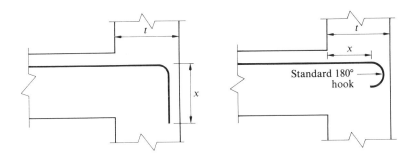

Fig. 5-P.8 Development Lengths

determine: (a) Required ℓ_d; (b) possible ℓ_e; (c) Length x.

5-15 given: Figure 5-P.9 with $t = $ _____ (8, 10, or 14 in.), $a = $ _____
(0.9, 1, 1.1, or 1.2 k/ft), $b = $ _____ (8, 10, 12, or 14 ft).

design: Section with steel cutoff points.
present: Stress sheet.

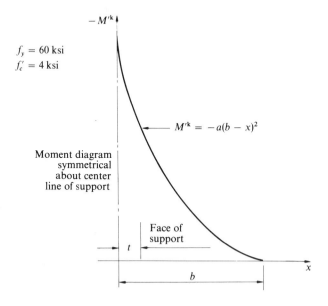

Fig. 5-P.9   Negative Moment Support

REFERENCES

1. *Bethlehem Construction Fasteners*, Bethlehem Steel Export Corporation, 25 Broadway, New York, N.Y. 10004, pp. 12, 11, 8, 4.

2. *Building Code Requirements for Reinforced Concrete* (ACI 318-71), American Concrete Institute, Detroit, Michigan 48219 (1971).

3. *Commentary on Building Code Requirements for Reinforced Concrete* (ACI 318-71), American Concrete Institute, Detroit, Michigan 48219 (1971).

CHAPTER 6

INCREASED CAPACITY FOR FLEXURAL MEMBERS

The capacity of a flexural member is usually controlled by bending, deflection, or shear. It is difficult to find economical materials with higher moduli of elasticity than currently in use. So both bending and deflection are controlled almost entirely by the moment of inertia, which may be increased directly or, in some cases, by use of the floor as flanges for T-beam action. Shear capacity can be increased in several ways.

This chapter is divided into four parts which treat shear; increasing I directly; use of the floor for T-beam action; and deflection of reinforced concrete beams.

PART A
SHEAR

This discussion is divided between steel and reinforced concrete because phenomena differ so. Excess shear in steel members actually causes shear failures. But excess shear in reinforced concrete members usually results in a diagonal tension (principal normal stress) failure.

6-1 SHEAR IN STEEL MEMBERS

Plastic design of a weightless beam (A36), shown in Fig. 6-1.1, requires a W16 x 40 to be adequate for bending. The shear at the left reaction and over the left three feet of the beam, $160(19)^2[3(3) + 19]/(22)^3 = 151.9$ kips,† exceeds the allowable (19.8 x 0.307 x $16.0 = 97.3$ kips) by 54.6 kips.

---

† This computation is for the reaction of an *elastic* fixed end beam. At ultimate load, the reaction of Pb/L is actually only 138.2 kips.

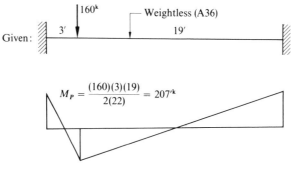

Given:

$$M_P = \frac{(160)(3)(19)}{2(22)} = 207'^k$$

Load factor already included in 160-kip load

Fig. 6-1.1   High shear

Shear capacity can be increased by using (1) a larger beam, (2) web doublers, or (3) diagonal stiffeners.

The lightest beam with adequate section modulus and web area is a W21 x 55. It has an allowable shear of 19.8 x 0.375 x 20.80 = 154.4 kips. Weight is increased 15 lb/ft for a total of 330 lb, or 37.5%.

The second alternative is to slot-weld a plate over the left three feet of web. This web doubler goes from the toe of one fillet to the toe of the other, the T dimension in *Steel Construction*, which is $13\frac{3}{4}$ in. Thus, 13.75 times the unknown web doubler thickness t, multiplied by the allowable stress of 19.8 ksi is equated to excess shear of 54.6 kips. Resultant t is 0.201 in., so a $\frac{7}{32}$ in. plate is adequate. (Dependent upon locality, a $\frac{1}{4}$ in. plate may have to be used; see page 1–108 of *Steel Construction*.) The increase in weight (13.75 in. x 7/32 in. x 3 ft) is only 30.7 lb, or 3.5%.

A third possibility is the use of diagonal stiffeners in tension (Fig. 6-1.2). Tensile force in the stiffener is found, by statics, to be 54.6 kips excess shear

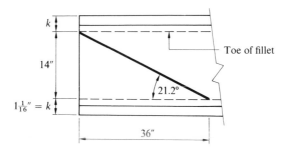

Fig. 6-1.2   Diagonal stiffener

divided by sin 21.2°. This yields a tensile force of 151 kips. At 36 ksi, the area of plates is (151/36) 4.20 in.². This is satisfied by two PL $\frac{3}{4}$ x 3 x 39 in. weighing 49.7 lb, or an increase in weight of 5.7%.

6-1.1 Comments. From the above three possibilities, web doublers appear to be the best choice. This is particularly true if fixity is obtained by framing into a column. (See sec. 5-8 for web stiffeners required in such a member.) Because it is likely that bearing stiffeners (discussed in Part B) are necessary, however, use of diagonal stiffeners might prove more economical.

Similar problems occur, although infrequently, in allowable stress design. For such members, a more common problem arises in the selection of bearing plates to distribute concentrated loads and reactions. For heavier loads, both bearing and intermediate stiffeners are often necessary, as discussed in Part B.

6-1.2 Bearing Plates. Design of bearing plates can be illustrated using a simply supported (weightless) A36 beam spanning 20 ft. A total load of 100 kips at mid-span produces a maximum moment of 500 ft-kips and maximum shear of 50 kips. A W30 x 99 is selected as most economical, with a web thickness of 0.522 in. and k of $1\frac{7}{16}$ in. (*Steel Construction*, pages 1–28 and 1–12).

By AISC Specification 1.5.1.3.5, allowable stress at the toe of the fillet is 0.75 F_y or 27 ksi. It is reasonable to assume that the load is distributed through the member at 45° (see Fig. 6-1.3), so effective length under the load is $\ell + 2k$ but only $\ell + k$ at the supports, where ℓ is the length of the bearing plate. Thus at the mid-span

$$\frac{P}{\sigma_{all}} = \frac{100}{27} = 0.522(\ell + 2\tfrac{7}{8})$$

so that minimum ℓ is 4.22 in. At the supports,

$$\frac{P}{\sigma_{all}} = \frac{50}{27} = 0.522(\ell + 1\tfrac{7}{16})$$

so that minimum ℓ is 2.11 in.

If the end bearing plate rests on a concrete pedestal, AISC 1.5.5 specifies allowable bearing as 0.25f'_c. If f'_c = 4000 psi, bearing plate area must be at

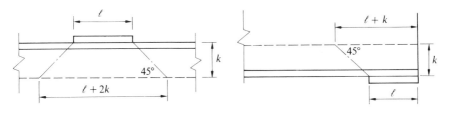

Fig. 6-1.3   Bearing Plates

least $(50^k/1^k/in.^2)$ 50 in.^{2}. Based upon the $10\frac{1}{2}$-in. flange width of a W30 x 99, welding ease, and overall economy, it seems reasonable to make both types of bearing plates 5 in. x 10 in.

6-2 SHEAR IN REINFORCED CONCRETE

When the term *shear* is used alone with reference to reinforced concrete, it usually means *shear as a measure of diagonal tension*. By contrast, true shear, usually referred to as *punching shear*, is critical at the intersection of columns and footings or floors. The following discussion is limited to shear as a measure of diagonal tension.

A Mohr's circle representation for stresses in a member subject to 1000 psi compression and, in succession, no shear, 200 psi shear, and 500 psi shear is shown in Fig. 6-2.1. From the sketches it is apparent that principal tensile stress, acting on a plane inclined to the beam's neutral axis, is related to the amount of shear stress present. Direct computation of this diagonal tension by Mohr's circle is not feasible, however, because reinforced concrete is not homogeneous, isotropic, or continuous, and it cracks. At the face of an open crack, steel is elongated while concrete stress is zero. Moving away from the face of the crack, concrete stress increases due to transfer from reinforcing by bond. Because of these and other difficulties, a simplified rather than a completely rational theory is used to develop an equation which produces reasonable results.

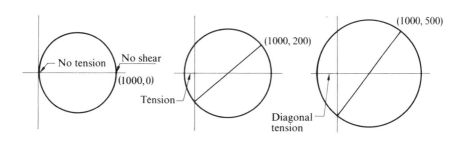

Fig. 6-2.1 Shear as a measure of diagonal tension

6-2.1 Shear as a Measure of Diagonal Tension. Figure 6-2.2 depicts a segmental length of reinforced concrete beam, corresponding to Fig. 5-2.8 used to develop the old bond stress equation. Changes in bending moment over incremental length dx are expressed as either Vdx or $\Delta T(jd)$, so that

$$\Delta T = Vdx/jd. \qquad (6\text{-}2.1)$$

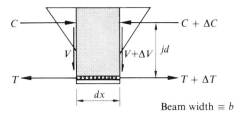

Fig. 6-2.2 Segmental beam length

However, ΔT may also be represented by the horizontal shear force as

$$\Delta T = vbdx. \tag{6-2.2}$$

Equating these two expressions for ΔT, we have

$$v = V/bjd, \tag{6-2.3}$$

the equation used for many years by ACI. Due to gross assumptions and comparative consistency of j, however, the equation was simplified in ACI-63 (ACI-63, formulas 12-1 and 17-1) to $v = V/bd$. In the 1971 ACI Code, formula 11-3 states

$$v_u = \frac{V_u}{\phi b_w d}, \tag{6-2.4}$$

where $v_u \equiv$ nominal shear stress, $b \equiv$ beam width, $d \equiv$ distance from extreme fiber in compression to centroid of tensile steel, $\phi \equiv$ capacity reduction factor = 0.85, and $V_u \equiv$ total applied design shear (including load factors).

A simplified expression for allowable stresses that takes into account an average j is $v_c = 2\sqrt{f'_c}$. Occasionally ACI formula 11-4 can be used to advantage:

$$v_c = 1.9 \sqrt{f'_c} + 2500 \, p_w \, V_u d/M_u \leq 3.5 \sqrt{f'_c}$$

with $V_u d/M_u$ limited to unity. When these stress allowances are exceeded, web reinforcement is required.

6-2.2 Web Reinforcement. Typical web reinforcement is pictured in Fig. 6-2.3, and includes a standard stirrup, a double (or W) stirrup, and the tie. These stirrups and ties may be used perpendicular to the neutral axis of the beam (vertical stirrups) or inclined at some angle (inclined stirrups). If vertical, they are normally looped over compression bars (top bars) for anchorage. If the stirrups are inclined by more than 45°, they must be welded to tension steel in order to keep them from slipping lengthwise along the beam (in case of bond failure).

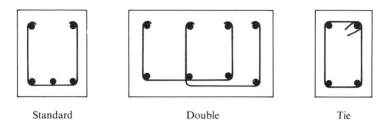

Standard   Double   Tie

Fig. 6-2.3 Typical stirrups as web reinforcement

Sometimes W stirrups are constructed from one piece rather than two. Often this is objectionable because it makes more difficult the placement of longitudinal reinforcing bars.

Ties are being used more frequently to increase ductility (ACI, Appendix A), to keep compression bars from buckling (ACI 7.12.5) and when stress reversals or torques exist (ACI 7.12.6), etc. Since ties are made of light steel (#3 or #4 bars), the top is sometimes left open so that longitudinal steel may be placed more easily, with final bends made in the field. This practice is no longer permitted unless shown on the plans or approved by the engineer.

Bent bars are also used to resist shear. Such bars are simply longitudinal steel reinforcement which is no longer needed to resist bending stresses.

Figure 6-2.4 defines and uses the nomenclature to derive formulas for spacing stirrups or bent bars. Stirrups (or bent bars) are shown inclined at some arbitrary angle β. The crack, however, is assumed to be at 45°, representing the worst possible condition. Crack length, defined as L', is related to β and stirrup spacing s by the Law of Sines:

$$L' = s \sin \beta / \sin (135 - \beta) = s \sin \beta / \cos 45 (\cos \beta + \sin \beta). \quad (6\text{-}2.5)$$

It is assumed that web reinforcement carries only shear in excess of that which concentrate alone could take. Therefore it can be seen that

$$V_u = V_c + V', \quad (6\text{-}2.6)$$

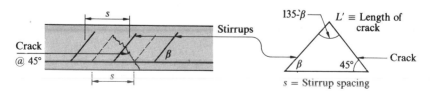

Fig. 6-2.4 Stirrup and crack geometry

where V_u ≡ total shear; V_c ≡ shear carried by concrete ($v_c bd$); and V' ≡ shear carried by stirrups.

Similarly,

$$v_u = v_c + v', \tag{6-2.7}$$

where

v_u ≡ total shear stress (if all were carried by concrete) = $V_u/bd\phi$,

v_c ≡ allowable shear stress in concrete (V_c/bd),

v' ≡ shear stress, in excess of that allowed on concrete, which must be carried by web reinforcement = $V'/bd\phi = v_u - v_c$.

Equations (6-2.6) and (6-2.7) disguise an anomaly which exists because of the capacity reduction factor ϕ. Using the latter equation and corresponding definitions, it is evident that Eq. (6-2.7) converts to $V_u = \phi V_c + V'$ rather than to Eq. (6-2.6). To facilitate computations, adhere strictly to the code, and to remain conservative, the author prefers to calculate V_c as $V_u - V'$ *without* dividing by ϕ. See, for example, step 8 in sec. 6-2.3.

Across a particular crack, the vertical component (only) of shearing force, $bL'v' \cos 45°$, is carried by web reinforcement. With L' expressed as in Eq. (6-2.5), rearranging gives

$$F_c = \frac{V's \sin \beta}{d(\sin \beta + \cos \beta)\phi}, \tag{6-2.8}$$

where F_c ≡ the vertical component of total force across the crack.

Total available resistance from steel crossing the crack is

$$R_s = A_v f_y \sin \beta, \tag{6-2.9}$$

where R_s ≡ vertical component of resistance of steel web reinforcement, A_v ≡ area of steel web reinforcement, and f_y ≡ yield stress in steel web reinforcement.

Equating F_c and R_s from Eqs. (6-2.8) and (6-2.9), we have $A_v = (v'bs)/[f_y(\cos \beta + \sin \beta)]$, which is ACI 11-14 when expressed as

$$A_v = \frac{(v_u - v_c)b_w s}{f_y(\sin \alpha + \cos \alpha)}. \tag{6-2.10}$$

If the web reinforcment is vertical, $\beta = \alpha = 90°$ and

$$A_v = V's/f_y d\phi = (v_u - v_c)b_w s/f_y, \tag{6-2.11}$$

which corresponds to ACI formula 11-13.

Inclined stirrups are expensive to place and are not used very often. Bending of longitudinal bars is quite common; and when shears are quite high they may be used to help resist shearing stresses. In many offices, however, designers use vertical stirrups to carry shear, with bent bars simply considered as an additional safety factor. Because of the preponderant use of vertical stirrups, only their spacing is discussed. This technique, of course, applies as well and almost directly to the spacing of inclined stirrups and bent bars.

6-2.3 Placement of Vertical Stirrups. Effective stirrup area is total steel crossing a crack, which for the standard U stirrups is two bars (for standard W stirrups, four bars). Minimum practical spacing for stirrups is about three inches and economical spacing of bars is accomplished by multiples of 3 in. (occasionally 4 in.). Although use of 3-in. multiples may necessitate more stirrups, overall economy probably is enhanced by reduced labor costs. Chalk marks can be made rapidly on the forms at 2-ft intervals and the stirrups placed (and inspected) by eye. (The method still applies for other spacing schemes.) A correct stirrup size, normally a #3, #4, or #5 bar can be determined by using in Eq. (6-2.11) a value of 3 in. for spacing.

An efficient approach for placement of stirrups in a uniformly loaded beam, 16-ft span, $b = 12$ in., and $d = 30$ in., is shown by the steps given below. Maximum ultimate shear is 108 kips, allowable shear stress v_c (ultimate) is $2\sqrt{f_c'} = 109.5$ psi, and allowable stirrup stress f_y is 40 ksi ($\phi = 0.85$ and $f_c' = 3000$ psi).

1. Determine $v_u = V_u/bd\phi$: $v = 108,000/(12)(30)(0.85)$
                                      $= 352.9$ psi.

2. Determine $v' = v_u - v_c$: $v' = 352.9 - 109.5 = 243.4$ psi.

3. Assume minimum spacing: say, $s = 3$ in. minimum.

4. Find $A_{sR} = (v_u - v_c)bs/f_y$: $A_{sR} = (243.4)(12)(3)/40000$
                                      $= 0.219$ in.^{2}.

5. Select stirrups: Use #3 U stirrups $2(0.11)$
                                      $= 0.22 > 0.219$.

6. With stirrup size known, use Eq. $s = 3$ in. 6 in. 9 in. 12 in. 15 in.
   (6-2.11) to obtain V' with $V' =$ $V' = 75.0$ 37.5 25.0 18.7 15 kips
   $(0.022)(40)(30)(0.85)/s$ to $s_{max} = d/2$.
   [By ACI 11.6.3, $s_{max} = d/4$ if
   $v_u - v_c > 4\sqrt{f_c'}$.]

7. Sketch shear diagram (only half because of symmetry) to scale (see Fig. 6-2.5).

8. Then solve for $V_c = V - V'$; $V_c = 108.0 - 74.5 = 33.5$ kips.

9. Compute maximum spacing as controlled by minimum area of shear reinforcement required; see ACI formula 11-1 and in this text sec. 5-10.4. $A_v = 50\ b_w s/f_y$ or $s_{max} = (0.22)(40,000)/50(12) = 14\frac{2}{3}$ in. Stirrups are required at this maximum spacing until $v_u \leq v_c/2$, or $V_c < 16.7$ kips.

10. Scale V_c downward from the top of the shear diagram and draw a line that represents the amount of shear that remains to be carried by stirrups. V_c is taken off the top so that the vertical scale is left intact for measuring V'.

11. Draw broken horizontal lines that correspond to V' computed for various spacings. Indicate maximum spacings permitted *between* broken horizontal lines.

12. Where horizontal broken lines intersect the V' curve, draw vertical broken lines. Maximum permitted spacings may also be shown between these lines.

13. Space the bars reasonably and to conform to minimum requirements.

14. Add extra stirrups at maximum spacing to conform with requirements of ACI 11.1.1(d).

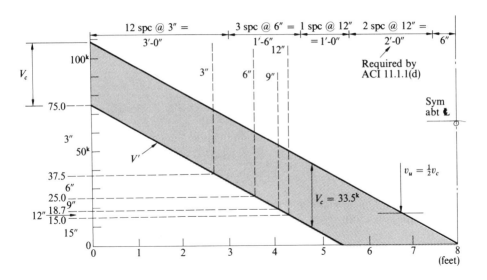

Fig. 6-2.5   Placement of stirrups

The preceding example did not take advantage of ACI 11.2.2 which states that maximum design shear shall be taken at d from the face of the support. Assuming 6 in. from the beam centerline to the face of the support, Fig. 6-2.6 shows the effect on required stirrup spacing. In this case, the V' curve is made horizontal from the point $b + d = 36$ in. from the centerline of the support. Although used for a uniformly loaded beam, this technique is even more powerful for members that have complex shear diagrams.

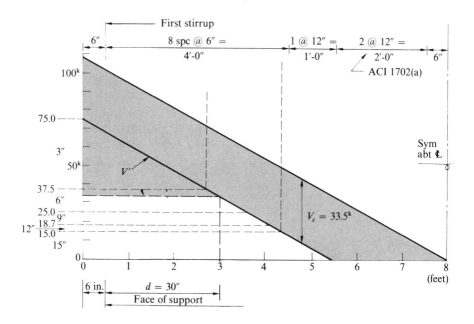

Fig. 6-2.6 Maximum shear at *d* for face of support

If the total shear exceeds $8\sqrt{f_c'}$ according to ACI 11.6.4, web reinforcement cannot be used to carry excess shear and it is necessary to increase beam dimensions.

In cases where shear is critical, a more detailed analysis is afforded for computing allowable v_c by ACI formula 11.4:

$$v_c = 1.9\sqrt{f_c'} + 2500\, p_w \frac{V_u d}{M_u} \leq 3.5\sqrt{f_c'}, \qquad (6\text{-}2.12)$$

where $v_c \equiv$ nominal permissible shear stress carried by concrete; $p_w \equiv A_s/b_w d$; $V_u \equiv$ total shear of section (including load factors); $M_u \equiv$ total bending moment of section (inch-pounds); and $V_u d/M_u \leq 1.0$. However, because of the complexity of computation little advantage is taken of this provision except in cases where design is facilitated by an electronic computer.

PART B
INCREASING THE MOMENT OF INERTIA DIRECTLY

The moment of inertia I is increased most directly by increasing dimensions. In reinforced concrete structures, however, an attempt is usually made to restrict overall dimensions and increase the amount of reinforcing steel used.

When using steel sections, maximum sizes are limited by those available unless special sections are fabricated. These special sections are called *plate girders* and are "built up" from (1) plates welded together (welded plate girder) or (2) plates and angles bolted or riveted together. *Cover plates* are also used to increase the moments of inertia of rolled sections without materially increasing overall depth.

6-3 REINFORCED CONCRETE BEAMS

In previous design and analysis of reinforced concrete beams by strength methods, q was limited (1) to approach balanced WSD design proportions, (2) to eliminate excessive deflection or related calculations, and (3) to facilitate placement of reinforcing bars. As pointed out in secs. 2-5.11 and 2-5.12, q actually may be higher and still be in compliance with code requirements. This indicates that effective moment of inertia can be increased by using more positive steel.

Such an approach is not possible in the alternate WSD method because an upper limit on the amount of positive steel is inherent in formulations. In such cases, to increase effective I without increasing overall dimensions, doubly reinforced beams are used. Such beams, with tensile reinforcing in one face and compressive in the other, are used also in stress design.

6-3.1 Increasing the Amount of Positive Reinforcement. Available beam space presents a major difficulty in the use of more reinforcing. To overcome the problem, bars may be collected together in small *bundles.*

Effect on web width can be illustrated by placing 12 in.^{2} of steel furnished by 12-#9 bars. Reference to Fig. D-3 shows that minimum web width for ten bars is 25.5 in. For 12 bars, the width is calculated as follows:

$$
\begin{array}{llll}
\text{12 times bar diameter} & = & 12 \times 1\tfrac{1}{8} = & 13.5 \\
\text{2-\#3 stirrup diameters} & = & 2 \times \tfrac{3}{8} = & 0.75 \\
1\tfrac{1}{2}\text{-in. cover, both sides} & = & 2 \times 1\tfrac{1}{2} = & 3.0 \\
\text{11 spaces between bars} & = & 11 \times 1\tfrac{1}{8} = & \underline{12.38} \ (\text{for } \tfrac{3}{4}\text{-in. maximum aggregate}) \\
 & & & 29.63 \text{ in.}
\end{array}
$$

The result is rounded upward to a minimum width of 30 in.

If bars are placed in bundles, the required web widths are reduced appreciably: three bundles of 4-#9 bars require 15.5 in. and four bundles of 3-#9 bars require 19.0 in.

To facilitate the choice of steel, charts showing several possibilities are available from PCA. Consider again the requirement of 12 in.². Some possibilities, assuming $\frac{3}{4}$ in. maximum aggregate, are:

| | Minimum | |
Bars	web	How placed
3-#18	16.0	One layer
12-#9	19.0	Four bundles
12-#9	16.5	Two layers
12-#9	15.5	Three bundles
12-#9	12.0	Three layers

Use of bars in more than one layer reduces the effective d and, therefore, increases the amount of steel required. As more bars are collected into one bundle, or as bar sizes increase, allowable bond is reduced and more development length is required. With these factors known, a suitable choice usually can be made rather easily.

When increasing the amount of positive steel above about $q = 0.18$: (1) do not exceed upper limit established by $p \leq 0.75p_b$; (2) be sure deflections are not excessive; (3) provide sufficient development length for reinforcing.

6-3.2 Analysis by Strength Method of Beams with Both Tension and Compression Reinforcement. The concept of analysis by the strength method of doubly reinforced beams is illustrated in Fig. 6-3.1. The doubly reinforced beam is replaced by a singly reinforced beam and a *couple* of steel. Related to stress diagrams, compressive force is composed of two components, one that represents concrete and the other steel. Static equilibrium is maintained when tensile forces equal the sum of these two compressive force. For ease, this tensile force is also broken into two components. Equivalent to C_s, T_2 implies that steel area of the T_2 component is exactly equal to steel area in compression only if $f'_s = f_y$. The remaining steel establishes T_1 which, in turn, establishes the magnitude of C_c and hence the distance a of Whitney's stress block. These relationships are found by combining the two couples so that $M_u = \phi[T_1 jd + T_2(d - d')]$, or

$$M_u = \phi[(A_s - A'_s)f_y(d - a/2) + A'_s f'_s(d - d')], \qquad (6\text{-}3.1)$$

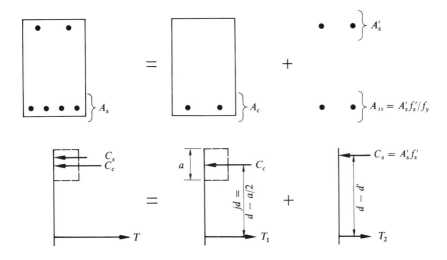

Fig. 6-3.1   Equivalent beams and stress diagrams

where bending moment is in inch-pound units if f_s' is in psi.

This moment may be expressed in foot-kip units as

$$M_u'^k = K_u F + \phi A_s' f_s'(\text{ksi})(d - d')/12 \qquad (6\text{-}3.2)$$

or

$$M_u'^k = M_{cs} + M_{ss}, \qquad (6\text{-}3.3)$$

where $M_{cs} \equiv$ moment of concrete in the top and steel in the bottom (ft-kips); $M_{ss} \equiv$ moment of compressive steel acting with corresponding tensile steel, that is, the moment of steel in the top and steel in the bottom (ft-kips); $f_s' \equiv$ stress in compressive steel which is often expressed as f_y (see sec. 6-3.3); and K_u is based on q, which is based on p, and computed as:

$$p = \frac{\text{area of tensile steel } not \text{ used to balance compressive steel}}{bd}. \qquad (6\text{-}3.3\text{a})$$

In general, Eqs. (6-3.2), (6-3.3), and (6-3.4) provide more economical structures if $f_s' = f_y$ and p is below its absolute maximum of $\frac{3}{4}p_b$. Simple means for exploring such possibilities are developed in the following sections.

6-3.3 Evaluating f_s'.   Because the strength method is based on idealized ultimate strength under factored loads (those that include load factors), extreme compressive strain in concrete is defined as 0.003 in./in., while tensile steel concurrently is at yield stress. At balanced design, $E_s = f_y/E_s$, but for other conditions steel strain increases without a concurrent change in stress. For these reasons, compressive steel strain can be determined only if c (often called kd), the distance from extreme compressive fiber to neutral axis, is known.

For design purposes, it is most convenient to work with (1) $A'_s \equiv$ total compressive steel at f'_s and (2) $A_s \equiv$ total tensile steel at f_y. The latter has two components: $A_c \equiv$ tensile steel offsetting compressive concrete, and $A_{ts} \equiv$ tensile steel offsetting compressive steel $= A'_s f'_s / f_y$.

By the use of these definitions, $A_c = A_s - A_{ts} = A_s - A'_s f'_s / f_y$, or

$$A_c = (A_s f_y - A'_s f'_s)/f_y. \qquad (6\text{-}3.4)$$

As for a singly reinforced beam $C_c = T_1$ (Fig. 6-3.1) or $0.85 f'_c ab = A_c f_y$. Replacement of A_c by Eq. (6-3.4) and defining $a \equiv \beta_1 c$ (ACI 10.2.7) yields the rearrangement:

$$kd = c = \frac{A_s f_y - A'_s f'_s}{0.85 f'_c b \beta_1}, \qquad (6\text{-}3.5)$$

where $\beta_1 = 0.85 - \langle f'_c - 4000 \rangle 0.00005$. Here $\langle f'_c - 4000 \rangle = 0$ when calculated to be less than or equal to 0.

From the strain triangles of Fig. 6-3.2, $\epsilon'_s{:}\epsilon_c{::}c\text{-}d'{:}c$. By Hooke's law, $\epsilon'_s = f'_s/29{,}000{,}000$ and ϵ_c is set at 0.003. These substitutions and rearrangement yield

$$f'_s = 87{,}000 \, (1 - d'/c). \qquad (6\text{-}3.6)$$

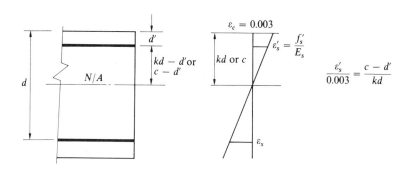

Fig. 6-3.2   Strain relationships

Substitution of the expression for c from Eq. (6-3.5) into Eq. (6-3.6) shows that

$$f'_s = 87{,}000 \left(1 - \frac{d'(0.85) f'_c b \beta_1}{A_s f_y - A'_s f'_s} \right). \qquad (6\text{-}3.7)$$

Of course, if the right-hand side of this equation yields a value either equal to or in excess of f_y, then $f_s' = f_y$. Conversely, values of limiting parameters can be obtained if (1) the equality sign is replaced by $\leq$ and (2) f_s' is replaced by f_y. Thus, if

$$A_s - A_s' < \left[\frac{0.85\beta_1 f_c' 87,000}{f_y(87,000 - f_y)} \right] bd', \qquad (6\text{-}3.8)$$

then $f_s' < f_y$.

The bracketed parameters of Eq. (6-3.8) are fractions of material strengths and are evaluated in Table 6-3.1. Also, actual steel areas can be replaced by steel ratios if both sides of Eq. (6-3.8) are divided by bd. In summary, if

$$A_s - A_s' \geq (\text{coef}) \, bd' \qquad (6\text{-}3.9)$$

or

$$p - p' \geq (\text{coef}) \, d'/d \qquad (6\text{-}3.10)$$

then

$$f_s' = f_y.$$

Otherwise,

$$f_s' = \frac{A_s f_y + A_s' 87 \pm \sqrt{(A_s f_y + A_s' 87)^2 + 4A_s'(87)(0.85 d' f_c' b \beta_1 - A_s f_y)}}{2A_s'}, \qquad (6\text{-}3.11)$$

where f_s', f_y, f_c', and 87 are in kips per square inch. Equation (6-3.11) is obtained directly from Eq. (6-3.7) by use of the quadratic formula.

Table 6-3.1 Coefficients for determination if $f_c' \leq f_y$ from Eqs. (6-3.8), (6-3.9), and (6-3.10) [figures in parentheses are values of $\bar{p}_b$ to be used with Eqs. (6-3.16) and (6-3.17)]

		f_c' (ksi)		
		3	4	5
	40	0.1003(0.0371)	0.1337(0.0495)	0.1573(0.0582)
f_y (ksi)	50	0.1019(0.0275)	0.1359(0.0367)	0.1599(0.0432)
	60	0.1164(0.0214)	0.1552(0.0285)	0.1826(0.0335)

6-3.4 Limiting Steel Ratio. Balanced reinforcement ratio p_b for doubly reinforced beams is based upon definitions of tensile stress blocks. In general terms, $T = A_s f_y = pbdf_y$ and is composed of two components: (1) $T_c = A_c f_y = p_c bd f_y$, that part which balances the compressive stress block resultant for concrete, and (2) $T_s = A_{ts}f_y = A'_s f'_s/f_y = p'bdf'_s/f_y$, that which balances the compressive steel. Thus

$$T = T_c + T_s. \tag{6-3.12}$$

A balanced ratio for beams with compressive reinforcement is defined when $p_c = \overline{p}_b$(balanced ratio for the same beam if there were no compressive reinforcement). Therefore

$$p_b bdf_y = \overline{p}_b bdf_y + p'bdf'_s/f_y \tag{6-3.13}$$

or

$$p_b = \overline{p}_b + p'f'_s/f_y. \tag{6-3.14}$$

By ACI 10.3.2, p is limited to three-quarters of this value, so

$$p \le (\tfrac{3}{4})(\overline{p}_b + p'f'_s/f_y), \tag{6-3.15}$$

where

$p \equiv A_s/bd$ ($A_s \equiv$ total tensile steel reinforcement);

$p_b \equiv$ same as p, except limited to the single value obtained which insures that $p_c = \overline{p}_b$

$\overline{p}_b \equiv A_b/bd$ ($A_b \equiv$ tensile steel required for balanced design if *no* compressive steel is used);

$p' \equiv A'_s/bd$ ($A'_s \equiv$ actual area of compressive steel used).

In most situations, total compressive steel denoted by A'_s is sought, so that a more useful form is

$$p' \ge \left(\frac{4p}{3} - \overline{p}_b\right) f_y/f'_c \tag{6-3.16}$$

or

$$A'_s \ge \left(\frac{4}{3}A_s - A_b\right) f_y/f'_c \tag{6-3.17}$$

where $\overline{p}_b$ can be read from Table 6-3.1, which was computed by Eq. (2-5.27), and $A_b = \overline{p}_b bd$.

6-3.5 Member Capacity versus A'_s. To illustrate some interesting relationships, curves are plotted in Fig. 6-3.3 for members with the following parameters: $M_u = 843.3$ ft-kips, $f_y = 60$ ksi, $f'_c = 4$ ksi, $d' = 2\frac{1}{2}$ in., $b = 12$ in., $d = 30$ in., and $F = 0.9$.

As plotted, solutions are obtained by *assuming* $f'_s = f_y$ for all cases, which is not theoretically correct. The procedure shown in Table 6-3.2 is:

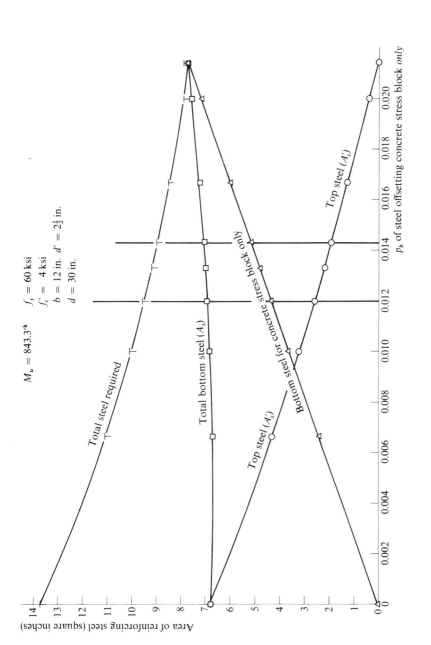

Fig. 6-3.3 Effects due to change in amount of compressive steel used

Table 6-3.2 Computations for Fig. 6-3.3

p	K_u	M_{cs}	M_{ss}	A'_s	A_c	A_s	Total A
0.0214	937	843.3	0	0	7.704	7.704	7.704
0.02	889	800.1	43.2	0.349	7.200	7.549	7.898
0.0167	767	690.3	153.0	1.236	6.012	7.248	8.484
0.0143	673.1	605.8	237.5	1.919	5.148	7.067	8.986
0.01333	635	571.5	271.8	2.196	4.800	6.996	9.192
0.012	579.2	521.3	322.0	2.602	4.320	6.922	9.524
0.01	492	442.8	400.5	3.236	3.600	6.836	10.072
0.00667	339	305.1	538.2	4.349	2.400	6.749	11.098
0	0	0	843.3	6.815	0	6.815	13.630

1. Assume p.
2. Read K_u for this p from Fig. D-6 and record.
3. Multiply this K_u by F of 0.9 to obtain M_{cs}.
4. M_{ss} is recorded as $843.3 - M_{cs}$.
5. With f'_s assumed equal to f_y, $M_{ss} = 0.9 f_y A'_s (d - d')/12$, so $A'_s = M_{ss}/123.75$.
6. A_c is calculated as pbd or $360p$.
7. A_s is the sum of $A'_s + A_c$.
8. Finally, total reinforcing used is $A'_s + A_s$. Two special values are used: $p = p_b/2 = 0.0143$ and $p = 0.012$ corresponding to $q = 0.18$.

To find true moments based on actual f'_s, Eq. (6-3.11) is specialized (by use of $f_y = 60$, $d' = 2.5$, $f'_c = 4$, $b = 12$, and $\beta_1 = 0.85$) to

$$f'_s = \frac{(60A_s + 87A'_s) - \sqrt{(60A_s + 87A'_s)^2 + 348A'_s(86.7 - 60A_s)}}{2A'_s}. \tag{6-3.18}$$

Subsequently, the ultimate moment may be found in Table 6-3.3 by the following steps:

1. Record A_s.
2. Record A'_s.
3. Compute the common factor $(60A_s + 87A'_s)$.
4. On line 4, compute value under the radical.
5. Take square root of previous value.
6. Subtract that result from the common factor to obtain the numerator for line 6.

Table 6-3.3 Check Computations for Fig. 6-3.3

1. A_s	6.815	6.749	6.922	7.067
2. A'_s	6.815	4.349	2.602	1.919
3. $60A_s + 87A'_s$	1001.8	783.3	641.7	591.0
4. $(60A_s + 87A'_s)^2 + 348A'_s(86.7 - 60A_s)$	239477	131922	114207	123983
5. Square root of line 4	489.3	363.2	338.0	352.1
6. Line 3 − line 5	512.5	420.1	303.7	238.9
7. Line 6 divided by 2 A'_s = f'_s (ksi)	37.6	48.3	58.4	62.2 > 60
8. $M_{ss} = 2.0625\,A_s f'_s$	528.5	433.2	313.2	
9. $p_c = (A_s - A'_s f'_s/60)/360$	0.0071	0.0090	0.0122	
10. K_u	350	448	590	
11. $M_{cs} = 0.9\,K_u$	315	403.2	531	
12. M_u = line 8 + line 11 = $M_{ss} + M_{cs}$	843.5	836.4	844.2	
13. $p_c = A_c/bd$ ($A_c = A_s - A'_s$)	0	0.0067	0.0120	0.0143

7. Obtain f_s' by dividing by $2A_s'$.

8. $M_{ss} = 0.9A_s'f_s'(d - d')/12 = 2.0625A_s'f_s'$.

9. Division of both sides of Eq. (6-3.4) by bd yields the equation for p_c (line 9).

10. Use p_c to obtain K_u from Fig. D-6.

11. $M_{cs} = FK_u = 0.9K_u$.

12. $M_u = M_{ss} + M_{cs}$.

In all cases, $p \leq \frac{3}{4}p_b$ because [see Eq. (6-3.17)] $4A_s/3$ is always less than A_b, that is, no compression steel is needed to satisfy ACI 10.3.2.

6-3.6 Optimum Practical Design. Figure 6-3.3 illustrates for a single case what seems to be on the conservative side of correct for most cases: (1) Tension steel *cannot* be reduced appreciably by the addition of compressive steel. (2) Moment capacity *cannot* be increased appreciably by the addition of compressive steel *only*.

Three typical design situations occur:

1. Steel occurs in the compression face because of requirements other than flexure. Such steel usually is disregarded in calculations because it is difficult to justify additional design time for relatively minor savings. Although f_s' may be less than f_y, balanced steel provisions are not violated.

2. Compressive steel is needed to reduce deflections, particularly those that occur over long time periods. For maximum overall economy, beam size should probably be sufficient so that A_s will provide more than $0.18q$ and less than $0.5p_b$. Compressive steel is then added so that A_s' remains at f_y and still satisfies ACI 10.3.2. If excess A_s' is used, it is necessary to compute f_s' and recalculate the minimum A_s' required. Note that as A_s' increases, f_s' decreases and *required* A_s' increases still more! This procedure is summarized in Table 6-3.4. It is often possible to reduce A_s somewhat, but time spent in such calculations does not always justify the savings. A direct procedure is discussed in connection with the third design situation.

3. Compressive steel is needed in conjunction with extra tension steel to carry loads which otherwise would require a singly reinforced beam exceeding $\frac{3}{4}p_b$. In general, this practice is not recommended because of high shear and steel placement difficulties, but the procedure is outlined in Table 6-3.5. Because of severe steel placement problems, p_c of $\frac{3}{4}p_b$ is probably more suitable than $\frac{1}{2}p_b$. Moment not carried by concrete and A_c is called M_{ss}. In step 4, c is related to a by β_1 and then to member parameters by equating T_c and C_c. When c is known, f_s' is computed without need for a quadratic. A_s' is computed for M_{ss} with actual f_s' known, and A_s is obtained readily. At this point, however, it is necessary to check against balanced design limitations. If A_s' is not large enough, it is, without need for additional calculations, simply increased to satisfy code requirements.

Table 6-3.4 Doubly Reinforced Beam Design Procedure

Compressive Steel Needed to Reduce Deflections:

1. Choose A_s so $q \geq 0.18$ and $p \leq 0.5\ p_b$. If impossible, see Table 6-3.5

2. Add compressive steel so

$$A'_s \leq A_s - \text{(coefficient from Table 6-3.1)} \times bd'$$

or

$$p' \leq p - \text{(coefficient from Table 6-3.1)} \times d'/d$$

to insure that $f'_s = f_y$

AND

$$A'_s \geq 4A_s/3 - A_b \text{ (parenthetical figures in Table 6-3.1 times } bd)$$

or

$$p' \geq 4p/3 - \overline{p}_b \text{ (taken from parenthetical figures of Table 6-3.1)}$$

to satisfy ACI 10.3.2.

3. If A'_s or p', actually used, is too large, then compute f'_s by Eq. (6-3.11); and make sure that

$$A'_s \geq (4A_s/3 - A_b)\ f_y/f'_c$$

or

$$p' \geq (4p/3 - \overline{p}_b)f_y/f'_c.$$

Table 6-3.5 Direct Procedure When Beam Size Crucially Limited

Given: $M_u^{'k}$, b, d, d', f_y, and f'_c.

1. Find $M_{cs} = K_u F$ for assumed q or p. (a) F from Fig. D-1 if more complete table unavailable; (b) K_u from Fig. D-5.

2. $A_c = p_c bd$.

←3. $M_{ss}^{'k} = M_u - M_{cs}$.

4. $c = a/\beta_1$

$\quad = A_c f_y/0.85 f'_c b\beta_1 \quad$ [see Eq. (6-3.5)]

$\quad = p_c df_y/0.85 f'_c \beta_1$.

5. $f'_s = 87(1 - d'/c)$ ksi [by Eq. (6-3.6)] .

→6. $A'_s = 12M_{ss}/0.9(d - d')f'_s$ [by Eq. (6-3.2)] .

7. $A_s = A_c + A'_s f'_s/f_y$.

8. $A'_s \geq 4A_s/3 - A_b$? If so, design is complete. If not, increase $A'_s \geq 4A_s/3 - A_b$ (same as $p' \geq 4p/3 - \overline{p}_b$, Table 6-3.1).

6-3.7 Example. Because of architectural limitations, it is impossible to make b greater than 12 in. or for overall depth to exceed 20 in. Design moment is 210 ft-kips (including load factors), $f_y = 60$ ksi, and $f'_c = 4$ ksi. To control deflections, it is decided to limit q to about 0.18 ($p_b = 0.012$). Assume that #3 ties and two layers of tension steel are used, d is estimated as follows: 20 in. $-$ $1\frac{1}{2}$-in. cover $-$ $\frac{3}{8}$-in. tie diameter $-$ $\frac{1}{2}$ in. between layers $-$ $\frac{1}{2}$ in. for one-half of a #8 bar diameter, or $d = 17.125$, say, $d = 17.0$. Similarly, d' is estimated as $1\frac{1}{2}$-in. cover $+$ $\frac{3}{8}$-in. tie diameter $+$ $\frac{1}{2}$ in. for one-half of a #8 bar diameter $=$ 2.375 in., say, $2\frac{1}{2}$ in. With $b = 12$ in. and $d = 17.0$ in., $F = 0.289$.

In accordance with the procedure of Table 6-3.5, find $K_u = 579.2$ using D-5. Then $M_{cs} = (579.2)(0.289) = 167.4$ ft-kips and $A_c = (0.012)(12)(17.0) = 2.45$ in.² and $M_{ss} = 210 - 167.4 = 42.6$ ft-kips. Use A_c to obtain $c = (2.45)(60)/0.85(4)(12)(0.85) = 4.24$ in. Thus (step 5 of Table 6-3.5), $f'_s = 87[1 - (2.5/4.24)]$ $= 35.7$ ksi. By use of M_{ss} in foot-kips and with f'_s known, $A'_s = 12(42.6)/0.9(17.0 - 2.5)35.7 = 1.10$ in.². Then $A_s = 2.45 + 1.10(35.7)/60 = 3.11$ in.². To insure that $p \leq \frac{3}{4}p_b$, A'_s must be $\geq 4(3.11)/3 - (0.0285)(12)(17)$, where 0.0285 is taken from Table 6-3.1. It is, and the design is complete.

For analysis of this same member, proceed as follows:

1. Check $A'_s \leq A_s - $ (coef)bd', that is, if $1.10 \leq 3.11 - (0.1552)(12)(2.5)$ this is not true (0.1552 from Table 6-3.1) so f'_s is not at yield.

2. Obtain f'_s by substitution of actual values in Eq. (6-3.11):

$$f'_s = \Big\{ (3.11)(60) + (1.10)(87)$$

$$- \sqrt{[(3.11)(60) + (1.10)(87)]^2 + 4(1.10)(87)\big[0.85(2.5)4(12)(0.85) - 3.11(60)\big]} \Big\}$$
$$/[2(1.10)],$$

or $f'_s = 35.77$ ksi.

3. Then obtain the moment carried by the steel couple, $M_{ss} = 0.9A'_s f'_s(d - d')/12 = 0.9(1.10)(35.77)(17.0 - 2.5)/12 = 42.8$ ft-kips.

4. With f'_s, A_s', and A_s known, find A_c, the amount of tensile steel used to balance concrete in compression, $A_c = A_s - A'_s f'_s/f_y = 3.11 - 1.1(35.77)/60 = 2.45$ in.².

5. Use any of the singly reinforced beam equations to obtain M_{cs}.

$$M_{cs} = 0.9A_c f_y(d/12)(1-0.59q), \quad [\text{see sec. 2-5.3}]$$

where $q = (A_c f_y)/bdf'_c$ or $q = 2.45(60)/(12)(17)(4) = 0.180$ and $M_{cs} = 0.9(2.45)(60)(17)[1-0.59(0.180)]/12 = 167.5$ ft-kips.

6. Find capacity as the sum of M_{ss} and M_{cs}:

$$M_u = M_{ss} + M_{cs} = 42.8 + 167.5 = 210.3 \text{ ft-kips.}$$

Analysis would be made with steel areas for bars actually chosen and with corresponding values of d and d'. Also, steel placement would have to be checked, as discussed in sec. 5-11.3. Remember that this procedure, as outlined in Table 6-3.5 is for only very special cases. The typical approach is that of Table 6-3.4.

6-3.8 Doubly Reinforced Beams (WSD). Design of doubly reinforced concrete beams by the alternate method (WSD) is not as direct as by the stress method. A similar approach, however, gives reasonable results.

Transformed area is the most direct means of analysis. According to ACI 8.10.1-4, compression steel is converted to an equivalent area of concrete by using $2n$ rather than the modular ratio n. This recognizes that creep in concrete makes actual compressive stresses about double the values computed by elastic analysis. Final compressive stresses, however, may not exceed allowable tensile stresses.

A typical transformed area analysis, shown in Fig. 6-3.4, is similar to the developments in Chap. 2. The neutral axis (kd) is located by taking moments of areas about its assumed position, with steel replaced by an equivalent area of concrete. Once kd is known, the total moment of inertia is found readily. Stresses are computed by Mc/I for the equivalent concrete section. Final steel stresses are found by multiplying equivalent concrete stresses by n or $2n$, as appropriate. In recognition of the fact that steel and concrete cannot occupy the same space, some designers use $2n$-1 rather than $2n$ to convert compression reinforcement.

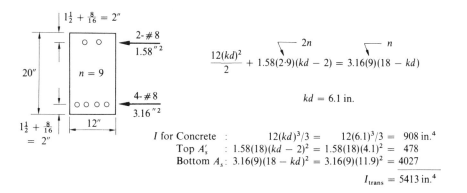

Fig. 6-3.4   Transformed area analysis for a doubly reinforced beam

6-4   STEEL BEAMS

When larger moments of inertia than those furnished by standard rolled sections are desired, the most economical solution may be to add cover plates. As bending moments become higher and spans longer, a more economical solution may be plate girders instead. For still longer spans, trusses, arches, and/or suspension systems often are used in preference to plate girders.

6-4.1 Cover Plates. To illustrate, a rolled A36 beam is designed, with a simple span of 100 ft, supporting a concentrated load of 50 kips located 40 ft from its right support. Deflection is not a design consideration. Overall depth is limited to 35 in. The beam is laterally supported throughout.

Without taking beam dead load into account, the maximum moment is 1200 ft-kips. A W33 x 141 is chosen as a trial section as it appears to be adequate and economical, being good for 896 ft-kips. Cover plates must carry the remaining bending moment (1200 + moment due to beam weight − 896) of about 500 ft-kips. (Fig. 6-4.1).

A W33 x 141 has a 33.31-in. depth, so a couple arm for cover plates could be assumed conservatively at, say, 34 in. Then $500 = (A_{PL})(24 \text{ ksi})(34)/12$ or $A_{PL} = 7.35 \text{ in.}^2$. Selection of a 10 in. x $\frac{3}{4}$ in. cover plate, top and bottom, yields the following:

	Weight (lb/ft)	I (in.^{4})	Z (in.^{3})	$M_{\text{allowable}}$ (ft-kips)
Beam alone	141	7,460	448	896
Cover plates	51	4,350	$= [7.5 \times 2 \times 17.03^2]$	
Beam and cover plates	192	11,810	678	1,356

The average weight is about $141 + \frac{1}{3}(51)$ or, say, 160 lb/ft, including weld material. Therefore, the maximum dead load bending moment at $0.6L$ is 192 ft-kips. These additional dead load moments are shown on Fig. 6-4.1, superimposed over moments due to the 50-kip load. Theoretical points of cutoff are also shown.

Maximum shear over the length of the cover-plated portion is at the right-hand theoretical cutoff point, that is, 26 ft from the right support, where the shear is $[30 + 24(0.16)]$ or 33.84 kips. Shear flow between the cover plate and beam is, therefore (see sec. A-4):

$$v = \frac{VQ}{I} = \frac{(33,840)(7.5)(16.66)}{11,810} = 360 \text{ lb/in.}$$

Use of a $\frac{5}{16}$-in. fillet weld (E60XX) on both sides of the plate (minimum for 15/16-in. flange, according to AISC Table 1.17.5), results in an allowable shear flow calculated as two sides times the number of one-sixteenths in the weld times the allowable, or $(2)(5)(800) = 8$ k/in. Intermittent fillet welds, $1\frac{1}{2}$ in.

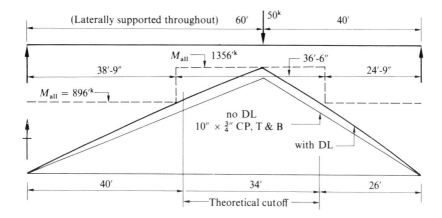

Fig. 6-4.1   Cover-plated beam

long (minimum, by AISC 1.17.8) on 12-in. centers are good for 1000 lb/in. equivalent. Spacing on 12-in. centers is the maximum permitted by AISC Specification 1.18.2.3 (plate thickness $\times\ 127/\sqrt{F_y} \leq 12$) and 1.18.3.1.

In accordance with AISC 1.10.4(2), cover plates are extended $(1.5a')$ 15 in. past their points of theoretical cutoff with $\frac{5}{16}$-in. welds continuous on both sides and across the end. This provides 40 in. of $\frac{5}{16}$-in fillet weld which is good for $40(5)(0.800) = 160$ kips. Actual plate force at the cutoff point is the plate stress times plate area, or

$$F_{PL} = \frac{My}{I}(A_{PL}) = \frac{MQ}{I} = \frac{(896)(17.03)(7.5)}{11,810} = 9.7 \text{ kips.}$$

Thus, 36 ft-6 in. cover plate, top and bottom, is adequate.

Total weight is $141(100) + 51(36.5) = 16,000$ lb. An alternate solution, without cover plates, is to use a W33 $\times$ 220 with a total moment of (264 dead + 1200 live), or 1464 ft-kips, and a weight of 22,000 lb. The cover-plated solution saves $(22,000 - 16,000)/100$, or 60 lb/ft and is more economical.

6-4.2 Welded Plate Girder. Figure 6-4.2 summarizes information needed to design a welded plate girder. To reduce complications, the uniform load of 1 k/ft is assumed to *include* weight of the plate girders itself. (Design equations used in this section are developed theoretically in Chap. 11.)

If ten engineers were given the design parameters of Fig. 6-4.2, in all likelihood, ten *different* welded plate girders would result, and they could all be "right." Probably each engineer would start with two similar assumptions: (1) An economical depth to length ratio is *about* 1 to 20, and (2) a "thin" web

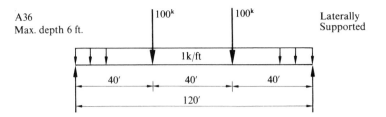

Fig. 6-4.2   Design parameters for a welded plate girder

requiring intermediate stiffeners is probably more economical than a web thick enough to act alone. It is clear, then, that the design detailed below is not the *only* satisfactory solution.

Maximum design moment is $100(40) + 1(120)^2/8 = 5800$ ft-kips. Maximum shear, at a support, is $100 + 1(120)/2 = 160$ kips. Because the plate girder is not a "compact" section, the allowable flexural stress (fully supported laterally) is 22 ksi. Therefore, the required section modulus is

$$Z_R = 12(5800)/22 = 3160 \text{ in.}^3.$$

A reasonable starting size can be obtained by using the tabulation for Dimensions and Properties of Welded Plate Girders (page 2–122 of *Steel Construction*). From this table, a plate girder with 22 in. **x** 2 in. flanges and a 66 in. **x** $\frac{1}{2}$ in. web has a section modulus of 3250 in.^{3} and a total depth of 70 in.

An alternate approach without the use of tables is $Af_s d \doteq M = Zf_s$, so $A = Z/d \doteq 3250/72 = 45$ in.^{2}. Full depth is used for d to compensate for not including the web in section modulus.

According to AISC Specification 1.10.2, slenderness ratio is limited by

$$\frac{h}{t} \leqq \frac{14,000}{\sqrt{F_y(F_y + 16.5)}} \leqq 320,$$

where $h \equiv$ clear distance between flanges and $t \equiv$ web thickness. Increasing the overall depth to the maximum allowable of 72 in., and assuming 2-in. thick cover plates, h is 68 in. Therefore, the minimum web thickness is $68/320 = 0.212$ in. Minimum practical plate thickness greater than 0.212 in. is $\frac{1}{4}$ in. The slenderness ratio for the $\frac{1}{4}$-in. plate is $68/(0.25)$ or 272. Increasing the web thickness to $\frac{5}{16}$ in. decreases the slenderness ratio to 218.

If shear stresses are sufficiently low, and if h/t is less than 260 (AISC Specification 1.10.5.3, paragraphs 1 and 2), no intermediate stiffeners are required. Otherwise, they must be used with a maximum permitted spacing of $3h$. Refer to Fig. 6-4.2 and it is obvious that very little shear exists over the middle 40 ft of span. Therefore, a decision must be made between (1) the use of $\frac{1}{4}$-in. plate and two sets of intermediate stiffeners in the middle 40 ft of span or (2) the use of $\frac{5}{16}$-in. plate and no intermediate stiffeners in the middle 40 ft.

Because shear in the end 40 ft of the span is high, stiffeners are required there with the choice of either thickness, so use a web plate of $\frac{1}{4}$ in. x 68 in.

To space intermediate stiffeners, formulas from AISC Specification 1.10.5.2 are used. These formulas, which are derived in Chap. 11, state that allowable average web shear in any panel between stiffeners must not exceed F_v, where

$$F_v = \frac{F_y}{2.89}\left[C_v + \frac{1 - C_v}{1.15\sqrt{1 + (a/h)^2}}\right] \qquad C_v < 1.0, \qquad (6\text{-}4.1)$$

or

$$F_v = \frac{F_y}{2.89}\ C_v \le 0.4\ F_y \qquad\qquad C_v > 1.0, \qquad (6\text{-}4.2)$$

where

$$C_v = \frac{45,000k}{F_y(h/t)^2} \qquad\qquad C_v < 0.8,$$

or

$$C_v = \frac{190}{h/t}\sqrt{k/F_y} \qquad\qquad C_v > 0.8, \qquad (6\text{-}4.3)$$

and

$$k = 400 + \frac{5.34}{(a/h)^2} \qquad\qquad a/h < 1.0,$$

or

$$k = 5.34 + \frac{4.00}{(a/h)^2} \qquad\qquad a/h > 1.0.$$

For any particular steel, these equations are functions of only slenderness ratio (h/t) and aspect ratio (a/h). Thus a tabular solution is developed readily as shown for A36 steel in Table 3–36 on page 5–95 of *Steel Construction*. At the same time, AISC formula 1.10–3 specifies the area of intermediate stiffeners as

$$A_{st} = \frac{1 - C_v}{2}\left[\frac{a}{h} - \frac{(a/h)^2}{\sqrt{1 + (a/h)^2}}\right]YDht, \qquad (6\text{-}4.4)$$

where $Y \equiv$ yield point web steel divided by yield point of stiffener steel; and $D \equiv 1.0$ for stiffeners furnished in pairs, $D \equiv 1.8$ for single angle stiffeners, or $D \equiv 2.4$ for single plate stiffeners. Therefore, required stiffener areas are also shown as a percentage of web area. From this table, with $h/t = 272$ and a shearing stress of $160/\frac{1}{4}(68) = 9.41$ ksi (reduction because of distance from support neglected), a reasonable aspect ratio a/h and area of stiffeners is obtained.

For $h/t = 260$, $F_v = 9.9$ and 9.4 for aspect ratios 0.7 and 0.8, respectively; 9.7 and 9.2 if $h/t = 280$. Interpolation between h/t-values yields approximately F_v-values of 9.8 and 9.3 for h/t of 272. Interpolation between aspect ratios for $F_v = 9.41$ suggests a conservative a/h of about 0.77. This limits maximum spacing to $0.77 \times 68 = 52.4$ in.

As a check,

$$k = 4.00 + 5.34/(0.77)^2 = 13.00,$$
$$C_v = \frac{45,000(13)}{36(272)^2} = 0.220 < 0.8,$$

and

$$F_v = \frac{36}{2.89} \left[0.220 + \frac{1 - 0.220}{1.15\sqrt{1 + (0.77)^2}} \right] = 9.43 \text{ ksi.}$$

In practice, such precision is seldom warranted because reasonable spacing of stiffeners, in general, is less than that permitted.

Area, by Eq. (6-4.4), is

$$A_{st} = \frac{1 - 0.220}{2} \left[0.77 - \frac{(0.77)^2}{\sqrt{1 + (0.77)^2}} \right] (1)(1)ht = 0.117 \, ht.$$

These values, shown as percentages in the tables, can be obtained rather easily by inspection. Total $A_{st} = 0.117 \times \frac{1}{4}(68) = 1.99$ in.^{2}.

Spacing at the end panel is covered by Supplement No. 2 to the *Specification for Design, Fabrication and Erection of Structural Steel for Building*, that is, the AISC Code, which requires that f_v (note that f_v is actual shear stress whereas F_v is allowable shear stress) not exceed $F_y(C_v)/2.89$. By use of this criterion and with f_v previously computed as 9.41 ksi,

$$9.41 = 36C_v/2.89,$$

or $C_v = 0.7554 < 0.8$. By the first of Eqs. (6-4.3),

$$0.7554 = \frac{45,000k}{36(272)^2} \quad \text{or} \quad k = 44.71.$$

Then, $44.71 = 4 + 5.34/(a/68)^2$, from which maximum end panel spacing a is 24.6 in.

For reasonable spacing:

1st space =	21 ins. =	1 ft 9 in.	< 24.6 in.
9 spaces at 51 in. (4 ft 3 in.)		= 38 ft 3 in. each	< 52.5 in.
2 spaces at 120 in. (10 ft 0 in.)		= 20 ft 0 in. each	< 3(68) in.
	To centerline —	60 ft 0 in.	

Stiffener dimensions are determined in the next section in conjunction with the design of bearing stiffeners. When it is known that an adequate web has reasonable stiffener spacing, however, it is appropriate first to determine final flange dimensions.

For a required section modulus computed as 3160 in.^{3} and a 68-in. deep web with 2-in. thick flange plates assumed, the required moment of inertia would be (3160 $\times$ 36) 113,800 in.^{4}. Moment of inertia of a 68 in. $\times \frac{1}{4}$ in. web plate is 6550 in.^{4} (or, see *Steel Construction*, p. 2–130). Flange plates, then, must have a moment of inertia of (113,800 $-$ 6550) $=$ 107,250 in.^{4}. By use of $2(A_{PL}y^2 + I_{PL})$ or $2[2w(35)^2 + w(2)^3/12]$ $=$ 107,250, w, the required plate width is determined as 21.88 in. Therefore reasonable flanges are made from PL 2 $\times$ 22. The width of the projecting flange is (22 in. $- \frac{1}{4}$ in.)/2 $=$ 10$\frac{7}{8}$ in. This provides a width-thickness ratio much less than 15.8 as permitted by AISC Specification 1.9.1.2, so the design is complete except for stiffeners.

6-4.3 Design of Stiffeners. Bearing stiffeners at supports are designed for total reactions of 160 kips; bearing stiffeners under concentrated loads are designed for only 100 kips. These stiffeners act as columns, assuming ℓ as $3h/4$, in accordance with AISC 1.10.5.1.

Cross-sectional dimensions of bearing stiffeners are based on a $12t$ length of web for end bearing and a $25t$ length of web at interior points. Additional area is provided by the portion of stiffeners which bears on the flange outside the flange-to-web weld.

Allowable bearing stress is 0.9 F_y in accordance with AISC Specification 1.5.1.5.1 because Specification 1.10.5.1 requires "close bearing against the flange." Minimum bearing area is, therefore, (160/0.9F_y $=$ 160/33) 4.85 in.^{2}, for end bearing stiffeners and 3 in.^{2} at interior points.

Bearing stiffeners require an area of at least 4.85 in.^{2} to satisfy bearing stresses, a width-to-thickness ratio of 15.8 to preclude local buckling, and sufficient area not to be overstressed as a column using an effective length of $3h/4 = 51$ in. Two PL $\frac{7}{16} \times 5\frac{3}{4}$ satisfy the first two criteria. The third criterion is checked by rounding to PL $\frac{7}{16} \times 6$.

Moment of interia is figured about the longitudinal axis of the girder, with both I and A including a part of the web as specified by AISC 1.10.5.1. F_a is read from AISC Table 1-36 based on ℓ/r. Allowable stiffener load P is this effective web and stiffener area times F_a. Calculations are shown in Fig. 6-4.3. Although this is adequate for points under loads, reactions exceed permissible values at the supports. By ratio, $\frac{9}{16}$-in. plates would probably be adequate (actually good for 156.5 kips).

By previous calculation, required area of the intermediate stiffeners is 0.117 ht or (0.117 $\times$ 68 $\times \frac{1}{4}$) 1.99 in.^{2}. AISC Specification 1.10.5.4 requires these to have a minimum moment of inertia, with respect to an axis parallel to the web, of at least $(h/50)^4$ or $(68/50)^4 = 3.43$ in.^{4}.

Based on these criteria and a minimum width to thickness ratio of 15.8, use 2 PL $\frac{1}{4}$ × 4 in pairs for $A = (2)(\frac{1}{4})(4) = 2$ in.² and $I = 2(\frac{1}{4})(4)^3/3 = 10.6$ in.⁴.

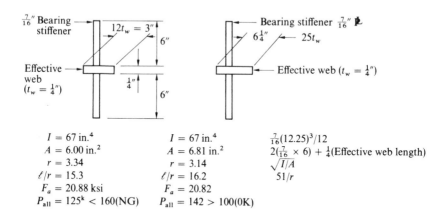

$$I = 67 \text{ in.}^4 \qquad I = 67 \text{ in.}^4 \qquad \tfrac{7}{16}(12.25)^3/12$$
$$A = 6.00 \text{ in.}^2 \qquad A = 6.81 \text{ in.}^2 \qquad 2(\tfrac{7}{16} \times 6) + \tfrac{1}{4}(\text{Effective web length})$$
$$r = 3.34 \qquad r = 3.14 \qquad \sqrt{I/A}$$
$$\ell/r = 15.3 \qquad \ell/r = 16.2 \qquad 51/r$$
$$F_a = 20.88 \text{ ksi} \qquad F_a = 20.82$$
$$P_{\text{all}} = 125^k < 160(\text{NG}) \qquad P_{\text{all}} = 142 > 100(\text{OK})$$

Fig. 6-4.3   Bearing stiffeners

6-4.4 Plate Girder—Final Design. The final design of a plate girder is shown in Fig. 6-4.4 and the final analysis in Fig. 6-4.6. Flange plates are reduced to 22 in. × 1 in. in the end 20 ft of the span, and have an allowable moment of 3080 ft-kips. Stiffeners are welded to the web for a total ($f_{vs} = h\sqrt{(F_y/340)^3}$, by AISC Specification 1.10.5.4) of 0.034h or 2.4 k/in., that is, 1.2 k/in. for each stiffener. Shear flow is calculated (VQ/I) between the web and flanges to determine the amount of fillet weld required.

If girders can be taken intact to the erection site, necessary splices can be accomplished by full penetration, full-strength welds without need for calculations. Otherwise, splices are made by using plates and (probably) high-strength bolts. Typical arrangements for flange splices are shown in Fig. 6-4.5. Net area of splice plates must provide sufficient capacity to carry the moment at the spliced point, in accordance with AISC 1.10.8.

Flange splices must be long enough to contain sufficient rows of high-strength bolts to carry the total stress block force from one flange, through the splice, to the other corresponding flange.

A typical shear splice connection is illustrated in Fig. 6-4.7; it is designed exactly as a discrete eccentric connection, considering connectors on *only* one side of the splice. Total shear V is maximum at the splice. Eccentricity e is the distance between centroids of connector patterns on opposite sides of the splice. Thus, design and analysis is as developed in Chap. 5.

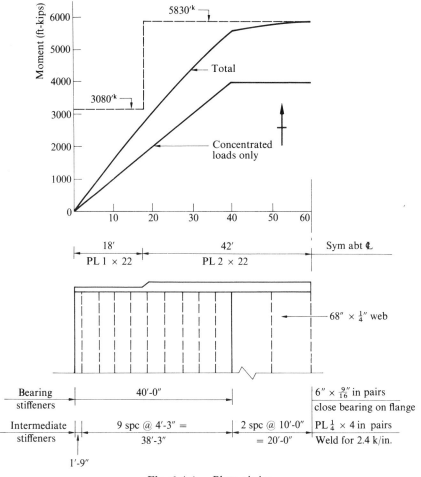

Fig. 6-4.4 Plate girder

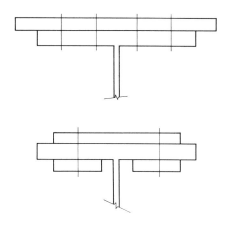

Fig. 6-4.5 Flange splice arrangements

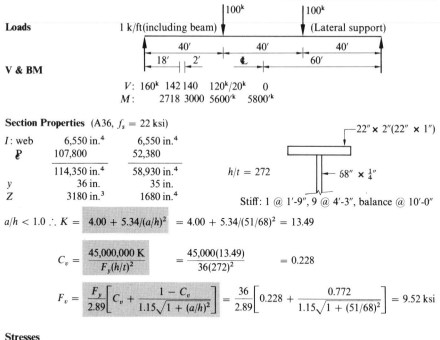

Loads — 1 k/ft(including beam) · 100^k · 100^k (Lateral support)

40′ | 40′ | 40′

18′ | 2′ | $\mathbb{C}$ | 60′

V & BM

V: 160^k 142 140 120^k/20^k 0
M: 2718 3000 5600$'^k$ 5800$'^k$

Section Properties (A36, $f_s = 22$ ksi)

$22'' \times 2''(22'' \times 1'')$

I: web	6,550 in.^{4}	6,550 in.^{4}
P	107,800	52,380
	114,350 in.^{4}	58,930 in.^{4}
y	36 in.	35 in.
Z	3180 in.^{3}	1680 in.^{4}

$h/t = 272$

$68'' \times \frac{1}{4}''$

Stiff: 1 @ 1′-9″, 9 @ 4′-3″, balance @ 10′-0″

$a/h < 1.0 \therefore K = 4.00 + 5.34/(a/h)^2 = 4.00 + 5.34/(51/68)^2 = 13.49$

$$C_v = \frac{45,000,000\ K}{F_y(h/t)^2} = \frac{45,000(13.49)}{36(272)^2} = 0.228$$

$$F_v = \frac{F_y}{2.89}\left[C_v + \frac{1 - C_v}{1.15\sqrt{1 + (a/h)^2}}\right] = \frac{36}{2.89}\left[0.228 + \frac{0.772}{1.15\sqrt{1 + (51/68)^2}}\right] = 9.52\ \text{ksi}$$

Stresses

$$f(60\ \text{ft}) = \frac{12(5800)}{3180} = 21.9\ \text{ksi} < 22.0\ \text{ksi} \therefore \text{OK}$$

$$f(18\ \text{ft}) = \frac{12(2718)}{1680} = 19.4\ \text{ksi} < 22.0\ \text{ksi} \therefore \text{OK}$$

$$f_v = \frac{160 - 1.75}{(\frac{1}{4})(68)} = 9.3\ \text{ksi} < 9.5\ \text{ksi} \therefore \text{OK}$$

Fig. 6-4.6 Welded plate girder analysis

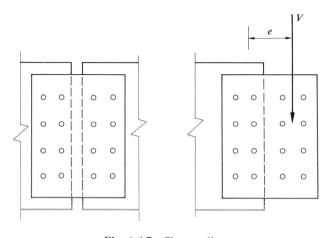

Fig. 6-4.7 Shear splice

PART C
USING THE FLOOR FOR T BEAM ACTION

Floors of reinforced concrete may often add section to supporting beams. If main slab steel parallels beam steel, ACI 8-7.5 specifies transverse reinforcement in the top of the slab. This transverse steel is designed to carry design loads, figured as for cantilevers, over the effective flange widths.

6-5 EFFECTIVE WIDTH OF T BEAMS

Supporting T beams may be made of reinforced concrete, steel, or prestressed concrete. Unless a large amount of additional tensile reinforcement is used in a floor, T beams are used to carry only positive bending moments since concrete alone is not good in tension.

The width of floor permitted to be used as part of a beam is specified in ACI 8.7 and AISC 1.11.1. Figure 6-5.1 illustrates three provisions: (1) effective overhang may not exceed eight times the slab thickness, (2) the same piece of floor may not be used as part of two different beams, and (3) effective width may not exceed one-quarter of the beam span length.

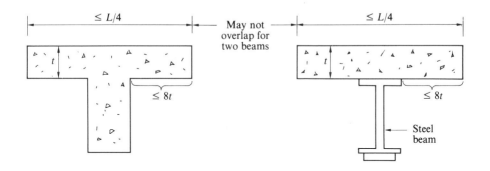

Fig. 6-5.1   Effective width of T beam

6-6 REINFORCED CONCRETE T BEAMS

Design and analysis of T beams is quite simple by both WSD and the strength method *if kd* or *c* is less than slab thickness *t*, which is almost always true. As apparent in Fig. 6-6.1, such a T beam acts exactly like a singly reinforced concrete beam and is so designed and analyzed because concrete in tension is neglected.

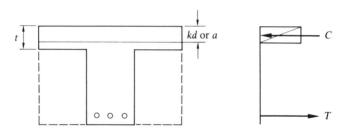

Fig. 6-6.1   T beam as a rectangular beam

When WSD is used, depth kd is first computed, and if less than t, rectangular beam analysis is appropriate. For the strength method, $kd \equiv c = a/\beta_1$ or

$$c = \frac{A_s f_y}{\beta_1 0.85 f'_c b} = \frac{pd f_y}{0.85 f'_c \beta_1} = \frac{1.18qd}{\beta_1}. \qquad (6\text{-}6.1)$$

If c is less than flange thickness, a member is analyzed as a singly reinforced beam.

Note, however, that in the calculation of shear as a measure of diagonal tension, b must be the width b' of the *stem*, and not flange width.

6-6.1 Neutral Axis within Flange.   As an example, consider a continuous beam that has a maximum positive moment of 194 ft-kips (including load factors). With an assumed effective flange width of, say, 4 ft, and d of 18 in., F is 1.30. Dividing this design moment of 194 by 1.30, K_u is obtained as 150. When curves of D-5 and 6 are used with $K_u = 150$ and $f'_c = 3000$, q is about 0.057. This q is apparently below the minimum allowed, however, for T-beam web width rather than flange width is used to obtain minimum permissible steel. Use Eq. (6-6.1) with $\beta_1 = 0.85$; distance to the neutral axis is

$$1.18(0.057)(18)/0.85 = 1.42 \text{ in.}$$

If flange thickness is greater than 1.42 in., then required steel is pbd or 0.057 $f'_c/f_y = 0.057(3/60) = 0.00285$. Then

$$A_{sR} = 0.00285(48)(18) = 2.47 \text{ in.}^2.$$

which obviously insures ductile behavior.

6-6.2 Neutral Axis Outside Flange.   When the neutral axis is not within the flange, design and analysis is only slightly more complex.

A basic equation is:

$$M_u = \phi \left[(A_s - A_{sf}) f_y (d - a/2) + A_{sf} f_y (d - 0.5t), \right. \tag{6-6.2}$$

where $A_{sf} \equiv 0.85(b - b')t f_c'/f_y$, $a \equiv (A_s - A_{sf}) f_y/0.85 f_c' b'$, $A_s \equiv$ amount of positive reinforcement (in.²), $f_y \equiv$ yield point of reinforcement, $d \equiv$ effective depth (top of flange to centroid of positive steel), $t \equiv$ flange thickness (in.), $f_c' \equiv$ ultimate strength of concrete, $b \equiv$ width of flange, and $b' \equiv$ width of stem.

Equation (6-6.2) is very similar to Eq. (6-3.1) for doubly reinforced beams except: (1) steel offsetting concrete in protruding flanges is called A_{sf} and (2) d' is replaced by the simpler dimension $t/2$, representing flange centroid. By equating C with T (see Fig. 6-6.2), A_{sf} is evaluated so that $0.85(b - b')t f_c' = A_{sf} f_y$ or $A_{sf} = 0.85(b - b')t f_c'/f_y$.

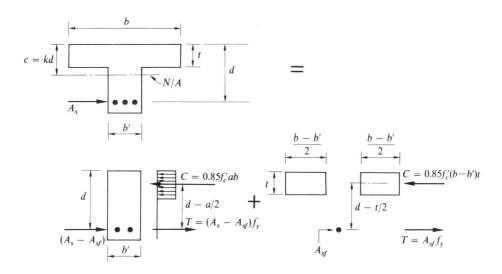

Fig. 6-6.2 True T beam—strength method

In accordance with the procedures of sec. 6-3.2, then,

$$M_u^{\prime k} = K_u F + \phi A_{sf} f_y (d - t/2)/12 \tag{6-6.3}$$

or

$$M_u^{\prime k} = M_{cs} + M_{ss}, \tag{6-3.3}$$

where K_u is based on a q, which is based on a p, computed as in Eq. (6-3.4), except altered to

$$p = \frac{\text{area of tensile steel } not \text{ used to balance equivalent compressive steel representing protruding part of flanges}}{bd}; \tag{6-6.4}$$

Typically, beam size is set by architectural considerations. For instance, the T beam of Fig. 6-6.3 must carry 50 ft-kips (including load factors) with $f_y = 60$ ksi and $f_c' = 4$ ksi. Design procedure is as follows.

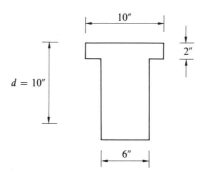

Fig. 6-6.3   T beam—neutral axis in stem

Step 1.

$$A_{sf} = \frac{0.85(b - b')tf_c'}{f_y} = \frac{0.85(10 - 6)2(4)}{60} = 0.453 \text{ in.}^2.$$

Step 2.

$$M_{ss} = \phi A_{sf} f_y (d - t/2)/12 = 0.9(0.453)60(10 - 2/2)/12 = 18.4 \text{ ft-kips.}$$

Step 3.

$$M_{cs} = M_u - M_{ss} = 50.0 - 18.4 = 31.6 \text{ ft-kips.}$$

Step 4.

$$F = 0.050 \qquad \text{for} \qquad b = 6 \text{ in., } d = 10 \text{ in.} \qquad \text{(by D-1).}$$

Step 5.

$$K_u = M_{cs}/F = 31.6/0.050 = 632.$$

Step 6.

$$p \doteq 0.0132 \qquad \text{(by D-6).}$$

Step 7.

$$A_s = pbd + A_{sf} = 0.0132(6)(10) + 0.453 = 1.245 \text{ in.}^2.$$

Usual checks to ensure that steel satisfies minimum and maximum requirements are fulfilled by visual inspection of K_u-versus-q or K_u-versus-p graphs in D-5 and D-6.

6-6.3 True T Beams, WSD. When WSD is used, occasionally kd is larger than the flange thickness; analysis can then be performed without undue difficulty by using transformed area. As an example, an algebraic equation for kd is illustrated in Fig. 6-6.4.

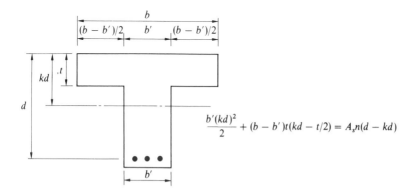

$$\frac{b'(kd)^2}{2} + (b - b')t(kd - t/2) = A_s n(d - kd)$$

Fig. 6-6.4   T beam—(WSD)

6-7 COMPOSITE CONSTRUCTION

Composite construction is a term applied when floors are used for T-beam action in conjunction with steel beams, prestressed concrete beams, or *precast* reinforced concrete beams. In this section, important and universal principles are illustrated by use of a combination of steel beams and reinforced concrete floors.

6-7.1 An Elastic Theory. To understand the difficulties associated with composite design, it is necessary to understand construction methodology. Two alternative methods are possible: (1) Shored construction means that beams are put in place with beam dead load carried by temporary supports, so that no deflection or stressing occurs. (2) Unshored construction uses no temporary supports. As placed, beams are subject to deflection and stress due to their own dead load.

In either method, when beams are in place securely, forms are erected to accept a concrete floor. In shored construction, all additional concrete and form load is carried by the temporary supports so that the beams remain free of stress. Conversely, in unshored construction, forms are supported by the beams themselves; these beams carry total dead load.

In either case, when the concrete has cured, formwork is removed and a composite beam exists. A shored beam carries the entire dead and live loads by composite action. But, an unshored beam already is carrying the total dead load itself so that the composite beam carries only the live load.

Stress diagrams for comparison of these two methods are shown in Fig. 6-7.1. For identical loads, total steel stresses are less for shored construction but slab stresses are higher.

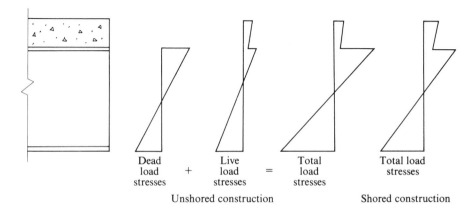

Dead load stresses + Live load stresses = Total load stresses   Total load stresses

Unshored construction   Shored construction

Fig. 6-7.1   Stress distribution—working load

Elastic analysis of beams erected by shored construction uses transformed area. But, in unshored construction, design is governed by summation of dead and live load *stresses*, where each is computed independently by using different section properties.

To insure that beams and slabs work together, shear connectors are provided. Load on a shear connector is computed by shear flow (VQ/I) between beam top and slab bottom (sec. A-4).

Based on the "Tentative Recommendations for the Design and Construction of Composite Beams and Girders for Buildings," [1] the AISC Specifications, among others, incorporate plastic theory within the framework of allowable stress design for simplicity and a more uniform factor of safety. Elastic analysis of composite sections is, however, still used in several codes.

6-7.2 Plastic Theory. Stress distribution for a composite beam at ultimate load is quite similar to that of a reinforced concrete beam, shown in Fig. 6-7.2. Such a shored composite beam can carry an ultimate load of 2.2 to 2.5 times its design (working) load. [1] An unshored beam has an even higher factor of safety. Compared to a typical 1.85 ratio for symmetrical rolled beams, a modified elastic analysis appears reasonable.

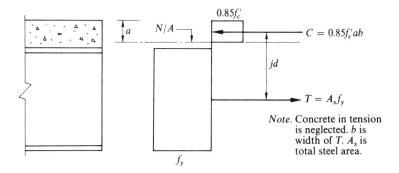

Fig. 6-7.2   Stress distribution—ultimate load

6-7.3 Modified Elastic Analysis. Based on this high factor of safey against ultimate failure, composite beams are designed to carry both dead and live loads, even when unshored construction is employed. But to safeguard against failure of composite beams, erection and actual working load stresses also are restricted.

Erection stresses, those in the beams before the floor has cured, must not exceed allowable working stresses for similar beams acting alone. Also, before the slab provides such support, particular care must be taken to preclude failure by lateral instability.

To develop AISC formula 1.11-2, a uniform overstress of 135% is permitted in steel of an unshored composite member (designed for both dead and live load) in recognition of the high factor of safey against ultimate failure. Expressed algebraically,

$$M_D/S_s + M_L/S_{tr} = 1.35(M_D + M_L)/S_{tr}, \qquad (6\text{-}7.1)$$

where $M_D \equiv$ dead load moment (in.-k), $M_L \equiv$ live load moment (in.-k), $S_s \equiv$ section modulus (measured to tension flange) of beam alone (in.³), and $S_{tr} \equiv$ section modulus (measured to tension flange) of beam or "transformed" composite beam (in.³).

The left side of Eq. (6-7.1) defines true stresses and the right side expresses code design provisions: (1) both dead and live load moments are carried by the composite beam and (2) allowable stress is increased by 135% above allowable working stresses.

Multiplication of Eq. (6-7.1) by S_{tr}, yields

$$S_{tr}M_D/S_s + M_L = 1.35(M_D + M_L), \tag{6-7.2}$$

which reduces to AISC 1.11-2,

$$S_{tr} = (1.35 + 0.35M_L/M_D)S_s. \tag{6-7.3}$$

By similar plastic consideration, shear connectors are usually spaced uniformly rather than in proportion to shear flow, as indicated by VQ/I. These shear connectors are designed to take one-half (reducing to working stress) the maximum ultimate stress block resultant of Fig. 6-7.2, i.e., V_h is equal to the minimum of $C/2$ or $T/2$. This is expressed by AISC formulas 1.11.3 and 1.11.4 as

$$V_h = \text{min of } 0.85f'_c A_c/2 \text{ or } A_s F_y/2, \tag{6-7.4}$$

where $V_h \equiv$ total force on shear connectors between maximum positive moment and zero moment, $f'_c \equiv$ ultimate strength of concrete, $A_c \equiv$ actual effective area of concrete flange (specified strength), $A_s \equiv$ cross-sectional area of steel, and $F_y \equiv$ yield strength of steel.

Allowable horizontal shear loads (q) per shear connector are tabulated in Table 1.11.4 of the AISC Specifications. The number of shear connectors is then determined as $N_1 = V_h/q$.

Note carefully that the tabulated q values are *not* to be used with shear flow (VQ/I) as errors greater than 200% could occur. The q values can be used *only* in conjunction with Eq. (6-7.4) and (6-7.5).

In some cases, generally when maximum moment does *not* occur under a concentrated load nearest to a simple support, AISC Formula 1.11-6 applies:

$$N_2 = N_1 \frac{[(M\beta/M_{max}) - 1]}{\beta - 1}, \tag{6-7.5}$$

where, for this particular configuration, $N_2 \equiv$ number of connectors required between this concentrated load and the nearest simple support; $N_1 \equiv$ number of connectors required between the point of maximum moment and the same simple support; $M \equiv$ moment at the concentrated load, less than M_{max}; and $\beta \equiv S_{tr}/S_s$.

6-7.4 Composite Design. The composite design of steel beams and concrete decks is greatly facilitated by three types of tables, provided in *Steel Construction* to be used with 4-, 4½-, 5-, and 5½-in. thick slabs: (1) *pp. 2-152 through 2-167*—beam and slab without cover plates, descending order of S_{tr}, section properties with full slab, slab one side only, and steel alone; (2) *pp. 2-168 through 2-191*—same as (1) except includes cover plates and order is by beam size; (3) *pp. 2-192 through 2-195*—composite beam selection tables ordered by S_{tr}. Modular ratio ($n = E_s/E_c$) is defined as 9 for all these tables.

To illustrate, say a composite beam is designed in accordance with the parameters of Fig. 6-7.3. Floor thickness has previously been set at 4 in. Design moments at point of concentrated load are:

$$M_{DL} = 0.4(20)(50 - 20)/2 = \qquad 120 \text{ ft-kips.}$$
$$M_{LL} = 0.2(20)(50 - 20)/2 = \quad 60.0$$
$$30 \times 30 \times 20/50 = 360.0 \quad \underline{420}$$
$$M_{TL} = \overline{540} \text{ ft-kips.}$$

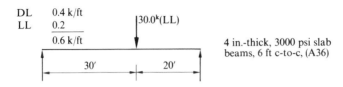

Fig. 6-7.3   Design parameters for a composite beam

Allowable steel stress is assumed to be 24 ksi, based on full lateral support, which is correct for the composite section. When the beam acts alone, however, it is necessary for the concrete forms to provide sufficient restraint for the top flange.

Effective flange width is computed as

$L/4$	$= 12 \times 50/4$	$= 150$ in.
c-to-c spacing	$= 6$ ft	$= 72$ in.
Beam width $+ 16 \times$ (4-in. slab)	$=$ width $+ 64$ in. $=$	72 in.

Beam flange width is unknown, but assumed at 8 in. If it were less than 8 in., total effective width would decrease; if it were greater than 8 in., effective width would remain unchanged at its minimum of 72 in. based on c-to-c spacing. Variations in width cause redesign in borderline cases.

Required section moduli are, therefore,

$$(S_{tr}) \equiv S_b(\text{composite}) = 12(540)/24 = 270 \text{ in.}^3$$

and

$$(S_s) \equiv S_b(\text{beam only}) = 12(120)/24 = 60 \text{ in.}^3.$$

By use of the table on page 2-192 of *Steel Construction* with $S_{tr} = 270$, choose a W18 × 45 with PL $1\frac{1}{2}$ × 6 (bottom only), and S_{tr} of 271 in.³. Turn to page 2-173 and read the section moduli of the beam alone as $S_s = 186$ in.³ and $S_{ts} = 94.1$ in.³. Because both values are greater than 60, the beam will not be over-stressed before the concrete cures. As a precaution, concrete stress also is checked. Beam flange width is 7.48 in., so effective width is $64 + 7.48 = 71.48$ in. and corresponds to that used in the table. By use of S_t and only the live load (because dead load does not stress slab) concrete stress is $f_c = 12(420)/(510)(9) = 1.0$ ksi < 1.35; well within the allowable. (S_t was found on p. 2-172.)

Finally, check for adequacy using Eq. (6-7.3) (bottom flange only): $(1.35 + 0.35 M_L/M_D)S_s = [1.35 + 0.35(420/120)] \, 186 = 479$. Because 479 is greater than S_{tr} of 271, the beam is adequate.

From Eq. (6-7.4), V_h is the smaller of

$$0.85(3 \text{ ksi})(4 \text{ in.})(71.48 \text{ in.})/2 = 365$$

or

$$13.2 \text{ in.}^2(36 \text{ ksi})/2 = 237.6.$$

From AISC Table 1.11.4 a $\frac{3}{4}$-in. diameter by 3-in. headed stud is good for 11.5 kips with $f_c' = 3$ ksi, and a C3 × 4.1 for $4.3w$ kips, where w is channel length. Such connectors must have at least 1-in. cover in all directions. Studs must be at least four diameters long, not greater than $\frac{7}{8}$-in. in diameter, spaced at least three diameters c-to-c, and if not loaded directly over the web must not have a diameter greater than two and one-half times the flange thickness (so the thin flange will not be torn out).

Using a 5-in. long C3 × 4.1, good for $4.3 × 5 = 21.5$ kips, $237.6/21.5 = 11$ are needed for *both* sides of the point of maximum moment. Spacing at 2 ft 8 in. left of the concentrated load with the first at 1 ft 4 in., and at 1 ft 10 in. right of the load with the first at 11 in., is adequate.

At an allowable of 11.5 kips each, $\frac{3}{4}$-in. diameter by 3-in. studs could be used as an alternate requiring $237.6/11.5 = 21$ both sides of concentrated load.

An analysis check is shown in Fig. 6-7.4. Moments are computed for both total and dead load at only maximum points and points of cover plate cutoff. More section property calculations are made than absolutely required, but this additional information is useful when making secondary checks of cover plate cutoffs, deflections, weld requirements, etc.

Stress checks provide answers to the following:

1. The beam is adequate when carrying all dead loads by itself.

2. Concrete is not overstressed when live loads are applied to the composite section.

3. Composite section is not overstressed by total loads and is adequate by AISC 1.11-2 [here, Eq. (6-7.3)].

4. Steel section carries shear easily.

Although very important, live load and long-term dead load deflections have been excluded in this analysis. These depend upon the transformed moment of inertia and are left as an exercise.

Loads (A36 steel, 3000 psi 4-in. thick slab, composite)

DL: 0.4 k/ft
LL: 0.2
TL: 0.6 k/ft $+30^k$ $+30^k$

30 ft ↓ 20 ft

V & BM   V: $0.6 \times 50/2 = 15^k$ 15
$30 \times 30/50 = 18$ 12
33^k Rt. support (max) $27^k(Lt)$

M: | 9 ft   30 ft   43 ft

$(27)(9) = 243$ $33(20) = 660$ $33(7) = 231$
$-0.6(9)^2/2 = -24$ $-0.6(20)^2/2 = -120$ $-0.6(7)^2/2 = -15$
Total $219'^k$ $540'^k$ $216'^k$

$\dfrac{0.4 \times (L - x)}{2}$ Dead $74'^k$ $125'^k$ $120'^k$ $60'^k$
@ 9 ft   @ 25 ft   @ 30 ft   @ 43 ft

Section Properties   W18 × 45 + PL1½ × 6 + 71.48 × 4 in. RC slab
$f_s = 24$ ksi, $f_v = 14.5$ ksi, $f_c = 1350$ psi

Effective slab width: ¼ × 50 = 12.5 ft
c-to-c = 6 ft
d(beam) = 17.86 in.   Beam w. + 16 × slab t = 71.48 in. ←———USE
$t_w = 0.335$ in.

	A	y	Ay	$Ay^2 + I_0$
Slab 4 × 71.48 ($n = 9$)	31.76	10.93	347.1	3794
				42
W 18 × 45	13.2	0	0	706
PL 1½ × 6 (bottom)	9.00	−9.68	−87.1	843

$\dfrac{71.48(4^3)}{(12)(9)}$

Slab + Beam: 45.00   7.71   347.1   4542
$y_{t(conc)} = 5.22$ $y_{t(st)} = 1.22$ $y_{b(st)} = 16.64$ −2677
$[s_t]Z_{t(conc)} = 357$ $Z_{t(st)} = 1529$ $[s_{tr}] Z_{b(st)} = 112$ 1865 in.^{4}

Slab + Beam + ⊥:   54.0   4.81   260   5385
$y_{t(conc)} = 8.12$ $y_{t(st)} = 4.12$ $y_{b(st)} = 15.24$ −1252
$[s_t]Z_{t(conc)} = 510$ $Z_{t(st)} = 1006$ $[s_{tr}] Z_{b(st)} = 272$ 4133 in.^{4}

Beam & ⊥:   22.2   −3.92   −87.1   1549
$y_{t(st)} = 12.85$ $y_{b(st)} = 6.51$ −342
$[s_{ts}] Z_{t(st)} = 94$ $[s_s] Z_{b(st)} = 185$ 1207 in.^{4}
Beam only: $y_{(st)} = 8.93$ $Z = 79$ 706 in.^{4}

Fig. 6-7.4 continued on next page

Fig. 6-7.4 (continued)

Stress Check   @ 9' & 43' w/o P @ 25' &/or 30' w/P

Dead: $\dfrac{12(74)}{79} = 11.3 < 24\,\text{ksi}(\text{T \& B})$ $\dfrac{12(125)}{94} = 16.0 < 24\,\text{ksi (top)}$

Beam only

Live only, conc: $\dfrac{12(219 - 74)}{35.7(9)} = 0.54 < 1.35\,\text{ksi}$ $\dfrac{12(540 - 120)}{510(9)} = 1.10 < 1.35\,\text{ksi}$

$0.45 f'_c$

D + L: $\dfrac{12(219)}{112} = 23.5 < 24\,\text{ksi(bot)}$ $\dfrac{12(540)}{272} = 23.8 < 24\,\text{ksi (bot)}$

$\left[1.35 + 0.35\left(\dfrac{145}{74}\right)\right]79 = 168 > 112(\text{OK})$ $\left[1.35 + 0.35\left(\dfrac{420}{120}\right)\right]185 = 476 > 272(\text{OK})$

Shear: @ Rt. support $f_v = \dfrac{33}{(17.86)(0.335)} = 5.5 < 14.5\,\text{ksi}$

Shear Connectors

$V_h = \dfrac{0.85(3)(71.48)(4)}{2} = 365\,\text{kips}$ $V_h = \dfrac{(13.2)36}{2} = 237.6\,\text{kips USE}$

5-in. wide $3[4.1 = 5(4.3) = 21.5\,\text{k}/\text{c}$ $\therefore$ USE $\dfrac{237.6}{21.5} = 11 \begin{array}{l} 0 \to 30\,\text{ft and} \\ 30\,\text{ft} \to 50\,\text{ft } also \end{array}$

Fig. 6-7.4   Analysis of composite section

Sometimes it is unnecessary or not feasible to obtain full composite action. In such cases, AISC formula 1.11-1 applies:

$$S_{\text{eff}} = S_s + \frac{V'_h}{V_h}(S_{tr} - S_s),\qquad (6\text{-}7.6)$$

where V'_h represents the amount of shear which connectors can carry and is less than V_h. Then, in basic formulas, S_{eff} is used rather than S_{tr}.

Composite action makes efficient use of materials and this reduces both cost and overall beam depths. These advantages are enhanced by the convenient tables of *Steel Construction* which make office design much more efficient and less costly in engineering time and money.

PART D
DEFLECTION OF REINFORCED CONCRETE

When thicknesses of beams or one-way slabs are less than specified by sec. 2-5.5, or when damage is likely if deflections are excessive, computations are required. Because concrete cracks, even complex calculations do not yield accuracy better than $\pm 20\%$. ACI-71, therefore, establishes an expression for determining effective moment of inertia in deflection computations:

$$I_e = \left(\frac{M_{cr}}{M_a}\right)^3 I_g + \left[1 - \left(\frac{M_{cr}}{M_a}\right)^3\right] I_{cr} \leq I_g, \qquad (6\text{-D.1})$$

$$\text{or } I_e = I_{cr} + (I_g - I_{cr})(M_{cr}/M_a)^3 \leq I_g, \qquad (6\text{-D.2})$$

where

$I_e \equiv$ effective moment of inertia (in.^{4}),

$I_g \equiv$ gross moment of inertia, $bh^3/12$ (in.^{4}),

$I_{cr} \equiv$ cracked moment of inertia, that is transformed moment of inertia (in.^{4}),

$M_{cr} \equiv$ cracking moment, $f_r I_g/y_t$, where $f_r \equiv 7.5 \sqrt{f'_c}$ for normal weight concrete and $y_t \equiv$ distance from extreme tension fiber to centroid of *transformed* section,

$M_a \equiv$ maximum moment (without load factors) occurring during the stage under consideration.

This equation and its terms are simple to use, except for I_{cr}. A conservative approach is to consider all members singly reinforced (recognizing that compression steel actually increases the value of I_{cr}). Then

1. Determine $n = E_s/E_c$.
2. Compute pn $= A_s n/bd$.
3. Obtain k from the curve of D-7.
4. $I_{cr} = b(kd)^3/3 + nA_s(d-kd)^2$.

Additional deflection that occurs with time can then be estimated as

$$\Delta_{\text{long term}} = \Delta_{\text{immediate}} \quad [2 - 1.2(A'_s/A_s)], \qquad (6\text{-D.3})$$

where the expression in brackets must be greater than or equal to 0.6.

Computed deflections must be limited by values of ACI-71, Table 9.5(b) which states, in general, that:

Deflection is limited to—	Due to—	When—	
$L/180$	Immediate live load	For flat roofs	Not likely to damage
$L/360$	Immediate live load	For floors and beams	adjacent structure
$L/480$	Part of load displacing	Roof or floor attached	
$L/240$	adjacent components	Roof or floor not attached.	

For continuous spans, or those with moment-restrained ends, averaged values from critical positive and negative moment sections may be used.

PROBLEMS

6-1 given: Beam of Fig. 6-P.1 with P = _____ (180, 200, or 220 kips) including load factors, L = _____ (24, 25, or 26 ft), a = _____ (0.1, 0.15, or 0.2).

    select: Most economical W-beam using bending only and (a) plastic design, (b) allowable stress design.

    adjust: for shear by (a) using web doublers, (b) using diagonal stiffeners, (c) using a larger section. Compare economy of the three solutions.

    design: Bearing plate under concentrated load using allowable stress designed section.

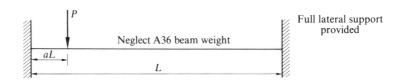

Fig. 6-P.1   Plastic design—high shear

6-2 given: Shear diagram of Fig. 6-P.2 with b = _____ (10, 12, or 14 in.), d = _____ (24, 28, 30, or 32 in.), f'_c = _____ (3000, 4000, or 5000 psi), f_y = _____ (40, 50, or 60 ksi).

    design: Stirrups and spacings.

    present: Stress sheet.

6-3 given: Shear diagram of the interior span of a continuous beam load by a uniform load of Fig. 6-P.3 with t_R = _____ (6, 7, or 8 in.), t_L = _____ (8, 10, or 12 in.), V_L = _____ (140, 180, or 214 kips), V_R = _____ (0.4, 0.5, 0.6, or 0.7), f'_c = _____ 3000, 4000, or 5000 psi), f_y = _____ (40, 50, or 60 ksi).

    select: (a) b and d so maximum shear to right of the span can be carried without stirrups, *OR* (b) b and d so maximum shear can just be carried if #4 W stirrups are used, *OR* (c) b and d so maximum shear to right of the span *could be* just carried by #3 U stirrups at 6-in. spacing.

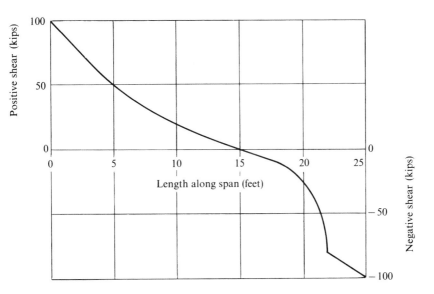

Fig. 6-P.2 Shear diagram

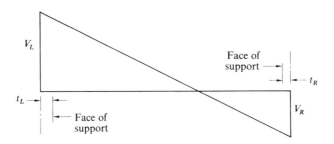

Fig. 6-P.3 Shear diagram—interior span

design: Stirrups and spacings.

6-4 given: $f'_c = $ _____ (3000, 4000, or 5000 psi), $f_y = $ _____ (40, 50, or 60 ksi), b limited to a maximum of _____ (10, 12, 14, or 16 in.), h limited to a maximum of _____ (20, 24, 28, or 32 in.), and *final q* limited to 0.18.

find: M_u for a singly reinforced beam using strength method if $q \equiv 0.15$.

design: Doubly reinforced beam to carry _____ (1.5, 1.7, or 1.9) times M_u with *final q* ≤ 0.18.

present: Analysis proof.

determine: (a) Transformed *I* of beam proven. (b) Maximum allowable WSD moment. (c) WSD stresses due to this moment in (1) concrete, (2) tensile steel, (3) compressive steel.

6-5 given: Beam span _____ [(a), (b), (c), or (d) of Fig. 6-P.4] with $P =$ _____ (40, 60, or 80 kips), $L =$ _____ (70, 85, or 90 ft), $a =$ _____ (0.1, 0.13, or 0.17), $b =$ _____ (0.4, 0.43, 0.47, or 0.5), $f_y =$ _____ (36 or 50 ksi).

select: Most economical rolled section. (If none is large enough, so state.)

design: Cover-plated beam with the basic section having less than or equal to $\frac{2}{3}$ the *I* of a uniform selection.

present: Analysis proof and stress sheet.

Full lateral support provided

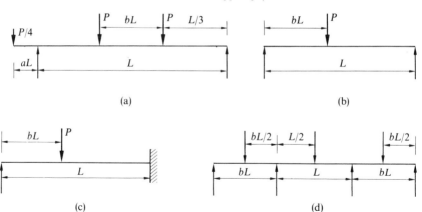

(a)   (b)

(c)   (d)

Fig. 6-P.4 Beam spans

6-6 given: Plate girder _____ [(a) or (b) of 6-P.5] with $f_y =$ _____ (36 or 50 ksi), $P =$ _____ (120, 140, or 160 kips), $L =$ _____ (120, 124, or 132 ft), $a =$ _____ (0.25, $\frac{1}{3}$, or 0.35), $b =$ _____ (*a*, 0.25, $\frac{1}{3}$, or 0.35).

design: Welded plate girder including (a) plate cutoff points, (b) stiffeners, (c) splice.

present: Complete analysis proof, and a stress sheet.

Full lateral support provided

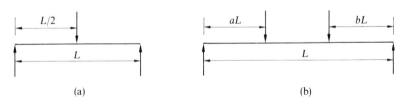

(a)   (b)

Fig. 6-P.5 Plate girders

6-7 given: T beam of Fig. 6-P.6 (strength design) with $b =$ _____ (10, 12, or 14 in.), $h =$ _____ (20, 26, or 32 in.), $f'_c =$ _____ (3000, 4000, or 5000 psi), $f_y =$ _____ (40, 50, or 60 ksi), $t =$ _____ (4, 5, or 6 in.), span length _____ (20, 24, or 32 ft), c-to-c spacing of beams _____ (18, 30, 60, or 90 in.).

determine: (a) M_u for beam acting alone with $q \leqq 0.18$.
  (b) M_u for beam acting as a T beam with $q \leqq 0.18$.
  (c) Using WSD analyze both (a) and (b) to find maximum working stress M permitted.

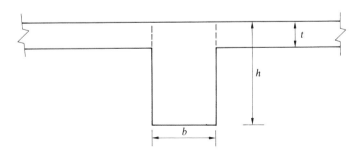

Fig. 6-P.6 T beam

6-8 given: Composite A36 steel beam with_____ (4-, $4\frac{1}{2}$-, 5-, $5\frac{1}{2}$-in.) thick, 150 lb/ft³, 3000 psi concrete slab of Fig. 6-P.7. $w_{LL} =$ _____ (20, 40, or 60 lb/ft²), $P_L =$ _____ (20, 40, or 60 kips), $L =$ _____ (40, 50, or 60 ft), $a =$ _____ (0.3, 0.4, or 0.5). c-to-c of beams is _____ ($5\frac{1}{2}$, 6, or 7 ft).

design: Composite beam including shear connectors.

present: Analysis proof and a stress sheet.

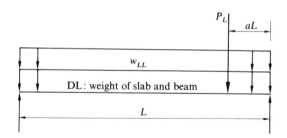

Fig. 6-P.7   Composite beam

6-9 given: Simple span length = _____ (30, 32, or 34 ft), b = _____
(12, 14, or 16 in.), h = 16 in., f_y = _____ (40, 50, or 60 ksi),
f'_c = _____ (3, 4, or 5 ksi).

determine: Maximum allowable uniform load considering flexure only.
_____ (40, 50, 60, or 70%) dead; balance, live load.

present: Analysis proof (using maximum allowable uniform load) including
(a) immediate dead load deflection, (b) immediate live load deflection, (c) long-term deflection after _____ (20, 25, or 30%) of
live load sustained for two years.

6-10 given: Same as problem 6-9 except _____ (4, $4\frac{1}{2}$, or 5 in.) thick slab
poured monolithically with beam. c-to-c of beams is_____ (3,
4, 5, or 6 ft).

determine: Same as in problem 6-9.

present: Same as in problem 6-9.

6-11 given: Cross section of Fig. 6-P.8 with b = _____ (10, 12, or 14 in.),
h = _____ (20, 21, 22, or 23 in.), n = _____ (9, 10, or 11).

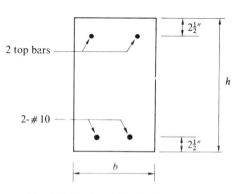

Fig. 6-P.8   Doubly reinforced beam

find: (a) Transformed moment of inertia without top bars. (b) Transformed moments of inertia as top bars increase by one size at a
time from 2 − _____ (#3, #4, #5, or #6) to 2 − _____ (#7,
#8, #9, or #10).

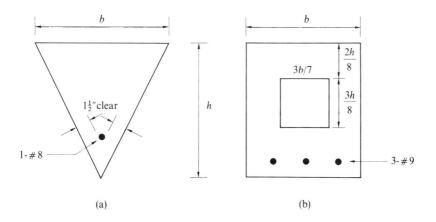

Fig. 6-P.9   Unusual cross sections

6-12 given: Cross section_____ [(a) or (b) of Fig. 6-P.9] *with* $b =$ _____
(14, 15, or 16 in.), $h =$ _____ (17, 18, or 19 in.), $f_c' =$ _____
(3, 4, or 5 ksi), $f_y =$ _____ (40, 50, or 60 ksi).

assume: Whitney's stress block applies, with uniform compressive concrete
strength $0.85 f_c'$.

determine: Maximum allowable M_u.

6.13 given: Beam_____ [(a), (b), or (c) of Fig. 6-P.10] *with* $L =$ _____
(18, 24, 28, 30, or 32 ft), $a =$ _____ (0.6, 0.7, or 0.8), $P =$
_____ (10, 11, or 12 kips) one-half live, one-half dead, $f_c' =$
_____ (3, 4, or 5 ksi), $f_y =$ _____ (40, 50, or 60 ksi). Uniform
dead load = _____ (0.9, 1.0, or 1.1 k/ft). Uniform live load =
_____ (1.2, 1.3, or 1.4 k/ft).

design: Beam, *completely.*

present: Analysis proof and stress sheet.

6-14 given: Same as problem 6-13, except_____ (4, 4½, or 5 in.) slab exists
with beams spaced at_____ (4, 5, or 6 ft).

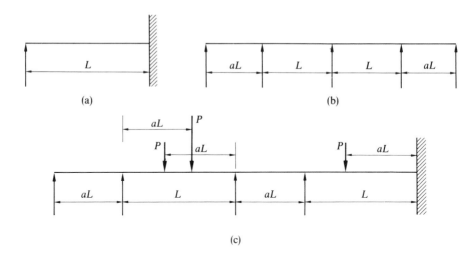

Fig. 6-P.10   Beam configurations

REFERENCES

1. "Tentative Recommendations for the Design and Construction of Composite Beams and Girders for Buildings," Progress Report of the Joint ASCE-ACI Committee on Composite Construction, ASCE, *Journal of the Structural Division*, ST12 (December, 1960), pp. 73–92.

CHAPTER 7

PRESTRESSED CONCRETE

Prestressed concrete has captured the imagination of engineers and architects alike. It is unique because it actively, rather than passively, resists superimposed loads. Standard beams need strength to carry themselves as well as external loads, but prestressed concrete has a "negative loading" even before it is erected. This negative load can be controlled to manage stresses due to all loads. In addition, prestressing can cause *all* of the concrete to work by eliminating tensile areas such as occur in reinforced concrete. Thus, prestressed concrete members can span longer distances and carry heavier loads than reinforced concrete beams of equal sizes.

Prestressing is sometimes applied to other materials, including steel, but the discussion that follows is limited to prestressed concrete.

7-1 BASIC CONCEPTS OF PRESTRESSING

Methods of manufacturing prestressed concrete are numerous; connections, types of tendons, and anchorages are constantly being revised and improved. Therefore, rather than consideration of various techniques, a graphical representation is utilized to demonstrate basic fundamentals and terms. Characteristics of *draping* are also discussed because this methodology directly affects design calculations. Spacing of tendons (strands, wires, cables), although seldom performed in a design office, is explained to augment an understanding of fabrication.

7-1.1 Pre-tension versus Post-tension. Two basic systems for prestressing concrete are (1) pre-tensioning and (2) post-tensioning. Both are shown in Fig. 7-1.1. Pre-tensioning is shown in part (a) of the figure. Step 1 shows two abutments with prestressed wire strands jacked taut between them, held in tension. Forms are placed (often before jacking). Then, concrete is poured while

PRE-TENSIONED   POST-TENSIONED

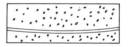

1. Strands jacked and held in
   tension between abutments.

1. Hollow cable, solid rubber cord,
   or other device placed in forms
   to leave longitudinal hole in
   concrete. Concrete poured.

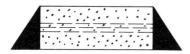

2. Forms placed, concrete poured,
   strands still held in tension.

2. After concrete sets, cable inserted
   in longitudinal holes. Often
   prestress cable is inside hollow sheath
   or other bond-breaking device in
   step 1 above.

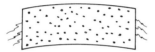

3. Wires cut after concrete sets.
   Beam deflects upward if
   prestress force is eccentric.
   Force transmitted from strands
   to concrete by bond.

3. Wires jacked against end of beam.
   Beam deflects upward if prestress
   force is eccentric. Wires
   mechanically secured against slip.
   Grouting is optional (preferred).

(a) Prestress force transferred
    to concrete by bond

(b) Prestress force transferred
    to concrete by end bearing

Fig. 7-1.1   Pre-Tension versus Post-Tension

tension is maintained on the wires (strands) (step 2). Step 3 shows that when the concrete has set, the wires are cut. If the prestress force is applied eccentrically (usually it is), as wires are cut the beam deflects upward. *This prestress force is transmitted from the wires (strands) directly to the concrete by bond*; this is the primary definition of pre-tensioned members.

In part (b) of Fig. 7-1.1, a post-tensioned member is shown. Step 1 shows a poured concrete beam with a hollow cable, a solid rubber cord, a coated cable (to prevent bond), or similar device imbedded in it. Although these devices may be placed straight, often they are constrained to produce a draped or parabolic profile. When the concrete has set, prestressed cable is inserted (if not already imbedded inside the devices) in the longitudinal holes left in the concrete (step 2). Then, as shown in step 3, cables are tensioned against the ends of the beam. If there is eccentricity, the beam will deflect upward as this prestressing force is applied. Wires are mechanically secured against slip at the ends of the beam, and prestressing force is transferred to the concrete by end bearing. The primary definition of post-tension concrete is that *the prestress force is transferred to the concrete (not by bond) by end bearing.*

7-1.2 Draped Cables. Another important consideration can be inferred from Fig. 7-1.1. At mid-span, it often is permissible for the centroid of the prestress force to act *below* the lower kern point. This is possible because the beam's own dead load becomes active concurrently with application of the prestressed force (either when the wire is cut in pre-tensioning or when the wires are jacked in post-tensioning). However, at beam ends where there is no dead load stress, a prestress force applied by straight cables could impose tensile stresses in the member. When cables are draped so that ends of the wires (strands, cables) go through the kern near beam ends, however, no tension exists in any concrete. This suggests that a more economical design may be obtained, without danger of tension, by the use of draped cables.

Many ingenious methods have been devised for prestressing. Even in the pre-tensioning method, modified draping is often secured by using mechanical *hold-downs.* The same effect has also been secured by the use of sheathing, grease, etc. to prevent part of the wires from developing their bond toward beam ends.

Prestressed concrete is still in its formative stage and many new techniques and innovations can be anticipated. This means that to obtain maximum economy, advantage must be taken of available techniques that are local to actual construction sites.

7-1.3 Spacing Strands. Figure 7-1.2 shows a sketch for a typical Type III AASHO girder.[1] It is required to space 33 strands to obtain an eccentricity e of 10.65 in. when distance between the member centroid and the beam bottom is 20.27 in. This means that the prestress steel must have its centroid acting $(20.27 - 10.65)$ or 9.62 in. above the base.

Solid circles on the figure represent a first estimate used to place 30 of the 33 strands. In the table (see sec. A-2), location of the centroid of these strands is computed as $(220/30) = 7.3$ in. above the base axis.

Next, two strands, represented by the open circles, are added a distance of 39 in. above beam bottom, adding 78 to Ny for a new total of 298. The centroid for the 32 strands $(298/32)$ is, therefore, 9.3 in.

Addition of one strand makes the total correct at 33. Required y is

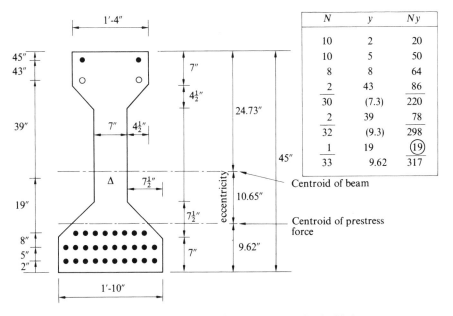

N	y	Ny
10	2	20
10	5	50
8	8	64
2	43	86
30	(7.3)	220
2	39	78
32	(9.3)	298
1	19	⑲
33	9.62	317

Fig. 7.1.2   Spacing Strands in Type III AASHO Girder

9.62 inches; therefore required $Ny = 33(9.62) = 317.46$. The difference between this and 298 is 19.46 which represents Ny to be furnished by the one remaining cable. This leads to a final y for this single strand, shown by the symbol Δ, of $19.46/1 = 19.46$ or, say, 19 in.

Other arrangements are possible; the primary criteria for spacing is the type of anchorage spacings available locally. Because the contractor and his fabricator usually have not been selected at design time, most engineering calculations simply state the amount of final prestress force required and its eccentricity. The engineer then may check shop drawings submitted by the fabricator to assure himself that spacing is satisfactory.

7-2 GEOMETRIC CONSIDERATIONS

Design of prestressed concrete beams is particularly dependent upon (1) loss of prestress (PR), generally time-related, and (2) geometric configuration. To delineate clearly between these two items, geometric configuration will be discussed as though any specified amount of prestressing can be obtained and maintained constant with time.

7-2.1 Solid Rectangular Section–Prestress Through Centroid. In this section a beam is designed in several ways for a live load of 1300 lb/ft over a simple span of 24 ft, with beam dimensions restricted to a 12-in. width and 20-in. overall depth. Table 7-2.1 shows preliminary computations for design of a beam that is assumed solid and rectangular. Stresses are obtained independently for live and dead loads, their total being 1670 lb/in.^{2}, compression at the top and tension at the bottom of the beam.

Table 7-2.1 Preliminary Computations for a Prestressed Beam Assumed Solid and Rectangular

---

Loads (24 ft Simple Span)

$$\text{LL:} \qquad = 1300 \text{ lb/ft}$$
$$\text{DL: } 12 \times 20 \times 150/144 = \underline{250}$$
$$\text{TL} = 1550 \text{ lb/ft}$$

M

$$\text{LL: } (\tfrac{1}{8})(24)^2(1.3) \quad = 93.5'^k$$
$$\text{DL: } (\tfrac{1}{8})(24)^2(0.25) \quad = \underline{18.0}$$
$$M_T = 111.5'^k$$

Section Properties

$$b = 12 \text{ in.} \qquad h = 20 \text{ in.} \qquad A = 240 \text{ in.}^2$$
$$I = \frac{12 \times 20^3}{12} = 8000 \text{ in.}^4 \qquad y_t = y_b = 10 \text{ in.}$$
$$Z_t = Z_b = \frac{8000}{10} = 800 \text{ in.}^3$$

Stresses

$$f_L = \frac{12 \times 93.5}{800} = 1.400 \text{ ksi} = 1400 \text{ psi}$$
$$f_D = \frac{12 \times 18.0}{800} = 0.270 \text{ ksi} = \underline{270 \text{ psi}}$$
$$f_T \text{(top and bottom)} = 1670 \text{ psi}$$

Figure 7-2.1 shows mid-span stress diagrams under various conditions of loading. Dead load is represented in stress diagram (a) as 270 psi (compression) at the top and -270 psi (tension) at the bottom. Compression is positive in accordance with the usual sign convention of reinforced concrete. Diagram (b) shows combined dead and live load stresses: 1670 psi compression (top) and 1670 psi tension (bottom).

A prestressing force is sought to eliminate all tension in the beam under the combined action of prestress force, dead load, and live load. At the beam bottom DL and LL stress is 1670 psi tension, requiring PR there to be 1670 psi compression. Because prestress force is applied through the centroid, it is uniform over the entire section as shown in (c).

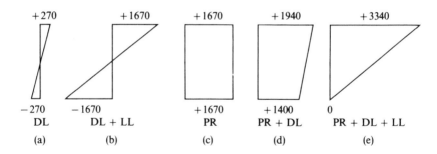

Fig. 7-2.1 Prestress Through Centroid (Mid-span)

In (d), stresses due to prestress and dead loads are added to yield 1940 psi compression at the top and 1400 psi compression at the bottom. Finally, in (e), live load stress is included to yield 3340 psi compression at the top of the beam and zero stress at the bottom.

Total prestressed force required is 400.8 kips, the volume of the stress block (12 × 20 × 1670), and the maximum required working stress of concrete is 3340 psi.

7-2.2 Solid Rectangular Section–Prestress Through Lower Kern Point. The kern of a section is that area through which a resultant tensile force can be applied without causing compression in the section (or through which a compressive resultant force can be applied without causing tension). Figure 7-2.2 shows a stress diagram of a 12-in. by 20-in. beam with a prestress force applied through the lower kern point. These diagrams are shown for the mid-span of the beam. Stress diagram (a) and (b) are unchanged from those of Fig. 7-2.1 because dead and live loads are identical. No prestress stress is necessary at the top of the beam because the top under gravity loads is always in compression. At the

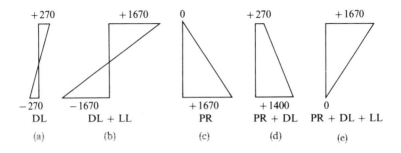

Fig. 7-2.2 Prestress Through Lower Kern Point (Mid-span)

bottom, however, a prestress stress of 1670 psi compression is required to offset the 1670 psi tension imposed by gravity loads. These two requirements establish a triangular prestress stress diagram, with zero stress at the top and 1670 psi at the bottom as in part (c) of the figure.

When prestress and dead load are added together, the beam is completely in compression, 270 psi at the top and 1400 psi at the bottom. (See part (d) of the figure.) Under full dead, live, and prestress loadings, stresses at the top are 1670 psi compression and zero at the base, as shown in (e).

To complete this comparison, it is necessary to compute both the total prestress force required and its eccentricity. This may be done by the use of the familiar

$$\sigma = P/A \pm My/I \quad \text{or} \quad \sigma = P/A \pm M/Z, \quad (7\text{-}2.1)$$

where $\sigma =$ stress at y from neutral axis; P is replaced by prestress force, PR; $A \equiv$ area of prestress member; $M \equiv$ (PR)e, prestress force times eccentricity; $y \equiv$ distance from centroid to point of stress investigation; $I \equiv$ total moment of inertia of prestressed member; and $Z \equiv$ section modulus $= I/y_{max}$. Thus, at the top fiber

$$0 = PR/240 - PR(e)\,10/8000$$

and at the bottom fiber

$$1.67 = PR/240 + PR(e)10/8000,$$

which yields PR $= 200.4$ kips at an eccentricity of $3\frac{1}{3}$ in.

The results may easily be verified by recalling that the kern of a rectangular section is one-sixth of the total depth below the centroid, and volume of the prestress stress block is $12 \times 20 \times 1.67/2 = 200.4$ kips.

Before proceeding, it is interesting to compare results of these two trials. By simply lowering the center of gravity of the strands (1) the required prestressing force has been reduced from 400.8 kips to 200.4 kips and (2) the required working strength of concrete has been reduced from 3340 psi to 1670 psi. What would happen if the prestressing force were lowered more?

7-2.3 Solid Rectangular Section—Prestress Below Lower Kern Point. In this third example, prestress force is applied sufficiently below the lower kern point of the section to produce 270 psi tension when the beam is unloaded otherwise. Of course, dead and live load stress diagrams (a) and (b) of Fig. 7-2.3 remain unchanged. As shown in Figure 7-2.3, the upper prestress stress, chosen as -270 psi at the top and $+1670$ at the bottom, results in no compression under full load, as in (e).

At the end of the beam, however, with zero bending moment, stress diagrams for prestress alone (c), prestress plus dead load (d), and prestress plus dead and live loads (e) are identical with one another, 270 psi tension in the top

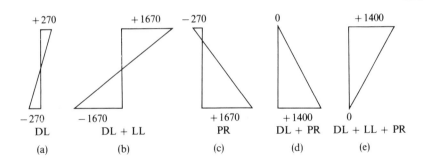

Fig. 7-2.3   Prestress Below Lower Kern Point (Mid-span)

and 1670 psi compression in the bottom. For this reason, cable should be draped or else mild steel is needed to carry these tension stresses. (The ACI Code specifies limits on such tension for various loading conditions.)

Prestress force and eccentricity are determined by equating the stress $(P/A \pm Pey/I)$ at the top of the beam to -270 psi (tension) and the stress at the bottom to $+1670$ psi.

Thus, using section modulus rather than y/I,

$$-270 = PR/240 - PR(e)/800$$

so that

$$1670 = PR/240 + PR(e)/800$$

$$PR = 168 \text{ kips and } e = 4.62 \text{ in.}$$

In much of the literature, the formula $\sigma = P/A \pm (Pe)y/I$ is altered by factoring P/A, to obtain $\sigma = P/A(1 \pm eyA/I)$ which, noting that radius of gyration r is $\sqrt{I/A}$, is rewritten in final form as

$$\sigma = P/A(1 \pm ey/r^2). \tag{7-2.2}$$

In Fig. 7-2.3, compressive stress due to prestress is greater than the compressive stress due to prestress plus any combination of loading. When comparing required concrete stress with previous trials, however, 1400 psi is used because a smaller factor of safety is applied when prestress acts alone.

7-2.4 Hollow Rectangular Section (or I)—Prestress Through Centroid. A solid rectangular beam yields interesting differences in required strengths brought about by simply relocating the centroid of prestress force. A second interesting observation is change in required material strengths due to decreased area. Size changes can be accomplished by making either a wide flange or a box beam, keeping within specified overall dimensions of 12 in. by 20 in. In general, a resulting section is still symmetrical.

Figure 7-2.4 shows computations for a reduced section that are comparable to those in Table 7-2.1 for a solid section. Section modulus is 16% smaller while member dead load is decreased by 45%. This results in reduction of final dead load stress; final live load stresses are increased.

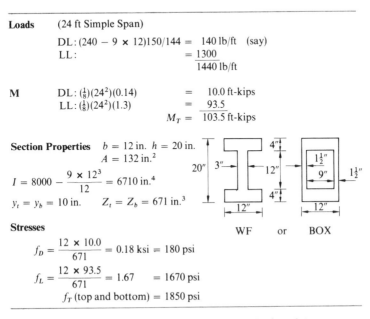

Loads	(24 ft Simple Span)	
	DL: $(240 - 9 \times 12)150/144 =$	140 lb/ft (say)
	LL: $\qquad = $	$\underline{1300}$
		1440 lb/ft

M	DL: $(\frac{1}{8})(24^2)(0.14)$	$=$	10.0 ft-kips
	LL: $(\frac{1}{8})(24^2)(1.3)$	$=$	$\underline{93.5}$
		$M_T =$	103.5 ft-kips

Section Properties $b = 12$ in. $h = 20$ in.
$A = 132$ in.^{2}

$$I = 8000 - \frac{9 \times 12^3}{12} = 6710 \text{ in.}^4$$

$y_t = y_b = 10$ in. $Z_t = Z_b = 671$ in.^{3}

Stresses

$$f_D = \frac{12 \times 10.0}{671} = 0.18 \text{ ksi} = 180 \text{ psi}$$

$$f_L = \frac{12 \times 93.5}{671} = 1.67 \quad = 1670 \text{ psi}$$

$$f_T \text{ (top and bottom)} = 1850 \text{ psi}$$

WF   or   BOX

Fig. 7-2.4   Preliminary Computations—Reduced Area

Prestress force is applied through the centroid of the beam of Fig. 7-2.4 in Fig. 7-2.5, which shows dead and live load stresses. A prestress force sufficient to develop 1850 psi compression is required to eliminate tension under any loading condition. Total prestress force is (132 × 1.85), or 244.2 kips, and required concrete working strength is 3700 psi.

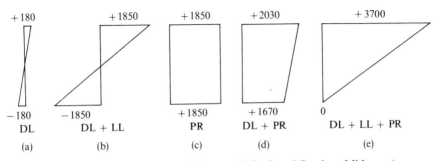

+180	+1850	+1850	+2030	+3700
-180	-1850	+1850	+1670	0
DL	DL + LL	PR	DL + PR	DL + LL + PR
(a)	(b)	(c)	(d)	(e)

Fig. 7-2.5   Prestress Through Centroid (Reduced Section, Mid-span)

Compared to a solid section, prestress force has been greatly reduced, from 400.8 kips to 244.2 kips, as required concrete working strength has increased from 3340 psi to 3700 psi.

7-2.5 Hollow Rectangular Section (or *I*)—Prestress Through Lower Kern Point. In Fig. 7-2.6, stresses are shown for a prestress force applied through the lower kern point.

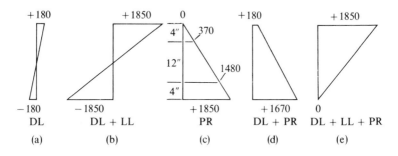

Fig. 7-2.6   Prestress Through Lower Kern Point (Reduced Section, Mid-span)

Prestress force and eccentricity are calculated by Eq. (7-2.1) or (7-2.2) as PR = 122.1 kips and $e = 5.08$ in. The required working strength of the concrete is 1850 psi.

Stresses noted as 370 and 1480 on the prestress stress diagram (c) are at the points where web and flanges meet and width changes. Such values, found by ratio and proportion, are useful if prestress force is computed by stress block volume. For instance:

Item	Area (in.^{2})	Average Stress (ksi)	Volume of Stress Block (kips)
Top flange	48	0.185	8.88
Web	36	0.925	33.30
Bottom flange	48	1.665	79.92
	132		122.1

7-2.6 Non-symmetrical Section. Figure 7-2.7 indicates an alternate shape of the same area as the WF or box section so that moments due to loads need not be recomputed. New section properties are calculated in a tabular form (sec. A-2), however, as well as dead and live load stresses.

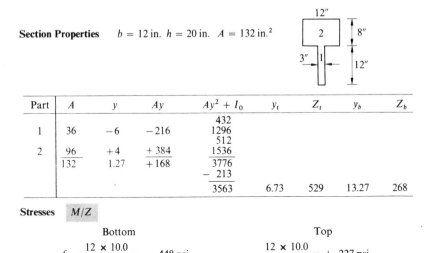

Section Properties $b = 12$ in. $h = 20$ in. $A = 132$ in.^{2}

Part	A	y	Ay	$Ay^2 + I_0$	y_t	Z_t	y_b	Z_b
1	36	−6	−216	432 1296 512				
2	96	+4	+384	1536				
	132	1.27	+168	3776 − 213				
				3563	6.73	529	13.27	268

Stresses M/Z

Bottom

$$f_D = \frac{12 \times 10.0}{268} = -448 \text{ psi}$$

$$f_L = \frac{12 \times 93.5}{268} = -4187 \text{ psi}$$

$$f_T(\text{bottom}) = -4635 \text{ psi}$$

Top

$$\frac{12 \times 10.0}{529} = +227 \text{ psi}$$

$$\frac{12 \times 93.5}{529} = +2121 \text{ psi}$$

$$f_T(\text{top}) = +2348 \text{ psi}$$

Fig. 7-2.7   Preliminary Computations—Nonsymmetrical Section

Stress diagrams with centroid of prestress force placed so that no compression exists at the top of the beam under a combination of prestress plus dead load is shown in Fig. 7-2.8. Required prestress force and eccentricity are computed as

$$-227 = PR/132 - (PR)e/529$$
$$4635 = PR/132 + (PR)e/268$$

to yield PR = 185.8 kips with an eccentricity of 4.65 in. Observe that top and bottom section moduli are different from this unsymmetrical beam.

This section has little advantage because it requires a greater working strength for concrete under dead load alone than when dead and live load are combined. But if it is inverted and used with a slab in composite construction, such a shape does have very real advantages.

7-2.7 Summary of Geometric Considerations. The results of the last six subsections are summarized in Table 7-2.2, and from these several conclusions can be inferred.

1. As the centroid of the prestress force is lowered: (a) required prestress force reduces rapidly; (b) required concrete strength reduces rapidly.

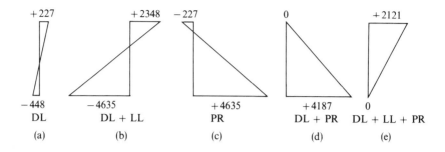

+227   +2348 −227   0   +2121

−448   −4635   +4635   +4187   0
DL   DL + LL   PR   DL + PR   DL + LL + PR

(a)   (b)   (c)   (d)   (e)

Fig. 7-2.8   Nonsymmetrical Section Stress Diagrams

Table 7-2.2   Geometric Effects

Shape (within 12 in. by 20 in.)	Area (in.²)	Eccentricity (in.)	Required PR (kips)	Strengths f_c (psi)
Solid	240	0	400.8	3340
Solid	240	3.33	200.4	1670
Solid	240	4.62	168	1400
WF or Box	132	0	244.2	3700
WF or Box	132	5.08	122.1	1850
T Inverted	132	4.65	185.8	4187 (DL)

2. As the area of the section is reduced: (a) required prestress force reduces rapidly; (b) required concrete strength increases slowly.

3. Careful choices of cross-sectional shape and area, combined with the *best* location of the centroid of prestress force, are mandatory for efficient design of prestress concrete.

7-3 PRESTRESS LOSSES

Section 18.6 of the ACI Code indicates sources of loss of prestress as (1) slip at anchorage, (2) elastic shortening of concrete, (3) creep of concrete, (4) shrinkage of concrete, (5) relaxation of steel stress, and (6) frictional losses. Before considering code provisions and recommendations, however, it is instructive to compute prestress losses theoretically.

7-3.1 Theoretical Losses of Prestress. As wires are cut and prestress force is transferred by bond to concrete, a loss in prestress occurs because the beam shortens elastically. Concrete strain is determined by dividing *stress in concrete at centroid of prestressing force* by *modulus of elasticity of concrete*. Because of bonding, steel strain is equal to concrete strain ($\epsilon_c = \epsilon_s$) and loss in the prestress steel is determined as the product of this strain times the modulus of elasticity of steel.

For post-tensioned concrete, prestress loss due to elastic shortening is variable because it is not usual for cables to be stressed simultaneously.

As discussed in Chapter 2, concrete is subject to creep under sustained loads. This creep is estimated to vary between 100% to 300% of the elastic strain, the smaller being for a wet climate and larger for a very dry climate. Thus concrete strain due to creep can be determined as a simple multiple of the strain due to elastic shortening under sustained loading (in general, PR + DL).

Concrete shrinkage varies from near zero to upward of 0.0005 in./in. A wetter climate and/or loading of the concrete a protracted time after curing results in less shrinkage. In drier climates and when loads come on concrete earlier shrinkage is higher.

The sum of concrete strains due to (1) elastic shortening, (2) creep, and (3) shrinkage, when multiplied by the modulus of elasticity of the steel, yields total losses from these three causes.

An additional loss due to the relaxation (plastic flow) of the steel under tension is estimated as between 2% and 8% of the initial prestressing force.

Losses due to anchorage slip and friction are highly dependent upon local construction materials and techniques. The ACI Code requires that frictional losses be computed by Eq. 18-1 or Eq. 18-2 with the wobble coefficient K and curvature coefficient μ based on experiment data verified during prestressing operations.

To illustrate the computation of prestress losses, the beam of Fig. 7-1.2 is used in conjunction with stress diagrams sketched in Fig. 7-3.1. Total required prestress force (PR), after losses, is 490 kips, area of steel is 3.52 in.^{2} (33 strands), ultimate strength of the concrete is 5000 psi, $E_c = 4.3 \times 10^6$ psi, and

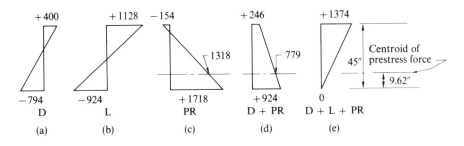

Fig. 7-3.1 Stress Diagrams for Type III AASHO Girder

$E_s = 27 \times 10^6$ psi. Stress diagrams for dead load (D), live load (L), prestress only (PR), dead plus prestress (D + PR), and total (D + L + PR) are shown. Stress in the concrete at the centroid of prestress force is shown for prestress only (PR) and prestress plus dead load (D + PR), 1318 psi and 779 psi, respectively, using simple ratio and proportion.

1. *Elastic shortening*, computed as f_c/E_c at the centroid of prestressing, is

$$\epsilon_{c(e)} = 1318/4,300,000 = 0.000306 \text{ in./in.}$$

2. *Creep of concrete* is based on a percentage of elastic strain under sustained load, that is, prestress plus dead load. Therefore

$$\epsilon_{c(e)D + PR} = 779/4,300,000 = 0.000181 \text{ in./in.}$$

Assuming a climate such that 200% is a reasonable ratio,

$$\epsilon_{c(c)} = 2(0.000181) = 0.000362 \text{ in./in.}$$

3. *Shrinkage* is estimated for the local climate as

$$\epsilon_{c(s)} = 0.000300 \text{ in./in.}$$

4. Therefore, total *effective concrete strain* is

$$\epsilon_c = 0.000306 + 0.000362 + 0.000300 = 0.000968 \text{ in./in.}$$

5. Based on this ϵ_c, loss in prestress ($\epsilon_s \times E_s$) is

$$0.000968 \times 27 \times 10^6 = 26,000 \text{ psi,}$$

or for the total area of prestress steel, 3.52(26) = 92 kips.

6. *Plastic flow* is assumed at 6%, so the loss is

$$0.06 \text{ PR} = 0.06(490) \doteq 30 \text{ kips.}$$

7. When summed, *total losses* are 92 + 30 = 122 kips or 122/3.52 = 34.6 ksi.

8. *Initial prestress force* must be the sum of 490 and the losses of, 122 kips or 612 kips.

9. *Loss* in percentage is (122/612) 100 $\doteq$ 20%.

7-3.2 Estimation of Prestress Losses. A number of calculations for various beams in various climates led to two lumped loss approaches. The earliest of these suggested that losses be estimated at about 20% of total prestress force. It is suggested in "Tentative Recommendations for Prestressed Concrete" [2] that losses of 35 ksi for pre-tensioning and 25 ksi for post-tensioning be used. This is also suggested by the Commentary on the ACI Code, which recommends study of a publication by ACI Committee 435. [3]

Interestingly, calculations of sec. 7-3.1 agree with both of these methods, 35 ksi or 20%, although for other beams agreement might not be so favorable.

(This does represent a real structure designed by the author in 1957 and built in 1958.)

An error in computation of losses does not appreciably affect the ultimate strength of the member. To obtain satisfactory service load behavior (with respect to deflection, camber, and cracking), however, losses must be reasonably estimated.

Because of the variety and complexity of variables, a simplified lump sum loss expressed in kips per square inch seems reasonable. In any particular locality, the value to be used should be checked against recommendations of local fabricators.

7-4 ADDITIONAL CONSIDERATIONS

Deflection, ultimate strength, shear, and time related load-stress-deflection considerations are discussed in this section.

7-4.1 Deflection. Deflection of prestressed concrete beams is determined as for beams of any other material. The only difficulty is that apparent modulus of elasticity of concrete varies with time (as it does with reinforced concrete members). Because prestress force is constant, deflections due to prestressing also vary with time.

Deflection at any given time is found by superimposing deflections due to dead loads, live loads, and prestress forces. Dead and live load deflections are obtained by standard procedures. Simplified equations for determining prestress deflections are developed in Fig. 7-4.1.

A constant eccentricity produces a bending moment diagram which is rectangular, with a magnitude of $-Pe$, equivalent to a simple weightless beam loaded only by end moments having magnitudes of $-Pe$. Thus deflection at mid-span due to such a loading is

$$\Delta_{max} = PeL^2/8EI. \tag{7-4.1}$$

When cables are draped in the form of a parabola, maximum mid-span moment is $-Pe_{max}$ which is analogous to the bending moment diagram obtained for a simple beam supporting a uniform load. Maximum moment due to a uniform load $(wL^2/8)$ equated to $-Pe$ produces an equivalent w of $8Pe/L^2$. With this substitution for w, the usual formula for maximum deflection of a uniformly loaded beam, $5wL^4/384EI$, is utilized to find the maximum deflection due to a parabolically draped cable:

$$\Delta_{max} = 5Pe_{max} L^2/48EI. \tag{7-4.2}$$

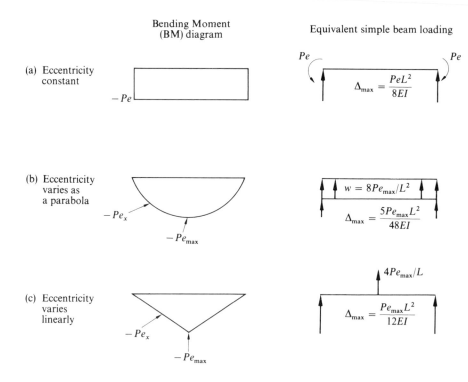

Fig. 7-4.1 Deflection Due to Prestress

When the cables vary linearly, it can, in a similar manner, readily be proved that the maximum deflection is

$$Pe_{max} L^2/12EI. \qquad (7-4.3)$$

7-4.2 Ultimate Flexural Strength (Design Strength). Flexural strength of prestressed members is computed by formulas of Chap. 2 with f_y replaced by f_{ps}. For bonded members,

$$f_{ps} = f_{pu}(1 - 0.5 p_p f_{pu}/f'_c), \qquad (7-4.4)$$

where f_{ps} ≡ calculated stress in prestressing steel at design load (psi) ; f_{pu} ≡ ultimate strength of prestressing steel (psi); $p_p \equiv A_{ps}/bd$, where A_{ps} ≡ area of prestressed reinforcement in tension zone, and f'_c ≡ specified compressive strength of concrete.

For unbonded members, conservative ACI formula 18-4 applies:

$$f_{ps} = f_{se} + 10,000 + f'_c/100 p_p \leq \begin{cases} f_{py} \\ f_{se} + 60,000 \end{cases} \qquad (7-4.5)$$

where $f_{se} \equiv$ effective streel prestress after losses (psi), and $f_{py} \equiv$ specified yield strength of prestressing steel (psi). These equations may be superseded if more information is available.

A strength check on prestress concrete is valid because prestressing forces have become nullified at the time of incipient failure. That is, steel is above the yield point and so prestress forces can no longer exist.

Conceptually, a problem is involved in determining d for strength checks of prestressed concrete beams. Prestressed steel usually is spaced over the entire area of the beam and there is some question as to whether steel in the compression zone (at ultimate load) should be used in these computations. The answer is on the safe side when all the steel is used, however, because any reduction in steel area is more than offset (in most cases) by an increased d. Also, as pointed out in "Tentative Recommendations" [2], these values have been substantiated by testing.

As for reinforced concrete beams, total amount of steel must be kept low enough to preclude sudden explosive type failures. This is not a serious problem in prestressed concrete except (1) when the amount of prestress force is small or (2) when prestress wires are used in conjunction with conventional reinforcement. "Tentative Recommendations" limit values of q to 0.3 and this limitation is maintained in ACI-71, 18.8.1, with q replaced by $\omega = pf_y/f_c'$. Load factors are the same for both prestress and reinforced concrete when the strength method is used. Elastic analysis without load factors is the basis for design at service loads.

7-4.3 Shear. ACI-71 Specifications are essentially unchanged from ACI-63. For a more detailed discussion, refer to the ACI-63 Commentary [4].

Area of stirrups is obtained by ACI formula 11-13, as discussed for non-prestressed members in sec. 6-2. It states that ultimate shear is carred by a combination of concrete at its ultimate allowable shear stress, and stirrups.

ACI Code formula 11-2 defines the minimum amount of web reinforcement, based on tests, as

$$A_v = \frac{A_{ps} f_{pu}}{80} \frac{s}{f_y} \frac{d}{d} \sqrt{\frac{d}{b_w}}, \tag{7-4.6}$$

where $f_{pu} =$ ultimate strength of prestressing steel, $f_y =$ yield strength of stirrups less than or equal to 60 ksi (to limit crack width); $s =$ stirrup spacing, $b_w =$ web width, $A_v =$ area of stirrups, and $A_{ps} =$ area of prestressed reinforcement in tension zone. Formula 11-1 can be used instead, but it is usually overly conservative.

As pointed out in sec. 6-2, shear actually is used as an indicator of the amount of diagonal tension present. Near points in a beam where shear is high and bending moments low, diagonal cracks occur due to principal tensile stresses, represented in the code by v_{cw}. When shear and bending moments are both high, flexural cracks due to axial extension are nearly vertical. Allowable shear representing these characteristics is denoted v_{ci}. Actual shear v_c carried by the concrete is, consequently, the lesser of v_{ci} or v_{cw}. In lieu of using either of these, conservative Code formula 11-10 may be used:

$$v_c = 0.6\sqrt{f'_c} + 700\, V_u d/M_u. \tag{7-4.7}$$

But $2\sqrt{f'_c} \leq v_c \leq 5\sqrt{f'_c}$ and $V_u d/M_u \leq 1$.

Derivation of v_{cw}. Code formula 11-12 is stated as

$$v_{cw} = (3.5\sqrt{f'_c} + 0.3 f_{pc}) + V_p/b_w d \tag{7-4.8}$$

and is a simplification of the classical principal stress equation. A solution by Mohr's circle (sec. A-1) is shown in Fig. 7-4.2. Thus

$$f_t = \sqrt{v_{cw}^2 + (f_{pc}/2)^2} - f_{pc}/2 \tag{7-4.9}$$

or

$$v_{cw} \equiv v_{act} = f_t\sqrt{1 + f_{pc}/f_t} \tag{7-4.10}$$

where, at the time of a diagonal tensile failure, $f_t \equiv$ tensile strength of concrete, $v_{act} \equiv$ actual shear stress of concrete, and $f_{pc} \equiv$ actual compressive stress of concrete due to prestress only. (For more precise definitions, refer to ACI Code, p. 35). The term f_{pc} does not include axial compression due to bending because of its low magnitude at points where the shear is maximum.

According to the ACI-63 Commentary, tests indicate that f_t is about equal to $4\sqrt{f'_c}$. To account for v_{cw} being a nominal shear stress rather than an actual maximum, however, f_t is replaced by $3.5\sqrt{f'_c}$. Substitution of this value of f_t in Eq. (7-4.10) yields

$$v_{act} = (3.5\sqrt{f'_c})\sqrt{1 + f_{pc}/3.5\sqrt{f'_c}} \tag{7-4.11}$$

A simplified approximate form of Eq. (7-4.11) was determined by numerical techniques to be

$$v_{act} = 3.5\sqrt{f'_c} + 0.3 f_{pc}. \tag{7-4.12}$$

Validity of this representation may be checked by plotting both equations evaluated with corresponding values of f'_c and f_{pc}.

Nominal shear stress v_{cw} is the difference between shear stress induced by loads and by prestress, or

$$v_{cw} = (V_{act} - V_p)/b_w d = v_{act} - V_p/b_w d, \tag{7-4.13}$$

where V_p is the component of shear induced by prestress.

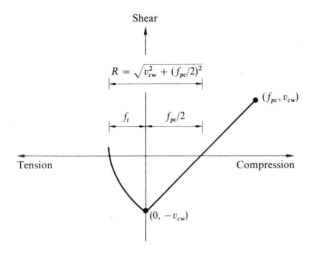

Fig. 7-4.2 Principal Tensile Stress by Mohr's Circle

Elimination of v_{act} between Eqs. (7-4.12) and (7-4.13) results in Code formula 11-12, given as Eq. (7-4.8) above.

Derivation of v_{ci} is based on the change in moment between two sections along a beam equated to the area of the shear diagram between end points (sec. A-3). Thus, assume that the bending moment at section 1 is M_{cr}, critical moment at which first cracking occurs, and M is the moment at section 2, located apart from section 1 by $d/2$:

$$M - M_{cr} = \left(\frac{V + V_{cr}}{2}\right)d \qquad (7\text{-}4.14)$$

Assume also that shear varies slowly so $(V + V_{cr})/2$ can be replaced by V; then

$$\frac{M}{V} - \frac{M_{cr}}{V} = \frac{d}{2} \qquad (7\text{-}4.15)$$

so that

$$V = \frac{M_{cr}}{(M/V) - (d/2)}. \qquad (7\text{-}4.16)$$

When flexural cracking occurs, M_{cr} is the moment due to applied loads and the corresponding stress is $M_{cr}y/I$. This stress is equal to (1) stress for the modulus of rupture of concrete, $6\sqrt{f_c'}$ [see ACI 18.4.2(b)]; (2) compressive concrete stress due to prestress f_{pc}; and (3) stress due to dead load. Therefore

$$M_{cr}y/I = 6\sqrt{f_c'} + f_{pc} - f_d \qquad (7\text{-}4.17)$$

which is the Code equation if both sides are multiplied by I/y.

Because dead load is usually uniform and because dead load section properties often are different from live load section properties (composite design), live and dead load shears are considered independently.

This means that Eq. (7-4.16) can be augmented for total shear V_{cr} due to applied and dead loads:

$$V_{cr} = \frac{M_{cr}}{(M_{max}/V) - (d/2)} + V_d. \tag{7-4.18}$$

The 1971 Code drops the term $d/2$, labels shear that occurs at the point of M_{max} by a subscripted parameter V_i, and changes the formulation from one for shear to one for stress by dividing throughout by $b_w d$. A final, experimentally determined factor of $0.6 \sqrt{f_c'}$ is added, and Code formula 11-11 is obtained as

$$v_{ci} = 0.6 \sqrt{f_c'} + \frac{V_d + \dfrac{V_i M_{cr}}{M_{max}}}{b_w d}, \tag{7-4.19}$$

where

$$M_{cr} = (I/y)(6 \sqrt{f_c'} + f_{pc} - f_d). \tag{7-4.20}$$

from Eq. (7-4.17).

When minimum A_v (allowed) and v_c (or its replacement as the minimum of v_{ci} or v_{cw}) are known, stirrup design is similar to that for a reinforced concrete member. However, spacing is limited to the minimum of $0.75h$ or 24 in. and critical shear is computed at $h/2$ rather than d from the face of the support.

In general, the definition of d is *distance from compression face to centroid of prestressing steel*, but in formulas for v_{ci} and v_{cw}, d is the greater of this value and $0.8h$.

7-4.4 Time Related Load-Stress-Deflection Considerations.

A prestressed beam typically is subjected to three independent types of loads: dead, live, and prestressed. Dead load components may be added onto the beam incrementally until their final (constant) value is reached. For example, dead load at time $t = 0$ (when prestress force first becomes effective) may be only the weight of the member itself, not changed until sometime later when a floor is placed upon it. Live load is not applied to the beam until it has been in place for a reasonable length of time. Prestress force varies with time, possibly being higher at some time before $t = 0$, that is, in cases of post-tensioning, prestress forces decrease at time $t = 0$ due to slip of anchors.

Corresponding to such variations, different factors of safety are needed. To account for these, the ACI Code recognizes three different sets of allowable stresses based on time:

At $t < 0$: Prestress tendons may be stressed to $0.80\,f_{pu}$ (ultimate strength of prestressing steel, in psi).

At $t = 0$: (time of initial prestress). Losses due to creep and shrinkage not having occurred; prestress tendons may not have stresses that exceed $0.70f_{pu}$. Concurrently, compression in concrete is limited to $0.60\,f'_{ci}$ (f'_{ci} is the ultimate concrete compressive strength at $t = 0$) and tension without bonded reinforcement is limited to $3\sqrt{f'_{ci}}$. For these calculations the loads involved are prestress and *applicable* components of dead load.

Example 1. These limits apply to a pre-tensioned member under prestress load, plus dead load due to beam weight, only if eccentricity causes the beam to deflect so as to actually carry the dead loads.

Example 2. These limits apply to a post-tensioned member under prestress and total dead load.

At $t > 0$: *All* losses have occurred; prestress tendons are considered, based on previous limits, to be adequate. Concurrently, compression in concrete is limited to $0.45\,f'_c$ and tension in the precompressed tension zone to $6\sqrt{f'_c}$ (in some environments, zero). At this time, the member is subject to prestress force, reduced by losses, plus full dead and live loads. If deflections based on cracked sections do not exceed those stated in ACI 9.5, concrete tension may be $12\sqrt{f'_c}$.

Continuing research and development may indicate that for certain applications these limitations are unnecessarily restrictive. In such cases, these values may be exceeded if analysis or experiment shows that structural performance is not impaired.

Because losses may accumulate for upward of two years, deflections should be computed carefully for appropriate loading conditions from $t = 0$ to $t = 2$ years so that undesirable positive or negative camber does not detract from structural integrity.

7-5 PLANT EFFICIENCY AND MAXIMUM ECONOMY

Plant efficiency and maximum economy are goals which cannot always be secured by consulting or design engineers. In the example which follows, however, a method is illustrated which can lead to efficient plant utilization thereby resulting in lower cost.

The problem is based on that explored earlier in this chapter, where span length was 24 ft, live loads were 1300 lb/ft, and limiting dimensions were 12 in. by 20 in. Maximum ultimate strength of the concrete is limited to 5000 psi. Because forms are expensive and limited in quantity, it is desirable to remove beams from them as soon as possible and to establish the value of f'_{ci} at 3000 psi. No tension is allowed under prestress only, in accordance with local code provisions.

Critical stresses are determined by comparing them with allowable stresses. Thus the stress at the top of the beam due to dead, live, and prestress is established as $0.45 f'_c$ or 2250 psi. Lower concrete stress at the time of initial prestressing is set by bottom stress due to live load, or $0.60 f'_{ci} = 1800$ psi, with prestress force at the top specified as zero. (Refer to the previous sketches of the stress diagrams if the reasoning behind this statement is unclear.)

Dead load moment from previous calculations is estimated at 10 ft-kips; when this is added to the live load moment of 93.5 ft-kips, a total moment of 103.5 ft-kips results.

Under full load with $M = 103.5$ ft-kips and with stress at 2250 psi, section modulus for the top of beam is determined from $\sigma = M/Z$ or $Z = M/\sigma$ as

$$Z_t = 12(103.5)/2250 = 552 \text{ in.}^3,$$

and for the bottom, using live load moment and initial allowable stress of concrete,

$$Z_b = 12(93.5)/1800 = 623 \text{ in.}^3.$$

By use of the relationships

$$Z_t = I/y_t \quad \text{and} \quad Z_b = I/y_b,$$

where I = moment of inertia (constant), y_t = distance from the centroid to the top of the beam, y_b = distance from the centroid to the bottom of the beam, Z_t = section modulus that refers to the top fiber, and Z_b = section modulus that refers to the bottom fiber, then $I = Z_t y_t = Z_b y_b$ or $552 y_t = 623 y_b$ so that $y_t = 623 y_b/552$ and $y_t + y_b = 20$ in. Simultaneous solutions for these equations are $y_t = 10.6$ in. and $y_b = 9.4$ in., so that $I = 5855$ in.[4]

In summary, a section is needed which will fit into a 12 in. by 20 in. space, with its centroid located 0.6 in. below mid-depth and a moment of inertia of 5855 in.[4]

The maximum area of all available space is 240 in.[2] with a corresponding I of 8000 in.[4] For a first trial, subtract a 6 in. by 14 in. section placed so that $\bar{y} = -0.6$. Computations are shown in Table 7-5.1, where

1. areas are summed to 156;
2. $\bar{y}$ is set at -0.6,
3. $\Sigma A\bar{y} = 156(-0.6) = -93.6$,
4. with base axis at the centroid of total area, Ay for the 6 x 14 hole must also be -93.6, so that
5. y for this 6 x 14 section $(-93.6/-84)$ is $+1.11$;
6. Ay^2 for the hole is $1.11(-93.6)$, about -105;
7. I_0 for the 6 x 14 hole is -1370;

Table 7-5.1   First Trial

Section (in.)	A	y	Ay	$Ay^2 + I_0$
12 × 20	240	0	0	8000
6 × 14 hole	− 84	+ 1.11	− 93.6	− 105
				− 1370
	156	− 0.6	− 93.6	6525
				− 55
				6470

8. total I, about mid-depth, is 6525, and

9. which, when transferred to the centroid, is 6470 in.[4].

Since 6470 is much greater than the required 5855, another trial is in order. A reasonable trial (several others not shown) is computed in Table 7-5.2 with a $7\frac{1}{4}$ in. by 15 in. hole removed from 12 in. by 20 in. gross dimensions.

Table 7-5.2   Last Trial

Section (in.)	A	y	Ay	$Ay^2 + I_0$
12 × 20	240	0	0	8000
$7\frac{1}{4}$ × 15	− 108.75	0.724	− 78.75	− 57
				− 2039
	131.25	− 0.6	− 78.75	5904
				− 47
				5857 $\doteq$ 5855 (OK)

This design is for maximum plant efficiency, but a more valid question asks whether this also leads to maximum beam economy. For instance, it might be desirable to use concrete with either higher or lower ultimate or initial strengths. In the prestressing yard, when many beams of the same type are being fabricated, it is worth while to prepare calculations for several values of f'_c and f'_{ci}, computing prestress forces required for all cases. From such data, curves can be plotted and the most economical section could be obtained to optimize use of the available materials and physical plant. As previously mentioned, true economy is dependent upon many criteria, and this illustrative example is limited in scope.

In summary, although prestress concrete members differ in many respects from conventional structural materials, design is still based upon fundamental concepts. When prestress is used for *continuous beams*, additional problems beyond the scope of this textbook are introduced. One fact is clear, however; the use of prestressed concrete will increase and the methods of manufacture will change and improve.

PROBLEMS

7-1 given: AASHO Type III girder as shown dimensioned in Fig. 7-1.2.

find: Placement of 22 strands so that $e =$ _____ (10.27 or 17.50 in.) measured downward from centroid of the section.

7-2 given: Simple prestressed beam of Fig. 7-P.1 with $L =$ _____ (38, 42, or 46 ft), $w_D =$ _____ (0, 0.8, 1.1, or 1.3 k/ft) plus beam itself, $w_L =$ _____ (0.8, 0.9, 1.2, or 1.4 k/ft).

find: Suitable cross section; show trials.

present: Analysis proof to include (a) prestress analysis—simply magnitude and eccentricity of *final* prestress force required, (b) ultimate strength analysis, (c) shear and stirrups, (d) deflections and camber.

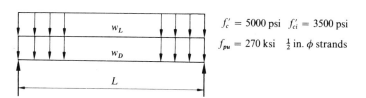

$f'_c = 5000$ psi $\quad f'_{ci} = 3500$ psi

$f_{pu} = 270$ ksi $\quad \frac{1}{2}$ in. ϕ strands

Fig. 7-P.1 Simple Prestressed Beam

7-3 given: The single tee of *PCI Design Handbook*, p. 3-40, as shown in Fig. 7-P.2. Designation _____ (10ST24, 10ST32, or 10ST44) with *corresponding h* = _____ (24, 32, or 44 in.) and *corresponding* span = _____ (66, 86, or 96 ft). Corresponding superimposed load = _____ (31, 29, or 67 lb/ft^2). One-half inch round strands are straight both sides of depressed point of midspan.

find: Required number of strands and eccentricity e_e at end and e_c at mid-span.

7-4 given: The inverted tee of Fig. 7-P.3, often used to support cross beams in precast construction. Typical designation is based on h; that is,

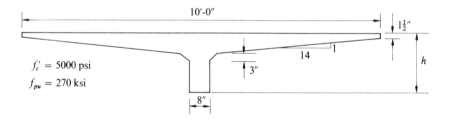

Fig. 7-P.2   Single Tee [PCI Design Handbook, page 3–40]

if $h = 20$ in., designation is 24IT20, where $h =$ _____ (12, 14, 18, or 28) and $h_1 =$ _____ (20, 22, 30, or 44). Span length = _____ (20, 24, or 28 ft). Uniform dead load = _____ (1.3, 1.4, or 1.5 k/ft). Uniform live load = _____ (1.5, 1.4, or 1.3 k/ft).

design: Straight strands required, and spacing.

present: Complete analysis proof.

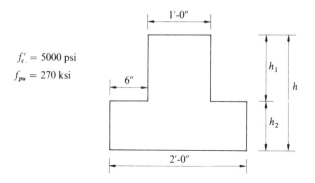

Fig. 7-P.3   Inverted Tee [PCI Design Handbook, page 3–48]

7-5 given: An AASHO-III girder (Fig. 7-1.2). Strands _____ (drape parabolically, have one hold-down, or have two hold-downs). $f'_c =$ _____ (5000, 6000, or 7000 psi), $f_{pu} =$ _____ (270,000 or 290,000 psi), $w_D =$ _____ (2, 3, or 4 k/ft), $w_L =$ _____ (2, 3, or 4 k/ft).

design: Girder, completely.

present: Complete analysis proof.

REFERENCES

1. *PCI Design Handbook—Precast and Prestressed Concrete*, Prestressed Concrete Institute, Chicago, Illinois (1971), pp. 3–49.

2. ACI-ASCE Committee 423, "Tentative Recommendations for Prestressed Concrete," *ACI Journal, Proceedings* V. 54, No. 7 (January 1958), pp. 545–578.

3. ACI Committee 435, "Deflections of Prestressed Concrete Members," *ACI Journal, Proceedings*, V. 61, No. 12 (December 1963), pp. 1697–1728.

4. *Commentary on Building Code Requirements for Reinforced Concrete* (ACI 318–71), American Concrete Institute, Detroit, Michigan (1971).

PLASTIC DESIGN

Plastic design theory originated from investigations of the true ultimate capacity of complex frames. As evidenced by preceding sections, however, this theory now is widely employed in design procedures which are tacitly linearly elastic. For this reason, concepts of a plastic hinge, shape factor, moment redistribution, etc., have already been introduced. To formalize these concepts into a unified design technique is the purpose of this chapter.

8-1 INTRODUCTION TO UPPER AND LOWER BOUND THEOREMS

A formal approach to plastic design is embodied in two theorems proven by H. J. Greenberg and William Prager[1]. Simplified by Beedle[2] these theorems are:

1. *Lower Bound Theorem*. "A load computed on the basis of an assumed equilibrium moment diagram in which the moments are not greater than the plastic moment is less than or at best equal to the true ultimate load."

2. *Upper Bound Theorem*. "A load computed on the basis of an assumed mechanism will always be greater than or at best equal to the true ultimate load." These theorems specify three conditions which must be satisfied at the precise time when the last plastic hinge forms:

1. The structure must be in static equilibrium.
2. The configuration must be, or approach being, a mechanism in whole or in part.
3. The bending moment at every place must be less than, or at most equal to, plastic moment capacity of any corresponding member (M_p).

A mechanism occurs when all or part of a structure can rotate without limit or increased resistance. For a simple span, one plastic hinge is sufficient, two are required for a propped beam, and three plastic hinges must form before a fixed end beam becomes a mechanism. A continuous beam or a frame becomes a mechanism when any integral member does.

These theorems provide two approaches to the plastic design of frames: the equilibrium and the mechanism method. In Table 8-1.1 these theorems are inverted and reworded for use with design related methods. The bending moment at any (every) point is signified by M_x, and the plastic moment capacity by M_p.

Table 8-1.1   Plastic Design Summary

Equilibrium Method

Based on   Inverse Upper Bound theorem (upper limit on Z_{reqd})

If an *equilibrium moment diagram* is assumed for a given loading condition, the section modulus required for the largest moment present will be equal to or *greater* than the actual required section modulus. In general, if a mechanism does exist, the correct answer is obtained when $M_x \leqq M_p$.

Procedure

1. Sketch bending moment diagram (often by parts).
2. Select system of points where M_p will form.
3. Check if $M_x \leqq M_p$ for each system.
4. M_p that satisfies 3, above, is the correct solution.

Mechanism Method

Based on   Inverse Lower Bound theorem (lower limit on Z_{reqd})

If a *mechanism* is assumed for a given loading condition, the section modulus required for the M_p obtained will be equal to or *less* than the actual required section modulus. In general, if all possibilities are tried, the correct answer is that yielding the largest M_p (at which time $M_x \leqq M_p$).

Procedure

1. Select possible mechanisms.
2. For each mechanism equate internal and external work; solve for M_p.
3. The correct result is that with the largest M_p (*if* all possible mechanisms have been utilized).

8-2   ILLUSTRATION OF BOUND THEOREMS

To illustrate these procedures, both methods are applied to (1) the same propped beam, (2) a propped beam with overhang, and (3) a propped beam with a variable moment of inertia.

8-2.1 The Equilibrium Method. Figure 8-2.1 shows a propped beam, assumed weightless, with concentrated loads specified as multiples of P and including load factors. Span length is 100 ft and EI is constant. Design is performed using procedural steps from Table 8-1.1.

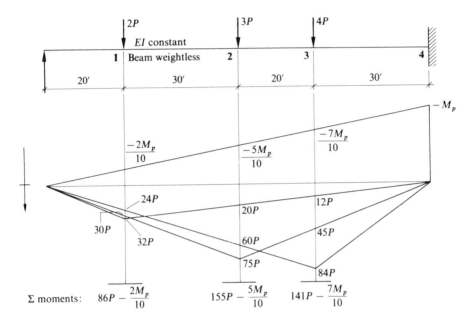

$$\Sigma \text{ moments:} \qquad 86P - \frac{2M_p}{10} \qquad 155P - \frac{5M_p}{10} \qquad 141P - \frac{7M_p}{10}$$

Fig. 8-2.1   Propped Beam, Equilibrium Method

Step 1. Sketch bending moment diagrams (by parts). In Fig. 8-2.1, bending moment diagrams are sketched by parts for each concentrated load and the end moment at point **4**, with positive moments plotted downward. An assumption, which experience tends to verify, is made that one plastic hinge will form at the fixed right support. A bending moment diagram for such an end moment is linear and intermediate values are simple to evaluate by geometry. Simple beam bending moment diagrams are drawn for each concentrated load with intermediate values also calculated by ratio and proportion. Finally, bending moments are summed at each load point, yielding expressions for corresponding composite values.

Step 2. Select system of points where M_p will form. A mechanism is formed by one additional plastic hinge. If it is assumed that it forms under the concentrated load at **1**, the composite bending moment at such point must equal M_p, or $86P - 2M_p/10 = M_p$, so that $M_p = 71.7P$.

Step 3. Check if $M_x \leq M_p$. When a value for M_p has been established, composite bending moments under the other loads are determined:

at 2: $155P - 5M_p/10 = 155.0P - 35.8P = 119.2P > 71.7P,\therefore NG;$
at 3: $141P - 7M_p/10 = 141.0P - 50.2P = 90.8P > 71.7P,\therefore NG.$

Because both values are greater than M_p of $71.7P$, this is not a correct solution. However, the plastic moment cannot exceed an upper bound of $119.2P$, as stated by the Inverse Upper Bound Theorem.

Step 4. Try additional points where M_p will form and check to see if $M_x \leqq M_p$ for each case. Computations for plastic hinges forming under loads **2** and **4** or **3** and **4** are shown in Table 8-2.1. From these calculations, choice of plastic hinges at points **3** and **4** does not yield a solution because M_x of $113.5P$ is greater than M_p of $83.0P$. However, $113.5P$ does represent a new (and lower) upper bound because it is less than the upper bound $119.2P$ obtained with plastic hinges at points **1** and **4**. It is evident that plastic hinges at points **2** and **4** represent a valid solution because at every location bending moments M_x are equal to or less than M_p.

A good estimate of correct plastic hinge formations can reduce the amount of computations required. Conversely, in very complex cases when the true mechanism condition is very difficult to locate, a reasonable design usually can be based on the lowest upper bound.

Table 8-2.1 M_x versus M_p

Position of M_p	2 and 4		3 and 4	
Equating Moments	$155P - 5M_p/10 = M_p$		$141P - 7M_p/10 = M_p$	
Plastic Moment	$M_p = 103.3P$		$M_p = 83.0P$	
Position of M_x	1	3	1	2
Simple Beam Moment	$86.0P$	$141.0P$	$86.0P$	$155.0P$
Moment due to M_p	$-20.6P$	$-72.3P$	$-16.6P$	$-41.5P$
Composite Bending Moment	$65.4P$	$68.7P$	$69.4P$	$113.5P$

8-2.2 The Mechanism Method. A solution for the beam of Fig. 8-2.1 can also be obtained directly by the mechanism method.

Step 1. Select possible mechanism. If plastic hinges form at points **1** and **4**, the mechanism condition, with elastic deformation disregarded, is as shown in Fig. 8-2.2. Angular rotations are considered to be small so that $\tan \theta \doteq \theta$. Positions for θ and Δ are usually chosen to simplify geometry; otherwise their locations are arbitrary. In this example, left support rotation is defined as θ and deflection under load $2P$ at point **1** as Δ. By geometry, the rotation at **4** is to 80 ft as θ is to 20 ft, so rotation at the right support is $\theta/4$, making rotation at **1** equal to $\theta + \theta/4$ or $5\theta/4$. Deflection at point **3** is to 30 ft as Δ is to 80 ft, or $3\Delta/8$. Because $\tan \theta \doteq \theta = \Delta/(L/5)$, it is seen that $\Delta = L\theta/5$.

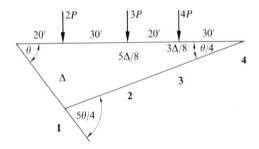

Fig. 8-2.2   Mechanism Condition with Plastic Hinges at **1** and **4**

Step 2a. Equate internal and external work; solve for M_p. (1) Internal work is performed as plastic hinges rotate and is equal to the product of *plastic moment of the hinge times angular rotation of the hinge.* Thus

$$M_{int} = M_p(5\theta/4) + M_p(\theta/4) = M_p(6\theta/4).$$

(2) External work is performed as loads move through a distance. Therefore, external work is equal to the summation of products of loads times their respective displacements:

$$M_{ext} = 2P(\Delta) + 3P(5\Delta/8) + 4P(3\Delta/8) = 43P\Delta/8.$$

(3) For a conservative system, $M_{int} = M_{ext}$ or

$$M_p(6\theta/4) = 43P\Delta/8 = 43PL\theta/40,$$

where Δ is replaced by $L\theta/5$.

When L is evaluated as 100 and the values of θ are canceled, the plastic moment is found to be $M_p = 71.7P$.

This is a lower bound on M_p and is correct only if it is the largest M_p found for any mechanism, and *all* have been tried. Alternately, this is the correct M_p if $M_x \leq M_p$ for this assumed mechanism condition.

Step 2b. Repeat for an assumed mechanism with plastic hinges at points **3** and **4** (depicted in Fig. 8-2.3) with θ and Δ redefined ($\Delta = 7\theta L/10$). Internal work is

$$W_{int} = M_p(10\theta/3) + M_p(7\theta/3) = M_p(17\theta/3),$$

and external work is

$$W_{ext} = 2P(2\Delta/7) + 3P(5\Delta/7) + 4P(\Delta) = 47P\Delta/7.$$

If internal and external work are equated, with Δ replaced by $7L\theta/10$, a new lower upper bound $M_p = 83.0P$ is found.

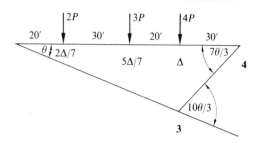

Fig. 8-2.3 Mechanism Condition with Plastic Hinges at **3** and **4**

Step 2c. Finally, with plastic hinges at points **2** and **4** the mechanism, as indicated in Fig. 8-2.4, is formed.

$$W_{int} = M_p(2\theta) + M_p(\theta) = M_p(3\theta),$$
$$W_{ext} = 2P(2\Delta/5) + 3P(\Delta) + 4P(3\Delta/5) = 31P\Delta/5.$$

If this result is equated, with Δ replaced by $L\theta/2$,

$$M_p(3\theta) = 31P\Delta/5 = 31PL\theta/10$$

so that $M_p = 103.3P.$

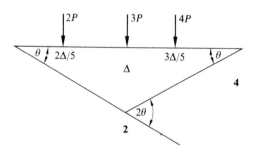

Fig. 8-2.4 Mechanism Condition with Plastic Hinges at **2** and **4**

Step 3. This latter M_p has the largest value and therefore is correct because all possible mechanisms have been tried. It represents the highest lower bound and agrees with the lowest upper bound obtained by the equilibrium method. For a complex design, members are sometimes sized for an M_p represented by a lowest upper bound when it is sufficiently close to a highest lower bound value.

8-2.3 Propped Beam with Overhang. This example emphasizes not only that internal energy is independent of direction of rotation but also that external energy is direction oriented. The equilibrium method applied to the propped beam of Fig. 8-2.5 shows that a mechanism form; with M_p either (1) at the right support or (2) at the fixed left support and under the load concurrently.

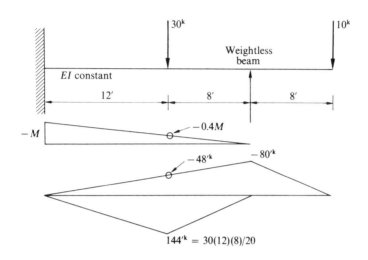

Fig. 8-2.5 Propped Beam with Overhang

The first condition is a correct solution if M_x is equal to or less than $M_p = 80$ at every x. Assume that M_p is of this maximum possible value, so as to lower as much as possible the positive bending moment of the point of load:

$$M_{max} = 144 - 48 - 0.4(80) = 64 \text{ ft-kips} < 80 \text{ ft-kips}$$

[where $0.4(80)$ represents $0.4M_p$]. Since no other moments are greater than 80, this is the solution and there is no need to check other conditions!

An alternate mechanism solution is sketched in Fig. 8-2.6. In the first case, $M_{ext} = 10\Delta$ and $M_{int} = M_p\theta$ so that $M_p = 80$ ft-kips (Δ replaced by 8θ). For the latter, $M_{ext} = 30\Delta - 10\Delta = 20\Delta$ (*note the signs!*) and $M_{int} = M_p(2\theta/3) + M_p(5\theta/3)$ (again, $\Delta = 8\theta$) so that

$$160\theta = M_p(7\theta/3) \qquad \text{or} \qquad M_p = 68.6 \text{ ft-kips.}$$

The largest M_p (when all mechanisms have been tried) represents the correct solution, according to the Inverse Lower Bound theorem, so design plastic moment is 80 ft-kips.

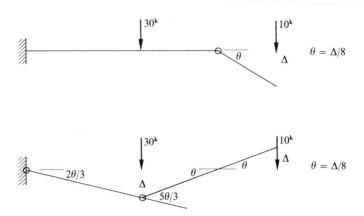

Fig. 8-2.6   Two Possible Mechanisms

8-2.4 Variable Section. Although variable sections do not occur commonly when plastic design is used, sometimes such variations are aesthetically desirable and/or lead to increased economy. To illustrate principles involved, design of a weightless fixed end beam subject to a single concentrated load (Fig. 8-2.7) is discussed. Such a solution is analogous to the design of one span of a continuous structure. Also shown are bending moment diagrams drawn by parts with summations leading to a composite bending moment diagram. Load includes the load factor of 1.7 and the member is laterally supported.

If EI is constant, $M_{p1} = M_{p2} = M_{p3}$, so $M_p = 525$ ft-kips. From page 2-17 of *Steel Construction* a W24 x 68 is chosen for a total of 3400 lb.

Alternately, a W21 x 55 with cover plates at the right support may be tried. Both M_{p1} and M_{p2} are established by member size as 378 ft-kips. At the load point, then, $M_{p2} = 378 = 1050 - 0.3(378) - 0.7M_{p3}$, so $M_{p3} = 798$ ft-kips. Thus cover plates must carry $(798 - 378)$ 420 ft-kips. By use of two PL$\frac{1}{2}$ x $13\frac{1}{2}$ cover plates, additional moment capacity computed as a couple (AF_y)(arm) is $(13\frac{1}{2})(\frac{1}{2})(20.80 + 0.50)(36/12) = 431$ ft-kips, where 20.80 is the depth of the W21 x 55 and 0.50 is two times one-half of a plate thickness. Required plate length is about $6\frac{1}{2}$ ft, making the total weight $50(55) + 2(6\frac{1}{2})(23) = 3050$ lb. This represents a savings of 7 lb/ft, a break-even point where welding costs nearly offset savings incurred by the reduction in amount of steel used.

A larger number of beam and cover plate combinations are possible for even this simple case, but unless weight is reduced by more than 7 lb/ft, constant section members, in general, are more economical. Iteration often is required to obtain an economical section if tapered beams are used.

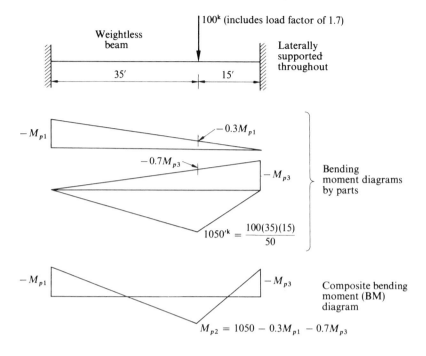

Fig. 8-2.7   Fixed End Beam

8-3   UNIFORM LOADING

Although common in practice, uniform loads increase the complexity of plastic design. Several methods have been suggested for handling such distributed loads, including graphical solutions and replacement by a series of equivalent concentrated loads. In this section a curve is developed which provides a more direct means of design for both constant and variable section members [3]. A second curve facilitates determination of cover plate lengths.

8-3.1 The Design Curve. Figure 8-3.1 shows a free body diagram for one span of a continuous structure subject to uniform load. The three plastic hinge moments required to produce a mechanism are called M_{p1} (left support), M_{p2} (interior of span), and M_{p3} (right support). To further simplify the problem, each plastic moment is expressed as a parameter, respectively, C_L, C_M, and C_R (left, mid, right) times simple beam bending moment $wL^2/8$. This derivation is based on two fundamentals: (1) Shear corresponding to M_{p2} must be zero if it is to be greater than any other bending moment close by; and (2) this bending moment is the plastic bending moment M_{p2}, expressed as $C_M wL^2/8$.

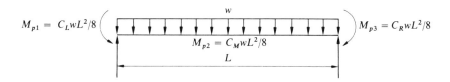

$M_{p1} = C_L wL^2/8$ $M_{p3} = C_R wL^2/8$

$M_{p2} = C_M wL^2/8$

Fig. 8-3.1   Free Body Diagram of One Span Uniformly Loaded

A general shear equation for the case in which x originates at the left support is

$$V_x = C_L wL/8 - C_R wL/8 + wL/2 - wx. \qquad (8\text{-}3.1)$$

With V_x set equal to zero, x distance to the point of maximum moment is expressed by

$$x = (C_L - C_R + 4)L/8. \qquad (8\text{-}3.2)$$

A general equation for bending moment is

$$M_x = -C_L wL^2/8 + C_L wLx/8 - C_R wLx/8 + wLx/2 - wx^2/2. \qquad (8\text{-}3.3)$$

When V_x is zero, M_x is defined by $C_M wL^2/8$ and Eq. (8-3.3) can be written as

$$C_M wL^2/8 = -C_L wL^2/8 + C_L wLx/8 - C_R wLx/8 + wLx/2 - wx^2/2$$

or

$$x^2/2 - Lx(C_L - C_R + 4)/8 + (C_M + C_L)L^2/8 = 0. \qquad (8\text{-}3.4)$$

By use of x from Eq. (8-3.2), Eq. (8-3.4) reduces to $x^2 = (C_M + C_L)L^2/4$ so that

$$x = L\sqrt{C_M + C_L}/2. \qquad (8\text{-}3.5)$$

By Eq. (8-3.2) C_R, C_L, and x are related and by Eq. (8-3.5) C_M and C_L are related also. The curve of Fig. 8-3.2 was constructed by using these relationships in the following manner: First, an assumed value for x is held constant while values for C_M ranging from 0.1 upward by tenths are used to find respective values of C_L. From each value of C_L and x, a corresponding value of C_R is then found. These points are plotted to obtain a curve for the assumed x. Then x is changed and the procedure is repeated.

Values of x are shown in terms of tenths of the span length L, and make 45° angles with axes. Values of C_M, varying from 0.0 to 0.9 are found along curved lines. On the vertical and horizontal axes, C_R and C_L, respectively, are plotted. The broken line labeled $C_L = C_M$ represents $M_{p1} = M_{p2}$. Another

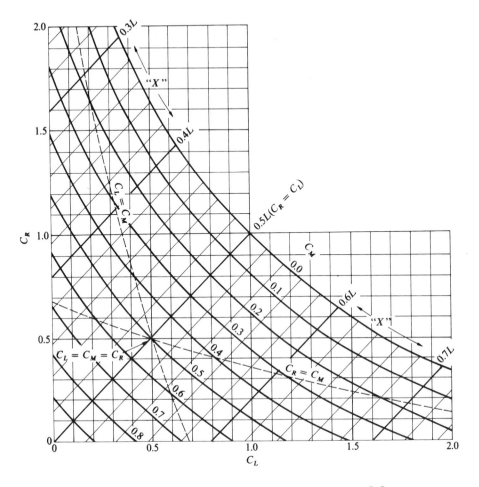

Fig. 8-3.2   Plastic Design Curve for Uniform Load [3]

broken line is drawn for the case when C_R is equal to C_M. Of course, C_L is equal to C_R when $x = 0.5L$. At the intersection of the two dashed lines is the point where $x = 0.5L$, $C_R = 0.5$, $C_M = 0.5$, and $C_L = 0.5$, representing a beam with a constant cross section. Use of Fig. 8-3.2 is illustrated in subsequent examples.

8-3.2 Bending Moment Curves for Uniform Loads. Figure 8-3.3 is a set of curves useful in determining theoretical points of cutoff for cover plates. To use these curves:

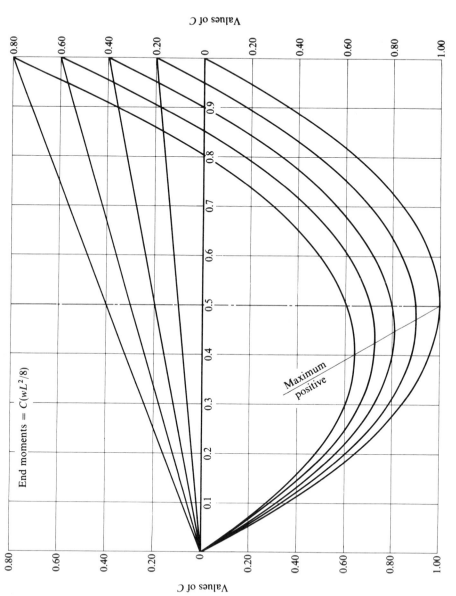

Fig. 8-3.3 Composite Uniform Bending Moment Curves with End Moments [3]

1. Plot the lower of C_L and C_R downward on the left axis and project a horizontal line that represents zero bending moment.

2. Plot the difference between the highest and lowest values of C_L and C_R upward on the right axis.

3. Using these curves as guides, sketch in the bending moment diagram.

The bending moment diagram is complete and to scale. Examples of its use are discussed in relation to Figs. 8-3.8 and 8-3.10.

8-3.3 Constant Section. A fixed-end A36 beam, laterally supported and loaded by a 2 k/ft working load is shown in Fig. 8-3.4. When EI is constant, $M_p = (wL^2/8)/2 = 3.4(40)^2/16 = 340$ ft-kips, thus requiring a W21 x 55.

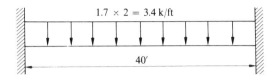

1.7 × 2 = 3.4 k/ft

40'

Fig. 8-3.4   Fixed-End A36 Beam with Uniform Load [3]

A chart solution of the problem, although trivial, is based on the fact that $C_L = C_M = C_R$ which establishes each of them as 0.5 and the location of the intermediate plastic hinge at $x = 0.5L$.

For completeness a check by the mechanism method is shown in Fig. 8-3.5. Internal work is $M_{\text{int}} = M_{p1}(\theta) + M_{p2}(2\theta) + M_{p3}(\theta) = M_p(4\theta)$, and external work is $M_{\text{ext}} = wL\Delta/2 = (3.4)(40)\Delta/2 = 3.4(40)^2\theta/4$. When internal and external work are equated, $M_p = 340$ ft-kips.

8-3.4 Mid-Span Cover Plate. Using the load and span of Fig. 8-3.4, a mid-span cover plate design can be based on the use of a W18 x 45 section (A36) with an allowable bending moment of 269 ft-kips. Simple beam bending moment, $wL^2/8$ is 680 ft-kips; therefore $C_L = C_R = 269/680 = 0.396$. A check of these values in Fig. 8-3.2 shows that $C_M = 0.604$ and $x = 0.5L$. Thus plastic bending moment at $L/2$ is $0.604(680) = 411$ ft-kips; cover plates are required to increase beam capacity by $(411 - 269)$ 142 ft-kips.

Results can be verified by either the statical or mechanism method. Use of the latter, in conjunction with the sketch of Fig. 8-3.5 shows that

$$W_{\text{int}} = M_{p1}\theta + M_{p2}(2\theta) + M_{p3}(\theta) = 269\theta + 411(2\theta) + 269\theta = 1360\theta$$

and

$$W_{\text{ext}} = wL(\Delta/2) = wL(L\theta/2)/2 = 1360\theta$$

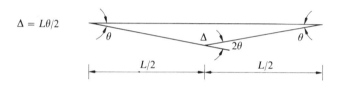

$\Delta = L\theta/2$

$L/2$ $L/2$

Fig. 8-3.5 Mechanism Method Check for Fixed-end A36 Beam

so that $W_{int} = W_{ext}$. Cover plate length can be determined by using Fig. 8-3.3.
Since $C_L = C_R = 0.396$, a horizontal base line ($M = 0$) is drawn from the left at $C = 0.396$. On the right axis, $C_L - C_R = 0$, so the lowest curve applies. A second horizontal line is drawn at a point 0.396 below the base axis which intersects the lower curve at $0.27L$ and $0.73L$, theoretical points of cutoff. Theoretical plate length is $(0.73 - 0.27)L = 0.46L = 18.4$ ft.

8-3.5 Three Unequal Plastic Bending Moments. Figure 8-3.4 could represent a free body diagram for one span of a continuous structure. It then might be appropriate to have a W16 x 50 extending through the left support and a cover-plated W16 x 45 extending through the right support, with maximum positive bending moment carried by the W16 x 45 acting alone.

With $wL^2/8 = 680$ ft-kips, corresponding values for the W16 x 45 are $M_{p2} = 246$ and $C_M = 246/680 = 0.362$; and for the W16 x 50, $M_{p1} = 275$ and $C_L = 275/680 = 0.404$. From Fig. 8-3.2 corresponding values are $x = 0.44L$ and $C_R = 0.89$, making $M_{p3} = 0.89(680) = 605$ ft-kips.

To check by the statical method, bending moment and shear are figured at $0.44L$ in Fig. 8-3.6. Left reaction is 59.76 kips, assuming that the values of M_{p1} and M_{p3} are correct. Then

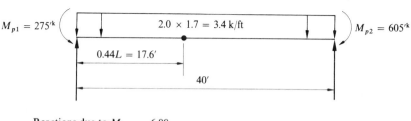

$M_{p1} = 275^{'k}$ $2.0 \times 1.7 = 3.4$ k/ft $M_{p2} = 605^{'k}$

$0.44L = 17.6'$

$40'$

Reactions due to M_{p1} 6.88
M_{p2} -15.12
w 68.00
          59.76^k

Fig. 8-3.6 Equilibrium Check [3]

$$V_{17.6} = 59.76 - 17.6(3.4) = 59.76 - 59.84 \doteq 0 \text{ (check)}$$

and

$$M_{17.6} = (59.76)(17.6) - 3.4(17.6)^2/2 - 275 = 250.0 \doteq 246.0 \text{ (check)}.$$

To check by the mechanism method drawing the beam as shown in Fig. 8-3.7 is required. Then

$$M_{\text{int}} = 275\theta + 246(1.785\theta) + 605(0.785\theta) = 1189\theta$$

and

$$M_{\text{ext}} = wL(\Delta/2) = 3.4(40)[0.44(40)\theta]/2 = 1196\theta,$$

so $M_{\text{int}} = M_{\text{ext}}$ (within one-half of 1%).

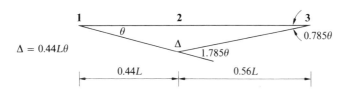

Fig. 8-3.7   Mechanism $M_{p1} \neq M_{p2} \pm M_{p3}$

Theoretical length of the cover plate is determined graphically in Fig. 8-3.8. A bending moment diagram is constructed according to the procedure of sec. 8-3.2. That is:

1. Because C_L is less than C_R, $C_L = 0.404$ is plotted downward on the left axis.
2. A horizontal line, representing the axis of zero bending moment, is drawn through this point.
3. $C_R - C_L = 0.89 - 0.404 = 0.48$ is then plotted upward on the right axis.
4. From this point, a bending moment curve is sketched as accurately as possible "by eye," using as guides the curves on the figure.
5. With the bending moment diagram complete, the C value, 0.362, for a W16 x 45 acting without cover plates is plotted upward from the base axis (axis of zero bending moment).
6. A horizontal line through this point intersecting the bending moment diagram at $0.86L$, signifies the theoretical point of cutoff.

8-3.6 Tapered Section. In some cases, often for architectural reasons, a tapered section is preferred. Imposing depth limitations between 12 and 24 in.

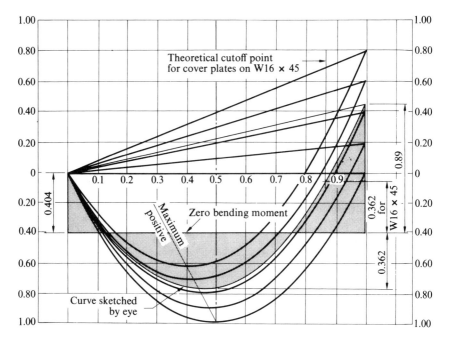

Fig. 8-3.8   Theoretical Point of Cutoff Determined by Bending Moment Curves [3]

for the beam of Fig. 8-3.4 it is decided to try a $\frac{3}{8}$-in. thick web plate and $\frac{1}{2}$-in. by 12 in. flanges. C values (plastic moment capacity for the cross section divided by maximum simple beam bending moment) and plastic moments are plotted versus depth in Fig. 8-3.9.

Although a reasonable first trial can be assumed by observing that a constant cross section has C equal to 0.5 at a corresponding depth of 20.2 in., the poorest first trial of $D_L = 12$ in. and $D_R = 24$ in. is illustrative of convergence.

A solution is best performed in tabular form (see Table 8-3.1). The first and third columns represent trial depths; the second and fourth columns represent C values as read from Fig. 8-3.9 for the trials. Using corresponding C_L and C_R values, Fig. 8-3.2 is used to determine $C_{M(\text{reqd})}$ (column 5), and x as a function of span length L (column 6). Column 7 represents the change in depth from left to right multiplied by x; this signifies the increase in depth over the leftmost section. Column 8 is the sum of columns 1 and 7 and indicates the depth of the section at xL. From Fig. 8-3.9, this depth yields the value of C_M shown in column 9 which is compared to the required value shown in column 5. In tabular form, then, a typical solution is as shown in Table 8-3.1. The C_M formulated is 0.484, less than the required C_M of 0.50, but the solution is accur-

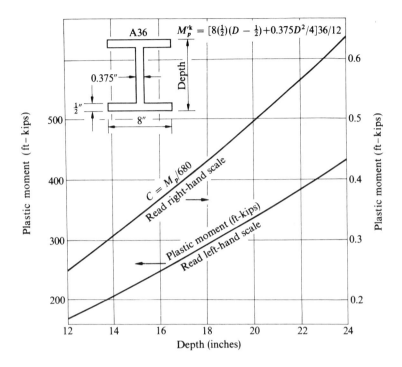

Fig. 8-3.9 Plastic moment capacity M_p and ratio of plastic moment capacity to simple beam bending moment, C, related to overall beam depth [3]

ate to within about 3%. Because 0.484 *is* less than 0.500, the OK shown in Table 8-3.1 is followed by a question mark. Would you accept this design or try again?

Validity of this solution is indicated in Fig. 8-3.10 where a bending moment diagram has been plotted in accordance with sec. 8-3.2. Allowable bending moment is plotted as a straight line from C_L of 0.367 to C_R of 0.634, both positive downward. This plot gives visual verification of the analytical solution.

Table 8-3.1   Tapered Section

(1) D_{left} (inches)	(2) C_L	(3) D_{right} (inches)	(4) C_R	(5) $C_{M(reqd)}$	(6) $x(L)$	(7) $x(D_R - D_L)$	(8) D_M	(9) C_M
12	0.253	24	0.634	0.57	0.455	5.45	17.45	0.42	NG
18	0.428	24	0.634	0.48	0.473	2.84	20.84	0.525	BIG
16	0.367	24	0.634	0.50	0.464	3.72	19.72	0.484	OK?

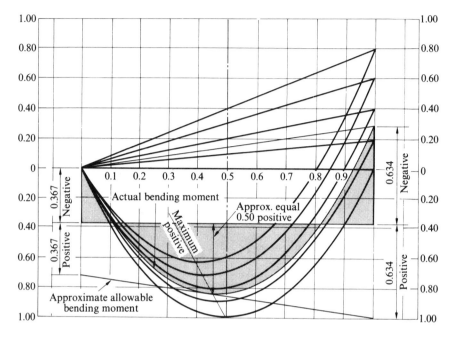

Fig. 8-3.10 Plastic Design for a Tapered Section; Actual versus Allowable Bending Moments Plotted [3]

The C values of a tapered section are not precisely linear (as drawn in Fig. 8-3.9) but the difference is minor. If webs are relatively thick or flanges vary in size, Fig. 8-3.3 can be used to verify a section by plotting actual allowable bending moments rather than a linear approximation.

8-4 CONTINUOUS SPANS

In addition to providing a more uniform factor of safety, a principal advantage of plastic design is the simplicity with which it is applied to redundant structures. This is true because a mechanism can be analyzed by statics alone. For design of continuous beams, the equilibrium method usually is easier than the mechanism method.

8-4.1 Mechanisms and Bending Moment Diagrams. A weightless, laterally supported, A36 continuous beam is to be designed for the loads shown in Fig. 8-4.1 Simple beam bending diagrams are sketched for each span and designated loads include load factors.

1. As a first trial mechanism, a plastic hinge is assumed to form at joint A, indicated by a solid circle at that point in Fig. 8-4.2. Supporting a cantilever,

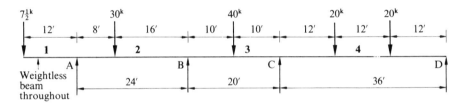

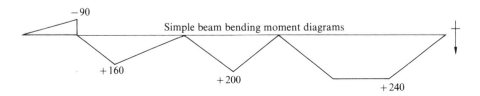

Fig. 8-4.1 Continuous Beam, Laterally Supported

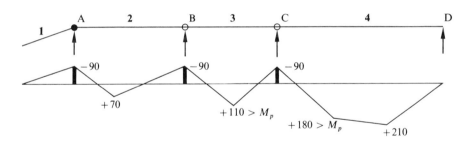

Fig. 8-4.2 First Trial Mechanism for Beam of Fig. 8-4.1

plastic moment is 90 ft-kips, symbolized by the heavy solid line on the bending moment diagram at the left support. At the next two joints, B and C, a moment is *assumed* also of 90 ft-kips, indicated by open circles on the mechanism diagrams and solid vertical lines on the bending moment diagram. These represent *possible* bending moments which satisfy equilibrium but do not exceed (or are equal to) the allowable plastic moment. When these end moments are related to the simple beam bending moment diagrams of Fig. 8-4.1, *possible* composite bending moment diagrams for the entire structure are completed. In spans **1** and **2**, all M_x are less than M_p, but in spans **3** and **4**, some M_x exceed M_p. By the Inverse Upper Bound theorem this design is inadequate, but an upper bound of 210 ft-kips has been established.

2. As a second trial mechanism, span **2** develops three hinges. This solution

is invalid because moment produced by the cantilever end is greater than M_p of 80 ft-kips (Fig. 8-4.3).

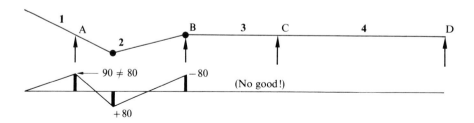

Fig. 8-4.3   Second Trial Mechanism for Beam of Fig. 8-4.1.

3. Figure 8-4.4, a third possible mechanism, shows three plastic hinges in span **3**, each developing 100 ft-kips of moment. At the cantilever support, bending moment cannot exceed 90 kips. At the right support, a hinge can support no bending moment. A composite bending moment diagram indicates that M_x is greater than M_p in span **4**, so a fourth solution must be tried.

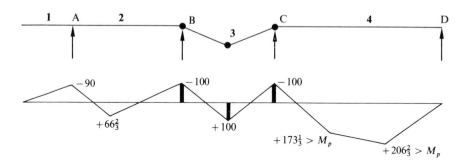

Fig. 8-4.4   Third Trial Mechanism for Beam of Fig. 8-4.1

4. An experienced engineer would, of course, check the fourth trial mechanism (Fig. 8-4.5) first because it appears critical upon even a cursory inspection.

Assuming a plastic hinge at the left support (joint C) of span **4** and under the rightmost load, the controlling plastic moment is determined by equating values from a mental picture of the bending moment diagram drawn by parts.

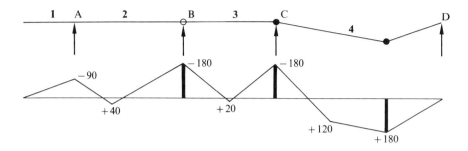

Fig. 8-4.5   Fourth Trial Mechanism for Beam of Fig. 8-4.1

Required plastic moment capacity is thus determined from $M_p = 240 - M_p/3$ to be 180 ft-kips.

This moment is also *assumed* as possible at joint B, while joint A maintains a 90 ft-kip value as required by statics. Completion of the composite bending moment diagram indicates that $M_x \leq M_p$ for all x; so trial 4 is the correct solution. During this process, the upper bound reduced from 210 to $206\frac{2}{3}$ to 180 and the lower bound increased from 90 to 100 to 180 ft-kips.

8-4.2 Beam Design. Design of continuous beams of constant cross section is almost trivial. A plastic moment (based on A36 steel) of 180 ft-kips is carried economically by a W18 x 35. When cover-plated sections lead to steel savings in excess of 7 lb/ft, they may prove to be more economical. In such a case, composite bending moment diagrams of trial mechanisms in which some M_x exceed M_p are very useful. For example, the third trial mechanism (Fig. 8-4.4) shows that for about 75% of the beam a section capable of sustaining a 100 ft-kip plastic moment is adequate; a W14 x 26 good for 120 ft-kips. The largest bending moment of Fig. 8-4.4 is $206\frac{2}{3}$, 86.7 ft-kips larger than the beam can sustain. Twenty-eight feet of PL $\frac{1}{2}$ x 4 (T & B) would suffice for a total weight of 381 + 2392 = 2773 as compared to 35 lb x 92 ft = 3220 lb for a W18 x 35 without cover plates, a savings of only $4\frac{1}{2}$ lb/ft.

In summary, the equilibrium method yields direct results for the design of continuous spans. If a constant section is used, often a proper mechanism can be determined by visual inspection and any solution is correct if all M_x are less than M_p. If cover-plated or tapered sections are desired, bending moment diagrams of conditions in which all M_x are not less than M_p are quite useful. Finally, a large number of solutions may be correct if validated only by considerations of safety, but the number of choices is reduced when overall economy is envisioned. Interestingly, it is probably impossible to say with certainty that any one design is absolutely the *best*.

8-5   ADDITIONAL CONSIDERATIONS

Design of concentrically and eccentrically loaded columns is discussed in Chaps. 3 and 10. Problems that involve high shear are solved in Chap. 6. Local buckling and lateral stability criteria are developed in Chap. 11.

Deflection of members designed by plastic analysis can be approximated by (1) ignoring elastic deformations, (2) computing deflection at the time of formation of the plastic hinge prior to collapse, and (3) dividing the result by the load factor. Such a deflection is *larger* than true elastic deflection. In case deflection as determined by this method is excessive, recourse must be made to the determination of deflections under elastic conditions.

Undoubtedly, use of plastic design and analysis will continue to increase (along with its companion "Yield Line Theory" as applied to reinforced concrete). Although difficult to forecast, it is anticipated that increased use of electronic computers could lead to a process of design by plastic theory and check by elastic analysis. Presently this procedure is fairly common in the aerospace industry and is similar in philosophy to checking prestressed beams by stress analysis after they have been designed elastically.

PROBLEMS

8-1 given: Propped beam of Fig. 8-P.1 with $f_y = 36$ ksi and $L = $ _____
(40, 60, 76.3, or 80 ft), $a = $ _____ (0.1 or 0.2), $b = $ _____
(0.2 or 0.3), $c = $ _____ (0.3 or 0.4), $P = $ _____ (30, 43.7,
69, or 80 kips) including L.F., $d = $ _____ (0.3, 0.4, or 0.7),
$e = $ _____ (0.9, 1.0 or 1.3).

determine: (a) Design moment by equilibrium method; (b) design moment by mechanism method; (c) choose suitable beam (if possible).

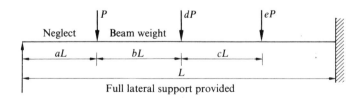

Fig. 8-P.1   Propped Beam

8-2 given: Beam _____ [(a) or (b) of Fig. 8-P.2] with, $f_y = $ _____
(36 or 50 ksi), $L = $ _____ (40, 42, or 44 ft), $a = $ _____

(0.05, 0.1, or 0.15), $b =$ _____ (0.2, 0.24, or $\frac{1}{3}$) $c =$ _____
(b, 0.2, 0.27, or $\frac{1}{3}$) $P =$ _____ (36, 40, or 42 kips).

design: Most economical rolled section.

present: Analysis proof.

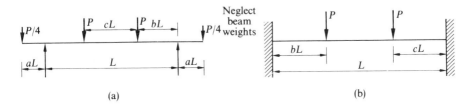

(a)   (b)

Fig. 8-P.2 Beams (Full Lateral Support Provided)

8-3 given: Same as in problem 8-2.

design: Cover-plated rolled section, with rolled section alone carrying about two-thirds of the maximum moment.

present: Analysis proof, including cover-plate lengths.

8-4 given: Beam of Fig. 8-P.3 which, for architectural reasons, is to be tapered. Top and bottom PL $\frac{1}{2}$ x 8. Web PL $\frac{3}{8}$ x variable. A36 steel with $P =$ _____ (40, 35, or 31 kips), $L =$ _____ (35, 40, or 45 ft), $a =$ _____ (0.3, 0.4, or 0.5). End A is to be approximately _____ ($\frac{1}{2}$, $\frac{2}{3}$, or $\frac{3}{4}$) as deep as end B.

design: Tapered beam.

present: Analysis proof.

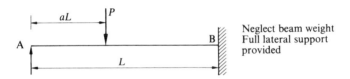

Fig. 8-P.3 Tapered Beam

8-5 given: Uniformly loaded beam of Fig. 8-P.4 with $w = $ _____ (0.9, 1.2, 1.4, or 1.7 k/ft) *not* including load factors or beam weight. $L = $ _____ (37, 39, 41, or 43 ft), $f_y = $ _____ (36 or 50 ksi).

design: (a) Economical rolled section. (b) Rolled section with _____ ($\frac{2}{3}$, $\frac{3}{4}$, or $\frac{4}{5}$) of I of the section determined in part (a) with either (1) cover plates in mid-span, or (2) cover plates over leftmost half ($\pm$) of beam.

present: Analysis proof.

Fig. 8-P.4   Uniform Load

8-6 given: Continuous beam _____ [(a), (b), (c), (d), (e), or (f) of Fig. 8-P.5] of A36 steel. $L = $ _____ (31.6, 33.7, 40, or 42 ft), $a = $ _____ (0.08, 0.11, or 0.14), $b = $ _____ (0.216, 0.232, 0.26, or $\frac{1}{3}$), $c = $ _____ (0.5, 0.52, 0.63 or $\frac{2}{3}$), $Q = $ _____ (18.4, 19.7, 21.3, or 24.3 kips), $P = $ _____ (37, 39.3, 41.6, or 42 kips), $w = $ _____ (1.4, 1.7, 2.1, or 2.3 k/ft). All loads are given without including load factor.

design: Continuous beam, (a) using constant section; (b) using _____ ($\frac{2}{3}$, $\frac{3}{4}$, or $\frac{4}{5}$) of section determined in part (a) and cover plates; (c) using constant section, *but* by the allowable stress design method.

present: Analysis proof.

8-7 given: Portal frame _____ [(a), (b), (c), or (d) of Fig. 8-P.6] with $H = $ _____ (10, 12, or 22 ft), $L = $ _____ (20, 24, or 30 ft), $P = $ _____ (8, 16, or 32 kips), $f_y = $ _____ (36 or 50 ksi).

design: Portal frame.

present: Analysis proof.

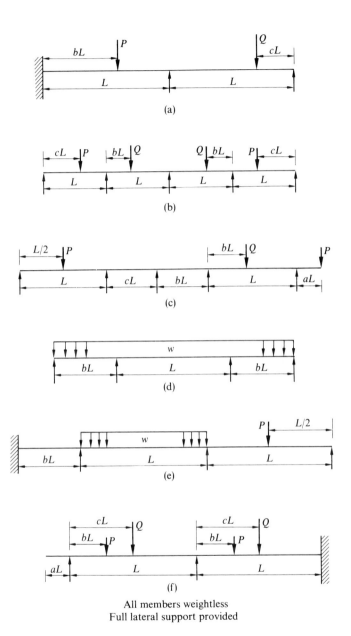

(a)

(b)

(c)

(d)

(e)

(f)

All members weightless
Full lateral support provided

Fig. 8-P.5   Continuous Beams

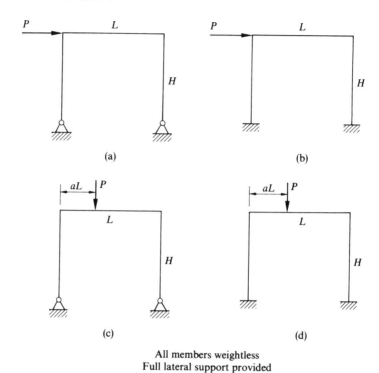

(a)   (b)

(c)   (d)

All members weightless
Full lateral support provided

Fig. 8-P.6   Portal Frames

REFERENCES

1. Greenberg, H. J. and W. Prager, "Limit Design of Beams and Frames," *Transactions*, ASCE, V. 117 (1952), 447.

2. Beedle, Lynn S., *Plastic Design of Steel Frames*, Wiley, New York (1958).

3. Hill, Louis A., Jr., and Garret N. Vanderplaats, "Design Curves for Plastic Design of Uniformly Loaded Steel Beams," *AISC Engineering Journal*, V. 5, No. 4 (October 1968), pp. 148–151.

CHAPTER 9

BIAXIAL BENDING AND TORSION

Simple bending theory is adequate for the analysis of beams only when the resultant of loads (and reactions): (1) lies in the plane containing one principal member axis, (2) passes through the member shear center, and (3) is perpendicular to the neutral axis. When the resultant of the load does not lie in the plane of one of the principal axes, biaxial bending occurs. As the term *biaxial bending* implies, a loading condition, in which all components are perpendicular to either neutral axis, can be resolved statically into two resultant loads, one that lies in the plane of the principal axis and the other in the plane of the minor axis. When the resultant of the load does not pass through the shear center, a member is subject to torsion.

Before loads which induce concurrent biaxial bending and torsion are considered, it is instructive to treat each concept independently. In this chapter, therefore, the ordering is significant and is as follows: (1) biaxial bending of mathematical models without torsion, (2) design of steel rolled sections subject to biaxial bending without torsion, (3) the concept of shear center, (4) St. Venant torsion, (5) nonuniform torsion, (6) analysis with combined torsion and bending, and (7) design.

9-1 BIAXIAL BENDING WITHOUT TORSION

In 1930, Hardy Cross [1] developed a general approach to stress at any point, and this has provided the basis of this approach. In Fig. 9-1.1, a generalized section is depicted, with two centroidal (X and Y) axes shown. On any elemental area of this section, an elemental force is represented by the product of average stress times differential area. Average stress is a function of three independent loading conditions: axial (constant a), bending moment about the Y axis (constant b times variable distance x), and bending moment about the X axis (constant c times the variable distance y). Thus

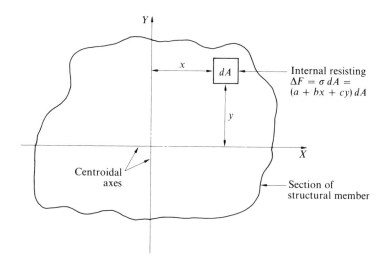

Fig. 9-1.1   Generalized Stress at a Point

$$\Delta F = \sigma dA = (a + bx + cy)dA, \qquad (9\text{-}1.1)$$

where $\Delta F \equiv$ incremental force, $\sigma \equiv$ average stress on the incremental area, $dA \equiv$ incremental area, x and $y \equiv$ coordinates measured perpendicular to principal axes, and a, b, and $c \equiv$ constants.

First moment of area about either principal axis is zero, and can be represented by the integral of first moments of differential areas as

$$\int y dA = 0 \qquad \text{and} \qquad \int x dA = 0.$$

These relationships make possible a simplified general solution by statics for constants a, b, and c.

The following convenient set of signs is used:

1. x measured to the right of the Y axis is positive;

2. y measured upward from the X axis is positive;

3. compressive stress is positive;

4. bending moments are by beam sign convention (sec. E-2).

In the upper right-hand quadrant, both x and y are positive. Compressive stress resisting applied bending moment can be visualized as upward from the plane in Fig. 9-1.1. This compressive stress is induced by a positive bending moment. Therefore, summing moments about the X axis (visualize right-hand rule) implies that the resisting moment is positive whereas applied bending moment is negative. To be in static equilibrium, their resultant must be zero, so

$$\Sigma M_{xx} = 0 \Rightarrow - M_x + \int \sigma y dA = 0,$$

or

$$M_x = a\int y\cancel{dA} + b\int y(x dA) + c\int y(y dA) = bI_{xy} + cI_x, \quad (9\text{-}1.2)$$

where the first moment of area vanishes and $I_x \equiv$ moment of inertia about centroidal X axis $= \int y^2 dA$; $I_y \equiv$ moment of inertia about centroidal Y axis $= \int x^2 dA$; and $I_{xy} \equiv$ product of inertia about centroidal axis $= \int xy dA$.

By a similar process:

$$M_y = bI_y + cI_{xy}. \quad (9\text{-}1.3)$$

Summation of forces along the Z axis (perpendicular to two principal axes) is also zero, to maintain static equilibrium. Therefore, $\Sigma F_z = 0 \Rightarrow - P + \int \sigma dA = 0$, so

$$P = \int \sigma dA = a\int dA + b\int x\cancel{dA} + c\int y\cancel{dA} = aA \quad (9\text{-}1.4)$$

or $a = P/A$ (where positive P is the external *compressive* force).

If each term of the equation for M_x is multiplied by I_y, and each term of the equation for M_y multiplied by I_{xy}, the difference between resulting equations is expressed as

$$M_x I_y - M_y I_{xy} = c(I_x I_y - I_{xy}^2)$$

or

$$c = \frac{(M_x I_y - M_y I_{xy})}{(I_x I_y - I_{xy}^2)}.$$

In a similar manner, b is evaluated:

$$b = \frac{(M_y I_x - M_x I_{xy})}{(I_x I_y - I_{xy}^2)}.$$

When these values of a, b, and c are substituted into the expression for average stress over an incremental area, the first general solution is obtained:

$$\sigma = \frac{P}{A} + \frac{M_y I_x - M_x I_{xy}}{I_x I_y - I_{xy}^2} x + \frac{M_x I_y - M_y I_{xy}}{I_x I_y - I_{xy}^2} y. \quad (9\text{-}1.5)$$

The complex form of this equation can be reduced by introducing primed terms:

$$I'_x \equiv I_x - I_{xy}\left(\frac{I_{xy}}{I_y}\right), \qquad M'_x \equiv M_x - M_y\left(\frac{I_{xy}}{I_y}\right),$$

$$I'_y \equiv I_y - I_{xy}\left(\frac{I_{xy}}{I_x}\right) \qquad M'_y \equiv M_y - M_x\left(\frac{I_{xy}}{I_x}\right)$$

Descriptions of the constants are now simplified:

$$a = \frac{P}{A}, \qquad b = \frac{M'_y}{I'_y}, \qquad c = \frac{M'_x}{I'_x},$$

so

$$\sigma = \frac{P}{A} + \left(\frac{M'_y}{I'_y}\right)x + \left(\frac{M'_x}{I'_x}\right)y. \qquad (9\text{-}1.6)$$

When centroidal axes are also principal axes, the product of inertia I_{xy} is zero, and primed values of Eq. (9–1.6) may replace their unprimed counterparts. Thus, with reference to principal axes:

$$\sigma = \frac{P}{A} + \frac{M_y x}{I_y} + \frac{M_x y}{I_x} \qquad (9\text{-}1.7)$$

Obviously, if no bending moments are present, $\sigma = P/A$. Also, if no axial force is present and bending is about only one principal axis (see also sec. A-3.3) the flexure formula is obtained:

$$\sigma = \frac{M_y x}{I_y} \qquad \text{or} \qquad \frac{Mc}{I}.$$

Note that these equations can be seriously in error when buckling is a consideration. Also, to be valid in this simple form, the resultant of all loads must pass through the shear center of the member.

Figure 9-1.2 shows a typical analysis that compares the use of the formulas derived above (see "Direct Procedure") with results obtained by resolution of loads, moments, and section properties to principal axes. Although answers are identical, developed in this section is somewhat simpler. (Required relationship between moments of inertia and products of inertia is reviewed in sec. A-1.)

If X and Y were principal axes, beam sign convention would show tension at A and compression at B and C due to $M_x = -120$ in.-kips. Conversely $M_y = +84$ in.-kips indicates tension at A and B with compression at C. To obtain this same stress pattern, using the "right-hand rule" to the right of the cross section, M_x would change to positive and M_y would remain positive. Although inappropriate for determining stresses, use of the right-hand rule for bending moments simplifies bending moment transformations to principal axes.

In both solutions, compression is positive and tension is negative. In Fig. 9-1.2, under "Section Properties," I_x, I_y, A, x, y, r, and $\tan \alpha$ may be read directly from page 1–59 of *Steel Construction*. Angle α is about 22°-50′, and $\tan [2(22°\text{-}50')] = 1.025$, so the product of inertia (see sec. A-1) is -9.2 in.^{4}.

Direct solution procedure involves (1) evaluation of the ratio I_{xy}/I_y and I_{xy}/I_x for use in (2) evaluation of I'_x, I'_y, M'_x, or M'_y; subsequently these values are

Moments $\quad M_x = -120$ in.-kips $\quad M_y = 84$ in.-kips
(Beam sign convention)

Section Properties $\quad \angle 6 \times 4 \times \frac{7}{8}$ A36

$$\sigma_{\text{allow}} = 24 \text{ ksi}$$

$I_x = 27.7$ in.^{4} $\qquad r_{ZY} = 0.857$ in.

$I_y = 9.75$ $\qquad\quad A = 7.98$ in.^{2}

$I_p = 37.45$ in.^{4} $\qquad I_{ZY} = Ar^2 =$
$$7.98(0.857)^2 = 5.86 \text{ in.}^4$$

$$I_{ZX} = I_p - I_{ZY} =$$
$$37.45 - 5.86 = 31.59 \text{ in.}^4$$

$$I_{xy} = \frac{I_y - I_x}{2}\tan 2\alpha = \frac{9.8 - 27.7}{2}(1.025) = -9.2 \text{ in.}^4$$

$$\left\{ \begin{array}{l} +120, +84 \text{ using} \\ \text{right-hand rule;} \\ \text{right part of section} \end{array} \right\}$$

$6.00''$

$\alpha = \tan^{-1} 0.421$

$y = 2.12''$

$x = 1.12''$

$4.00''$

Direct Procedure

$$I_{xy}/I_y = \frac{-9.2}{9.75} = -0.944 \qquad\qquad I_{xy}/I_x = \frac{-9.2}{27.7} = -0.332$$

I'_x:		27.7	I'_y:	9.75
$-I_{xy}(I_{xy}/I_y): -(-9.2)(-0.944) =$		-8.68	$-I_{xy}(I_{xy}/I_x): -(-9.2)(-0.332) =$	-3.05
$I'_x =$		19.02	$I'_y =$	6.70

M'_x:		-120	M'_y:	$+84$
$-M_y(I_{xy}/I_y): -(84)(-0.94) =$		$+79$	$-M_x(I_{xy}/I_x): -(-120)(-0.332) =$	-40
$M'_x =$		-41 in.-kips	$M'_y =$	$+44$ in.-kips

Stresses $\qquad \dfrac{M'_x}{I'_x} = \dfrac{-41}{19.02} = -2.15 \qquad \dfrac{M'_y}{I'_y} = \dfrac{+44}{6.70} = 6.57$

	A	B	C
x	-1.12	-1.12	2.88
y	3.88	-2.12	-2.12
$\left(\dfrac{M'_y}{I'_y}\right)x$	-7.4	-7.4	$+18.9$
$\left(\dfrac{M'_x}{I'_x}\right)y$	-8.3	$+4.6$	$+4.6$
σ	-15.7	-2.8	$+23.5 < 24$ ksi (OK)

Fig. 9.1.2 continued on next page

Fig. 9.1.2 continued

$$\sin \alpha = 0.388 \qquad \cos \alpha = 0.922$$

For principal axis solution

	A	B	C		
X	-1.12	-1.12	2.88	M_x	120
Y	3.88	-2.12	-2.12	M_y	84
$y \sin \alpha$	1.51	-0.82	-0.82		32.6
$x \cos \alpha$	-1.03	-1.03	2.66		110.6
zx	0.48	-1.85	1.84	M_{zx}	143.2
$y \cos \alpha$	3.58	-1.95	-1.95		77.4
$-x \sin \alpha$	0.44	0.44	-1.12		-46.6
zy	4.02	-1.51	-3.07	M_{zy}	30.8

Stresses

$$\frac{M_{zy}}{I_{zy}} = \frac{30.8}{5.86} = 5.26 \qquad \frac{M_{zx}}{I_{zx}} = \frac{143.2}{31.59} = 4.53$$

	A	B	C
x	0.48	-1.85	1.84
y	4.02	-1.51	-3.07
$\left(\dfrac{M_{zy}}{I_{zy}}\right)x$	$+2.5$	-9.7	$+9.6$
$-\left(\dfrac{M_{zx}}{I_{zx}}\right)y$	-18.2	$+6.8$	$+13.9$
σ	-15.7	-2.9	$+23.5 < 24.0$ ksi (OK)

Fig. 9-1.2   Typical Biaxial Bending Analysis

used to obtain M'_x/I'_x and M'_y/I'_y, which, when multiplied by respective "arms" (with correct signs) and summed, yield final stresses.

To work about principal axes, it is necessary to determine principal moments of inertia. When the radius of gyration $r = \sqrt{I/A}$ is known, I_{zy} is obtained. Then, based on the fact that the polar moment of inertia I_p is the sum of any two moments of inertia computed about mutually perpendicular axes, I_{zx} is obtained as the difference between I_p and I_{zy}.

Next, moments and arms must be transferred to the principal axes coordinate system, where the bending moment about the ZX principal axes,† $M_{zx} = M_y \sin \alpha + M_x \cos \alpha$ and $M_{zy} = M_y \cos \alpha - M_x \sin \alpha$; with the co-ordinate in the ZX principal axis direction, $zx = y \sin \alpha + x \cos \alpha$ and $zy = y \cos \alpha - x \sin \alpha$ and M_x and M_y obtain their signs by the right-hand rule at the right part of the section.

---

† Principal axes ZX and ZY are often denoted by single roman and Greek characters such as U and V or ξ and ζ.

By use of the flexure formula about both principal axes, stresses are obtained and summed. Although it is possible to establish a complete set of sign convention rules, it usually is easier and safer to visualize stresses as either tension or compression for each Mc/I calculation.

As a computational technique, the direct procedure (Fig. 9-1.2) appears superior in both speed and accuracy. Conversely, the principal axis solution seems to be related more readily to physical reality.

9-2 DESIGN AND BIAXIAL BENDING

Analysis of beams by generalized formulas is not, in general, appropriate for steel rolled sections even though they may be symmetrical about both axis. The reason for this is that allowable stresses usually differ due to reductions imposed by lateral stability and buckling criteria. In this section biaxial bending is considered without simultaneous axial loading, although the method with only slight modification applies when axial loads account for less than 15% of member capacity. Combined bending moment and axial force is discussed in Chap. 10.

9-2.1 Evolving a Design Technique. When $f_a/F_a \leq 0.15$, AISC formula 1.6-2 applies:

$$\frac{f_a}{F_a} + \frac{f_{bx}}{F_{bx}} + \frac{f_{by}}{F_{by}} \leq 1.0, \tag{9-2.1}$$

where $f_a \equiv$ axial compressive stress $= P/A$; $F_a \equiv$ allowable axial stress computed as for a column; $f_{bx} \equiv M_x/Z_x$, where $M_x \equiv$ the component of bending moment about the major axis and $Z_x \equiv$ the corresponding section modulus; $F_{bx} \equiv$ allowable bending stress about the X axis; and f_{by} and F_{by} are defined similarly to f_{bx} and F_{bx} except with respect to the minor axis.

Validity of Eq. (9-2.1) can be visualized if both f_a and f_{by} are 0, so that f_{bx}/F_{bx} less than or equal to 1 implies.

$$f_{bx} \leq F_{bx}. \tag{9-2.2}$$

This equation is of the more familiar form that shows actual stress limited by allowable stress.

If axial force is nonexistent or neglected, Eq. (9-2.1) can be rewritten, with f replaced by M/Z, as

$$\frac{M_x/Z_x}{F_{bx}} + \frac{M_y/Z_y}{F_{by}} \leq 1. \tag{9-2.3}$$

Evaluation of allowable stresses is dependent upon beam parameters which can be expressed as k_x and k_y so that (with C_b specified as 1.0 by AISC 1.5.1.4.6a)

$$\frac{M_x/Z_x}{k_x F_y} + \frac{M_y/Z_y}{k_y F_y} \leq 1 \tag{9-2.4}$$

or

$$\frac{M_x/F_y}{k_x Z_x} + \frac{M_y/F_y}{k_y Z_y} \leq 1, \tag{9-2.5}$$

where F_y = yield stress (ksi); k_x = less than 0.6 F_y but otherwise the greatest of

$$\frac{12(10^3)}{F_y L(d/A_f)} \tag{9-2.6}$$

or if

$$L/r_T \leq \sqrt{510{,}000/F_y}, \tag{9-2.7}$$

then

$$\left(\frac{2}{3} - \frac{F_y(L/r_T)^2}{1{,}530{,}000}\right), \tag{9-2.8}$$

but if

$$L/r_T \geqq \sqrt{510{,}000/F_y}, $$

then

$$\frac{170{,}000}{F_y(L/r_T)^2}; \tag{9-2.9}$$

and $k_y = 0.75$ if $b_f/2t_f$ is less than $52.2/\sqrt{F_y}$ or

$$0.933 - 0.0035\,(b_f/2t_f)\,\sqrt{F_y} \quad \text{if} \quad b_f/2t_f < 95/\sqrt{F_y}. \tag{9-2.10}$$

Equation (9-2.6) is AISC 1.5-7 divided by F_y, (9-2.8) is 1.5-6a, and (9-2.9) is 1.5-6b also divided by F_y. AISC 1.5-5b was added by Supplement No. 1, effective November 1, 1970, and is represented by Eqs. (9-2.10).

Equation (9-2.5) is linear and can be rewritten in terms of required section moduli as

$$\frac{Z_{xR}}{k_x Z_x} + \frac{Z_{yR}}{k_y Z_y} \leqq 1, \tag{9-2.11}$$

where $Z_{xR} = M_x/F_x$ and $Z_{yR} = M_y/F_y$. End points of this line are obtained when (1) $Z_{yR} = k_y Z_y$ and $Z_{xR} = 0$, and (2) when $Z_{xR} = k_x Z_x$ and $Z_{yR} = 0$.

If a point with coordinates (Z_{xR}, Z_{yR}) is plotted, then any line drawn from the end points $k_y Z_y$ and $k_x Z_x$ that passes through or beyond (Z_{xR}, Z_{yR}) satisfies Eq. (9-2.11). Conversely, if a line passes between (Z_{xR}, Z_{yR}) and the origin, Eq. (9-2.11) is not satisfied as its terms sum to greater than one.

In summary, to determine the most economical beam sufficient for given biaxial bending moments, use the following process:

1. Make a graph with axes Z_x and Z_y.
2. Plot point **1** at the intersection of $Z_{xR} = M_x/F_x$ and $Z_{yR} = M_y/F_y$.
3. Also plot point **2** where $Z'_{xR} = M_x/0.6F_y$ and $Z'_{yR} = M_y/0.75F_y$ intersect.
4. Try a section by obtaining Z_x and Z_y from *Steel Construction.*
   a) Lay a rule on the graph between end points (O, Z_y) and (Z_x, O).
   b) If the line falls inside point **2**, discard this choice and try another section. Otherwise, continue by computing $k_x Z_x$ and $k_y Z_y$.
   c) Plot end points $(0, k_y Z_y)$ and $(k_x Z_x, 0)$.
   d) If line falls inside point **1**, discard this choice and try another section.
   e) Otherwise this choice is either too heavy, in which case another section must be tried, or it is the final selection.

9-2.2 Example Design. Given a 32-ft simple span with $M_x = 246$ ft-kips (exclusive of beam dead load) and $M_y = 65.3$ ft-kips, design the most economical A36 rolled section. For the graph shown in Fig. 9-2.1:

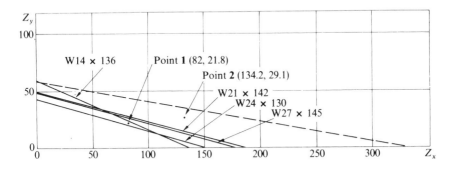

Fig. 9-2.1 Graphical Solution to a Design Involving Biaxial Bending

$$Z_{xR} = 12(246)/36 = 82 \quad \text{and} \quad Z_{yR} = 12(65.4)/36 = 21.8;$$
$$Z'_{xR} = 12(246)/22 = 134.2 \quad \text{and} \quad Z'_{yR} = 12(65.4)/27 = 29.1.$$

It is advantageous before proceeding with the design to systematize calculations for $k_x Z_x$ and $k_y Z_y$.

Equation (9-2.6) reduces to

$$\frac{12,000}{(36)\left[(12)(32)\right](d/A_f)} = \frac{0.868}{(d/A_f)}. \qquad (9\text{-}2.12)$$

If d/A_f is less than 1.42, $k_x = 0.6111$ and calculations are unnecessary. This limit is used rather than 0.60 because $22/36 = 0.6111$.

Equation (9-2.7) is evaluated as

$$\frac{12(32)}{r_T} = \sqrt{510,000/36} \qquad \text{or} \qquad r_T = 3.226. \qquad (9\text{-}2.13)$$

An r_T of 7.214 or greater (based on $L/r_T \leq \sqrt{102,000/F_y}$) signifies that $k_x = 0.6111$ and that further computations are not required.

Between $r_T = 3.226$ and 7.214, from Eq. (9-2.8),

$$k_x = \frac{2}{3} - \frac{36\left[(12)(32)\right]^2}{1,530,000\ r_T^2} = 0.6667 - \frac{3.47}{r_T^2}. \qquad (9\text{-}2.14)$$

If $r_T < 3.226$, then by Eq. (9-2.9),

$$k_x = \frac{170,000 r_T^2}{36\left[(12)(32)\right]^2} = 0.0320 r_T^2. \qquad (9\text{-}2.15)$$

When $b_f/2t_f$ (taken directly from *Steel Construction* for trial member) is less than $52.2/\sqrt{F_y}$, or 8.7, $k_y = 0.75$. Otherwise, if $b_f/2t_f < 95.0/\sqrt{F_y}$, or 15.83, then from Eq. (9-2.10)

$$k_y = 0.933 - 0.0035\left(\frac{b_f}{2t_f}\right)\sqrt{F_y} = 0.933 - 0.021\left(\frac{b_f}{2t_f}\right). \qquad (9.2.16)$$

Equations (9-2.13) through (9-2.16) contain five parameters which, as in Table 9-2.1, are read from *Steel Construction* for any trial member: Z_x, Z_y, r_T, d/A_f, and $b_f/2t_f$. Typical computations are also given in the table, where rows are arranged for slide-rule computations. Although unnecessary for calculator computations, such intermediate values are helpful in eliminating unnecessary computations for subsequent trials.

The object is to secure the lightest adequate section. From the graph it can be seen that the first trial for line (between 404,0 and 63.5,0) for a W27 x 145, passes beyond point **2**. Note that necessary parameters used in Table 9-2.1 are obtained by moving from left to right. Equation [B] is used rather than [A] because $3.226 < 3.72 < 7.214$, and yields a possible k_x of 0.416. Because $d/A_f > 1.42$, another possibility for k_x is, from Eq. [C], 0.441. Taken as the larger of 0.416 and 0.441, $k_x = 0.441$. With $b_f/2t_f < 8.7$, k_y is simply 0.75. Final values are obtained by multiplication, such as $k_x Z_x = (0.441)(404) = 178.2$. The line between coordinates (178.2, 0) and (0, 47.6) (labeled W27 x 145 in Fig. 9-2.1) passes beyond point **1** so the section is adequate.

Table 9-2.1   Biaxial Bending—Coordinate Computations

	First trial section W 27 × 145	Second trial section W 24 × 130	Third trial section W 21 × 142	Fourth trial section W 14 × 136
Z_x	404	332	317	216
Z_y	63.5	58.9	63	77
r_T	3.72	3.76	3.58	4.12
d/A_f	1.97	1.92	1.49	0.941
$b_f/2t_f$	7.16	7.78	6.00	6.93

Computations:

$r_T < 3.226$, then

[A] $0.032r_T^2$	—	—	—	—

$3.226 < r_T < 7.214$, then

$3.47/r_T^2$	0.251	0.245	0.271	
[B] $0.6667-3.47/r_T^2$	0.416	0.421	0.396	—

$d/A_f > 1.42$, then

[C $0.868/(d/A_f)$		0.441	0.452	0.583	—
k_x	k_x	0.441	0.452	0.583	0.6111

$8.7 < b_f/2t_f < 15.8$, then

$0.021(b_f/2t_f)$		—	—		
$0.933 - 0.021(b_f/2t_f)$					
k_y	k_y	0.75	0.75	0.75	0.75
$K_x Z_x$		178.2	150.1	184.8	132
$K_y Z_y$		47.6	44.2	47.2	57.8
Status		OK	NG	OK	OK

The next lightest W27, a W27 × 114, is ruled out by laying a straightedge between the coordinates (300, 0) and (0, 31.6) and observing that point **2** is beyond this line.

To try a deeper section, a W30 × 132 is the next possibility. All heavier sections weigh more than a W27 × 145 and are not tried. This deeper section also fails by inspection; that is, a line between (380, 0) and (0, 37.2) falls inside point **2**. By this same method a W33 × 141 and a W36 × 135 are excluded also. In summary, of beams ranging in depth from 27 in. to 36 in., the most economical and sufficient section is a W27 × 145.

Next, to try more shallow sections use of a W24 × 130, with $Z_x = 322$ and $Z_y = 58.9$, is suggested. A line formed from these properties going from (0, 332) to (58.9, 0) is shown by a broken line in Fig. 9-2.1. Further investigation is required and is shown as the second trial section in Table 9-2.1. Many engineers, however, would perform no calculations beyond listing parameters, but would simply observe inadequacy by quickly estimating k_x based on the k_x given for the W27 x 145. For completeness, this second trial is plotted in Fig. 9-2.1.

A third trial possibility, W21 x 142, is shown in Table 9-2.1. It represents a better solution since it weighs 3 lb/ft less than a W27 x 145. A W21 x 127 is ruled out by inspection. Neither a W18 x 114 nor a W16 x 96 is adequate, but a W14 x 136 might prove to be satisfactory. It is shown as the fourth trial section, with k_x set at 0.6111 without calculation because $d/A_f < 1.42$. No shallower section is adequate, so the final choice is the W14 x 136 if beam dead load is neglected.

To include dead load, point **1** must be moved to the right by 12 $M_D/36$, or by about $5\frac{1}{2}$ to 6 units. Therefore the W21 x 142 is selected and analyzed in the next section.

9-2.3 Analysis Proof. An analysis of this choice is made in Table 9-2.2. Computations are straightforward with but few exceptions:

1. F_v is not actually computed because it is so much less than 14.5.
2. Flange shear is multiplied by 3/2 because the flanges are rectangular.
3. Vertical and horizontal deflections are both within $L/360$ or 1.07 in.; however, when combined by the Pythagorean theorem, these deflections slightly exceed this limit.

9-2.4 Extensions to Method. As discussed in sec. 9-11.2, this technique can be applied with good results when only the top flange receives transverse load. In such cases, $2Z_{yR}$ is used as a coordinate rather than simply Z_{yR}.

Also, when small axial loads are present, several (Z_{yX}, Z_{yR}) points can be plotted by multiplying both values sequentially by 1.05, 1.10, and 1.15, corresponding to permissible f_a/F_a values of 0.0476, 0.0909, and 0.1304. Then by finding f_a and F_a, adequacy can be checked visually. When f_a/F_a exceeds 0.15 the method is not effective.

This design tool is recommended for efficient design of crane runway girders.

9-3 SHEAR CENTER

Shear center† of a beam is defined as the point in a cross section through which the resultant of internal transverse shearing stresses passes. *Elastic axis* is the locus of such shear centers taken along the member axis. Torsion is induced in

---

† *Center of twist*, often defined as the point about which the cross section of a bar in torsion rotates, is used by some authors, rather than *shear center* as in this definition.

Table 9-2.2   Analysis Proof—Biaxial Bending

Loads (32-ft Simple Span)

	Gravity:	LL	1.00 k/ft
		DL	0.92
		Beam W21 × 142	0.14
		Total	2.06 k/ft
	Lateral:	LL	0.51 k/ft

V & BM	**Gravity**	**Lateral**
V:	$2.06(32/2) = 33.0^{\mathrm{K}}$	$0.51(32/2) = 8.2^{\mathrm{K}}$
M:	$2.06(32^2/8) = 264'^{\mathrm{K}}$	$0.51(32^2/8) = 65.3'^{\mathrm{K}}$

Section Properties   **W21 × 142 A36** $E = 29{,}000$ ksi

$I_x = 3410$ in.^{4} $I_y = 414$ in.^{4} $A_w = (21.46)(0.659) = 14.14$ in.^{2}

$Z_x = 317$ in.^{3} $Z_y = 63.0$ in.^{3} $2A_f = 2(13.13)(1.095) = 28.8$ in.^{2}

$r_T = 3.58$ $d/A_f = 1.49$ $b_f/2t_f = 6.00$ $C_b = 1.0$ $F_v = 14.5$ ksi

$F'_{bx} = \left[\dfrac{2}{3} - \dfrac{36}{1530000}\left(\dfrac{12 \times 32}{3.58}\right)^2 \right] 36 = 14.3$ ksi

$\left. \begin{array}{l} \\ \\ \end{array} \right\}$ $\Delta = \dfrac{12 \times 32}{360} = 1.07$ in.

$F''_{bx} = \dfrac{12{,}000}{12 \times 32 \times 1.49}$ $= 21.0$ ksi

$\left. \begin{array}{l} \\ \\ \end{array} \right\}$ $F_{bx} = 21$ ksi $F_{by} = 27$ ksi

Stresses

$\dfrac{f_x}{F_{bx}} + \dfrac{f_y}{F_{by}} = \dfrac{12(264)/317}{21} + \dfrac{12(65.3)/63.0}{27} = 0.476 + 0.461 = 0.937 < 1$ (OK)

$f_v = \dfrac{33.0}{14.14} + \dfrac{3}{2}\left(\dfrac{8.2}{28.8}\right) \ll 14.5$ ksi $\therefore$ OK

Deflection

$\Delta_x = \dfrac{5(2.06)(32)^4(1728)}{384(29{,}000)3410} = 0.492$ in. $< 1.07 \therefore$ OK

$\Delta_y = \dfrac{5(0.51)(32)^4(1728)}{384(29{,}000)414} = 1.002$ in.

$\begin{aligned} \Delta_x^2 &= 0.242 \\ \Delta_y^2 &= \underline{1.004} \\ & 1.246 \end{aligned}$

$\Delta_{\max} = \sqrt{1.246} = 1.116$ in. $\therefore$ Reasonable

---

a beam when loads do not pass through the elastic axis. Thus the location of of shear center for structural members is necessary for adequate consideration of torsion.

Direct application of this definition is sufficient to locate shear center in many typical sections. In Fig. 9-3.1, for example, resultants of internal trans-

Fig. 9-3.1   Shear Center by Inspection

verse shearing forces can act at only the indicated points. Furthermore, in symmetrical and antisymmetrical sections (Fig. 9-3.2), the shear center is located at the centroidal axis. However, in shapes where symmetry does not make the shear center obvious by inspection it must be calculated.

Fig. 9-3.2   Shear Center by Symmetry and Antisymmetry

Figure 9-3.3 shows a typical channel section with the shear flow plotted from the calculation of VQ/I_x (sec. A-4). Summation of forces produced by shear flow (which varies linearly from 0 when $x = 0$ to a maximum of $Vtbh/2I_x$ when $x = b$) is computed for each component of the cross section. To maintain static equilibrium, summation of moments about any longitudinal axis is zero. Thus, internal shear force in the legs induces a torsional moment which is balanced by external shear V acting at a distance e from the web centerline:

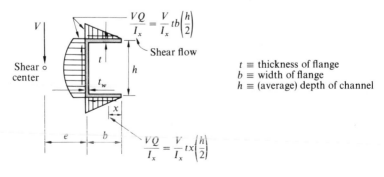

$$\frac{VQ}{I_x} = \frac{V}{I_x}tb\left(\frac{h}{2}\right)$$

Shear flow

$t \equiv$ thickness of flange
$b \equiv$ width of flange
$h \equiv$ (average) depth of channel

$$\frac{VQ}{I_x} = \frac{V}{I_x}tx\left(\frac{h}{2}\right)$$

Fig. 9-3.3   Shear Flow and Shear Center

$$Ve = \left(\frac{V}{I_x}\frac{t}{2}bh\right)\frac{b}{2}h. \tag{9-3.1}$$

A general equation for locating shear center for equal-legged channels is thus obtained:

$$e = \frac{tb^2 h^2}{4I_x}, \tag{9-3.2}$$

where t = thickness of outstanding legs, b = width of outstanding legs, and h = depth of channel minus one leg thickness. This general equation is often rearranged for special purposes. Typically, Seely and Smith[2] use

$$I_x = (t_w h^3/12) + [2(bt)(h/2)^2]$$

in Eq. (9-3.2), so that

$$e = \frac{tb^2 h^2}{4\left(\dfrac{t_w h^3}{12} + \dfrac{bth^2}{2}\right)} = \frac{b/2}{\dfrac{1}{6}\left(\dfrac{t_w h}{tb}\right) + 1} \tag{9-3.3}$$

$$e = \frac{b/2}{1 + \dfrac{1}{6}\dfrac{a_{web}}{a_{flg}}}, \tag{9-3.4}$$

where $t_w \equiv$ thickness of web, $a_{web} \equiv$ area of web, and $a_{flg} \equiv$ area of one flange.

Equation (9-3.4) may be no easier to use than (9-3.2), but it shows clearly that (a) when a_{web} approaches zero, e approaches a maximum of $b/2$, and (b) when a_{flg} approaches zero, e approaches a minimum of zero.

Popov [3] makes a similar equation by assuming that $t = t_w$, so Eq. (9-3.3) reduces to

$$e = \frac{b}{2 + h/3b}. \tag{9-3.5}$$

For unsymmetrical members the calculations are more complex, although the procedure is unmodified essentially. The steps are:

1. Locate principal axes.
2. Determine shear flow in elements of the section (VQ/I).
3. Equate moment produced by this internal shear flow to an externally applied load which is parallel to the major axis. Distance of the shear center from this major axis is equal to the arm required by the external load to maintain equilibrium.
4. Repeat step 3, but use the minor axis.
5. External shearing force V appears in both internal and external equations and cancels, thus providing a general solution.

Figure 9-3.4 shows these computations for a complex section. Part (a) is

	A	y	Ay	$Ay^2 + I_0$	x	Ax	$Ax^2 + I_0$	$Axy + I_0$
$(5)(\frac{1}{2})$	2.50	4.75	11.88	56.5	2.25	5.63	12.7	26.7
							5.2	
$(8\frac{1}{2})(\frac{1}{2})$	4.25	0.25	1.06	0.3	0	0	0	0
				25.5				
$(4)(1)$	4.00	-4.50	-18.00	81.0	-1.75	-7.00	12.3	31.5
							5.3	
	10.75	-0.47	-5.06	163.3	-0.13	-1.37	35.5	58.2
		$\bar{y}$		-2.4	$\bar{x}$		-0.2	-0.7
				$I_x = 160.9 \text{ in.}^3$			$I_y = 35.3 \text{ in.}^3$	$I_{xy} = 57.5 \text{ in.}^3$

(a) Moments of inertia and centroid

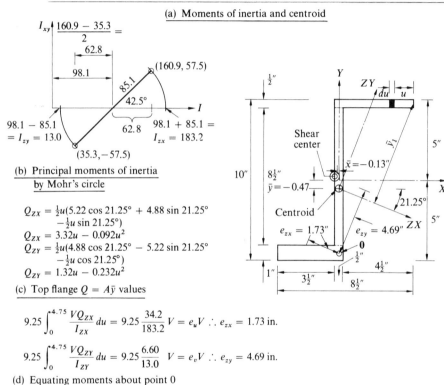

I_{xy} $\dfrac{160.9 - 35.3}{2} =$

62.8

98.1

$(160.9, 57.5)$

85.1

42.5°

I

$98.1 - 85.1 = I_{zy} = 13.0$

62.8

$98.1 + 85.1 = I_{zx} = 183.2$

$(35.3, -57.5)$

(b) Principal moments of inertia
     by Mohr's circle

$Q_{zx} = \frac{1}{2}u(5.22\cos 21.25° + 4.88\sin 21.25°$
$\qquad -\frac{1}{2}u\sin 21.25°)$
$Q_{zx} = 3.32u - 0.092u^2$
$Q_{zy} = \frac{1}{2}u(4.88\cos 21.25° - 5.22\sin 21.25°$
$\qquad -\frac{1}{2}u\cos 21.25°)$
$Q_{zy} = 1.32u - 0.232u^2$

(c) Top flange $Q = A\bar{y}$ values

$\frac{1}{2}''$

Y

ZY

du u

Shear center

$\bar{x} = -0.13''$

X

5″

$10''$ $8\frac{1}{2}''$
$\bar{y} = -0.47$

21.25°

Centroid

$e_{zx} = 1.73''$

$e_{zy} = 4.69''$ ZX

0

$\frac{1}{2}''$

5″

$1''$ $3\frac{1}{2}''$

$4\frac{1}{2}''$

$8\frac{1}{2}''$

$9.25\displaystyle\int_0^{4.75}\frac{VQ_{zx}}{I_{zx}}\,du = 9.25\,\frac{34.2}{183.2}\,V = e_uV \therefore e_{zx} = 1.73 \text{ in.}$

$9.25\displaystyle\int_0^{4.75}\frac{VQ_{zy}}{I_{zy}}\,du = 9.25\,\frac{6.60}{13.0}\,V = e_vV \therefore e_{zy} = 4.69 \text{ in.}$

(d) Equating moments about point 0

Fig. 9-3.4 Shear Center Computation

a tabular solution for I_x, I_y, I_{xy}, and the position of centroid. The X axis is midway between outer fibers and the Y axis bisects the web. In part (b) is a solution for principal moments of inertia by Mohr's circle (see sec. A-1). Coordinates (I_x, I_{xy}) (160.9, 57.5) and $(I_y, -I_{xy})$ (35.3, -57.5) are plotted; intersections of Mohr's circle and the I axis provide $I_{zx} = 183.2$ (max) and $I_{zy} = 13.0$ (min). The maximum moment of inertia is $\frac{1}{2}(42.5°)$ clockwise from

I_x, so all values of x and y are related to a 21.25° rotation clockwise in part (c) where the Q values are computed.

Area for $Q_{zx}(A\bar{y})$ is computed as flange depth ($\frac{1}{2}$ in.) times length u along the flange. Statical arm is found by transforming coordinates to the centroid of the effective area. This centroid is located at mid-depth of the flange. Vertical distance from centroid to mid-depth of flange is 5 in. (base X axis to extreme top fiber) less 0.25 in. (half of flange thickness) plus 0.47 in. ($\bar{y}$) distance between X axis and centroid) or 5.22 in. Horizontal distance from a point directly above the centroid to the end of the top flange is 5 in. (width of flange) less 0.25 in. (left edge of web to Y axis) plus 0.13 in. ($\bar{x}$ distance between Y axis and centroid) or 4.88 in. Horizontal distance from the end of the top flange to the centroid of effective area of length u is one-half u. These three distances have components parallel to the principal axes. In the figure, the distance $\bar{y}$, measured from the ZX principal axis, parallel to the ZY principal axis, to the centroid of the effective area is 5.22 cos 21.25° + 4.88 sin 21.25° − $(u/2)$ sin 21.25°.

Distances to the shear center are obtained by equating moments about a point located at the intersection of the centerlines of the bottom flange and web. Therefore, moment due to shear flow is arm (9.25 in.) times the force due to shear in the top flange. Force in the top flange is found by integrating $VQ \, du/I$, with Q expressed at any point u by either $Q_{zx} = 3.32u - 0.092u^2$ or $Q_{zy} = 1.32u - 0.232u^2$. To be conservative, maximum shear flow VQ/I is assumed to occur at mid-width of web so that limits of integration are zero to 4.75. Integration is required because shear flow does not vary linearly. In step (d) of Fig. 9-3.4, then, $\int_0^{4.75} Q_{zx} \, du = 34.2$. Because V occurs on both sides of the equations, coordinates to the shear center from point 0 are the only unknowns.

A similar process is illustrated for a semicircular ring cross section in Fig. 9-3.5. Thickness is denoted by t, radius by R, and θ is measured counterclockwise from a vertical position. Therefore, $ds = R d\theta$ and $\bar{y} = R \cos \theta$. For any particular point Q is found by integrating $A\bar{y}$ expressed as $(tds)R \cos \theta = tR^2 \cos \theta \, d\theta$ from 0 to θ radians. Thus, shear flow q varies as does sin θ. Equating moments about the origin, external shear V times unknown eccentricity e is equated to the integral from 0 to π of incremental shear flow moment $(qds)R$. Final simplification is obtained by expressing I as $\pi R^3 t/2$, yielding $e = 4R/\pi$.

More precise formulas have been developed to account for tapering flanges and fillets [4]. Shear center for standard rolled channels are tabulated with other properties for designing in *Steel Construction*.

9-4 ST. VENANT TORSION

Equations applicable to St. Venant torsion are obtained by specializing linearized equations from the theory of elasticity; a brief development with defini-

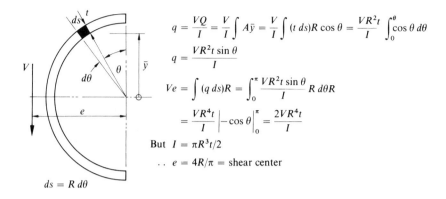

The equations shown in the figure:

$$q = \frac{VQ}{I} = \frac{V}{I}\int A\bar{y} = \frac{V}{I}\int (t\,ds)R\cos\theta = \frac{VR^2t}{I}\int_0^\theta \cos\theta\,d\theta$$

$$q = \frac{VR^2t\sin\theta}{I}$$

$$Ve = \int (q\,ds)R = \int_0^\pi \frac{VR^2t\sin\theta}{I}R\,d\theta R$$

$$= \frac{VR^4t}{I}\left|-\cos\theta\right|_0^\pi = \frac{2VR^4t}{I}$$

But $I = \pi R^3 t/2$

$\therefore\ e = 4R/\pi = $ shear center

$ds = R\,d\theta$

Fig. 9-3.5 Shear Center of Semicircular Ring

tion of symbols and sign convention is presented in Appendix B. *St. Venant torsion*, often called *pure* or *uniform* torsion, is based on assuming that cross sections along a member are free to warp out of plane. A simple experiment using a solid rectangular bar can illustrate how warping is caused by torsional moments. Obviously, if ends of such a rubber block are restrained to remain plane and parallel to each other, warping is resisted and beam-type stresses are induced. This latter type of torsion is discussed in sec. 9-9 and is called *non-uniform torsion*.

9-4.1 The Basic Equations of St. Venant Torsion. By visualizing a torsional model intuitive expressions can be obtained (as set forth by St. Venant[5]) for stresses. A first conclusion is that no normal stresses are present:

$$\sigma_x = \sigma_y = \sigma_z = 0. \tag{9-4.1}$$

Also, because the member is not twisted about either its vertical or horizontal axis,

$$\tau_{yz} = 0. \tag{9-4.2}$$

However, the two remaining shearing stress possibilities are not zero because of twisting about the longitudinal X axis. Magnitudes of those stresses are dependent upon their cross-sectional locations, as measured by coordinates y and z, so

$$\tau_{xy} = f_1(y, z) \quad \text{and} \quad \tau_{xz} = f_2(y, z). \tag{9-4.3}$$

Based upon these conclusions, classical equations for solving St. Venant torsion problems are developed by (1) reducing linearized three-dimensional equations from the theory of elasticity (summarized in Tables B-6.1 and B-6.2), and (2) obtaining a solution for the resulting simultaneous partial differential equations.

Equilibrium of x-direction forces is obtained by using stress relations expressed by Eqs. (9-4.1) and (9-4.2) to reduce the general equation for x-component equilibrium [Eq. (B-6.8)] to

$$\frac{\partial \tau_{yx}}{\partial y} + \frac{\partial \tau_{zx}}{\partial z} = 0. \tag{9-4.4}$$

By reduction of the remaining equilibrium equations [(Eqs. (B-6.9) and (B-6.10)], it is concluded that stresses do not vary along the length of a member:

$$\frac{\partial \tau_{xy}}{\partial x} = 0 \quad \text{and} \quad \frac{\partial \tau_{xz}}{\partial x} = 0. \tag{9-4.5}$$

A single compatibility equation is obtained by elimination. Terms which are functions of x are identically zero because variations are dependent upon only y and z. Without restraints, strains exist only with corresponding stresses, so

$$\epsilon_{xx} = \epsilon_{yy} = \epsilon_{zz} = \epsilon_{yz} = 0.$$

Thus the six compatibility equations of Fig. B-6.2 all vanish except Eq. (B-6.18) which reduces to:

$$\frac{\partial^2 \epsilon_{zx}}{\partial y^2} = \frac{\partial^2 \epsilon_{yx}}{\partial z \, \partial y}. \tag{9-4.6}$$

This equation can be rewritten as

$$\frac{\partial}{\partial y}\left(\frac{\partial \epsilon_{zx}}{\partial y} - \frac{\partial \epsilon_{yx}}{\partial z}\right) = 0 = \frac{\partial}{\partial y}\left(\frac{\partial \gamma_{zx}}{\partial y} - \frac{\partial \gamma_{yx}}{\partial z}\right), \tag{9-4.7}$$

which, when integrated with respect to y, yields

$$\frac{\partial \gamma_{zx}}{\partial y} - \frac{\partial \gamma_{yx}}{\partial z} = \text{constant} \equiv 2\theta, \tag{9-4.8}$$

where the constant in subsequent development will prove to be twice the angle of twist per unit length.

A function satisfying equilibrium and compatibility represents a solution. Therefore a solution function, commonly called a *stress function*, is assumed satisfying both equilibrium and compatibility. This unknown function is called Φ and is defined with relation to stresses as

$$\tau_{xy} = \frac{\partial \Phi}{\partial z} \quad \text{and} \quad \tau_{xz} = -\frac{\partial \Phi}{\partial y}. \tag{9-4.9}$$

Substitution of this definition into equilibrium Eq. (9-4.4) results in

$$\frac{\partial^2 \Phi}{\partial y \, \partial z} - \frac{\partial^2 \Phi}{\partial y \, \partial z} = 0, \tag{9-4.10}$$

which shows that defined Φ satisfies equilibrium. In order to satisfy compatibility, strains are equated to stresses expressed in terms of Φ [Eqs. (B-6.26) and (B-6.28)]:

$$\gamma_{xy} = \frac{(\partial \Phi / \partial z)}{G} \quad \text{and} \quad \gamma_{xz} = -\frac{(\partial \Phi / \partial y)}{G}, \tag{9-4.11}$$

so

$$\frac{\partial \gamma_{xy}}{\partial z} = \frac{1}{G}\left(\frac{\partial^2 \Phi}{\partial z^2}\right) \quad \text{and} \quad \frac{\partial \gamma_{xz}}{\partial y} = -\frac{1}{G}\left(\frac{\partial^2 \Phi}{\partial y^2}\right). \tag{9-4.12}$$

When these relations are used in modified compatibility Eq. (9-4.8), compatibility in terms of Φ is represented by

$$\frac{\partial^2 \Phi}{\partial y^2} + \frac{\partial^2 \Phi}{\partial z^2} = -2G\theta, \tag{9-4.13}$$

where $\Phi \equiv$ stress function which must satisfy Eq. (9-4.9) but is otherwise unknown, $\theta \equiv$ angle of twist *per unit length*, and $G \equiv$ modulus of rigidity. Because function Φ meets the assumptions of Eq. (9-4.9) and satisfies Eq. (9-4.13), it represents a solution. To be completely specified, Φ must also satisfy boundary conditions.

9-4.2 Torsional Moment. Torsional resisting moment is found by taking summation of moments in a cross section about the longitudinal X axis (Fig. 9-4.1) Both shearing stresses are functions of their y and z coordinates and each produces a differential force acting over differential area $dydz$. Summing moments about the longitudinal axis X for these differential forces, torsional resisting moment (represented by superscript t) about the X axis (represented by subscript x) shows

$$\Delta M_x^t = \tau_{xy} z - \tau_{xz} y,$$

or equating internal and external moments:

$$M_x^t = \iint_A (y\tau_{xz} - z\tau_{xy})dydz, \tag{9-4.14}$$

where the y coordinate is positive when measured above the Z axis, the z coordinate is positive when measured to the left of the Y axis, τ_{xz} is the shearing stress positive to the left, and τ_{xy} is the shearing stress positive upward.

In Fig. 9-4.1, differential elements are shown in each quadrant to simplify verification of signs used in Eq. (9-4.14). By use of Eq. (9-4.9), Eq. (9-4.14) is rewritten in terms of Φ as

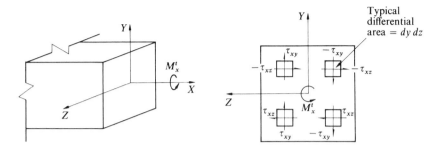

Fig. 9-4.1 Torsional Moment

$$M_x^t = \int\int_A \left(-y\frac{\partial\Phi}{\partial y} - z\frac{\partial\Phi}{\partial z} \right) dydz \qquad (9\text{-}4.15)$$

or in expanded form as

$$M_x^t = -\int dz \int y\frac{\partial\Phi}{\partial y}dy - \int dy \int z\frac{\partial\Phi}{\partial z}dz. \qquad (9\text{-}4.16)$$

Recall that by the method of integrating by parts,

$$\int_a^b udv = uv \Big|_a^b - \int_a^b vdu$$

with u defined as y so that $du = dy$, and with dv defined as $(\partial\Phi/\partial y)dy$ so that $v = \Phi$. Then the second integral of Eq. (9-4.16) is replaced by:

$$\int y\frac{\partial\Phi}{\partial y}dy = y\Phi_c - \int\Phi dy, \qquad (9\text{-}4.17)$$

where Φ is variable, but Φ_c is a constant obtained by evaluating the integral. Using the results of Eq. (9-4.17) the first term of Eq. (9-4.16) becomes

$$-\int dz \int y\frac{\partial\Phi}{\partial y}dy = -\int y\Phi_c dz + \int\int \Phi dydz,$$

where $\int ydz = A$, so that

$$M_x^t = -\Phi_c A + \int\int_A \Phi dydz - \Phi_c A + \int\int_A \Phi dydz$$

or

$$M_x^t = 2\int\int_A \Phi dA - 2\Phi_c A. \qquad (9\text{-}4.18)$$

Along boundaries of a cross section, Φ must be constant because shearing stresses (directives of Φ) are zero. It will be shown in sec. 9-4.4 that Φ_c must not only be constant, it must be zero, so

$$M_x^t = 2 \iint_A \Phi \, dA. \tag{9-4.19}$$

To complete this solution, it is helpful to use an analogy which relates torsional problems to membranes under uniform pressure. Therefore, the next section represents a temporary divergence from torsion to concentrate on a discussion of membranes.

9-4.3 Membrane Solution. A *membrane* is defined as a structural member which can carry loads by tension only. A familiar example of this type of configuration is a soap bubble. Figure 9-4.2 shows a cross section of a membrane (soap bubble) when loaded by uniform pressure q. Tensile force is constant throughout the membrane to maintain equilibrium, that is, $T_L = T_R \equiv T$.

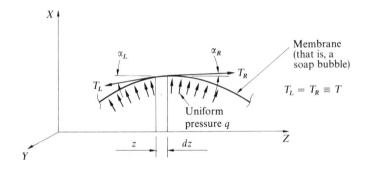

Fig. 9-4.2 Cross Section of Membrane

Difference in vertical components of force in the X direction due to membrane tension in the X–Z plane is determined from geometry of an incremental *dydz* element as

$$dF_x = T(-\sin \alpha_L + \sin \alpha_R)dy, \tag{9-4.20}$$

where α_L is the angle between a line parallel to the Z axis and membrane tensile force acting on left-hand edge of the element. Conversely, subscript R refers to the right-hand side of the element.

For small angles, $\sin \alpha \doteq \tan \alpha \doteq dx/dz = \partial x/\partial z$. Thus, Eq. (9-4.20) becomes

$$dF_x = T \left[- \left(\frac{\partial x}{\partial z}\right)_L + \left(\frac{\partial x}{\partial z}\right)_R \right] dy, \tag{9-4.21}$$

where $(\partial x/\partial z)_L \equiv$ slope at left-hand edge of element, $(\partial x/\partial z)_R \equiv$ slope at right-hand edge of element, and $dz \equiv \partial z =$ distance between left-hand and right-hand edges. A Taylor series expansion (see sec. E-4) of $(\partial x/\partial z)_L$ about the right R edge of the element results in

$$\left(\frac{\partial x}{\partial z}\right)_L = \left(\frac{\partial x}{\partial z}\right)_R + \left(\frac{\partial^2 x}{\partial z^2}\right)_R (-dz) + \left(\frac{\partial^3 x}{\partial z^3}\right)\left(\frac{dz^2}{2}\right) + \cdots . \quad (9\text{-}4.22)$$

If higher ordered terms are neglected, this expansion reduces the previous equation for dF_x to:

$$dF_x = T\left[\left(-\frac{\partial x}{\partial z}\right)_R + \left(\frac{\partial^2 x}{\partial z^2}\right)_R dz + \left(\frac{\partial x}{\partial z}\right)_R\right] dy \quad (9\text{-}4.23)$$

or, after cancellation of common terms,

$$dF_x = T\left(\frac{\partial^2 x}{\partial z^2}\right) dydz. \quad (9\text{-}4.24)$$

By a similar process differential vertical force dF_y produced by tensile forces parallel to the Y axis is obtained. Then the sum of all vertical forces leads to $dF_x + dF_y +$ elemental load $= 0$, or

$$T\left(\frac{\partial^2 x}{\partial z^2}\right) dydz + T\left(\frac{\partial^2 x}{\partial y^2}\right) dydz + qdydz = 0, \quad (9\text{-}4.25)$$

which simplifies to

$$\frac{\partial^2 x}{\partial z^2} + \frac{\partial^2 x}{\partial y^2} = -\frac{q}{T}, \quad (9\text{-}4.26)$$

where q is the uniform pressure on the membrane and T is the uniform membrane tension.

9-4.4 The Membrane Theory. Equations (9-4.13) and (9-4.26) represent identical mathematical formulations if

$$x \equiv \Phi, \qquad q \equiv 2\theta, \qquad \text{and} \qquad 1/T \equiv G. \quad (9\text{-}4.27)$$

In the use of these relationships it is seen that the stress function Φ is analogous to membrane displacement. By Eqs. (9-4.9), membrane slope along the Z axis is analogous to the torsional shear stress τ_{xy}, while the *negative* of membrane slope paralleling the Y axis is analogous to τ_{xz}. In conclusion, any member subject to torsion has shearing stresses proportional to slopes on a soap bubble that spans an opening which is similar (geometrical definition) in shape to the cross section of the actual member. In practice, for irregular cross sections, relative magnitudes of shearing stresses are found sometimes by optically measuring slopes on a membrane. [6]

If in Eq. (9-4.18) Φ is replaced by x [see Eq. (9-4.27)], the right-hand side would then represent volume under a membrane, composed to a *domed* ($\int\int\Phi dA$) and a *cylindrical* ($\Phi_c A$) part. No vertical displacement x exists along boundaries, so the cylindrical part must vanish and a reduced form, Eq. (9-4.19), is obtained. Torsional resisting moment, therefore, equals twice the volume encompassed by an analogous membrane and provides a basis for conversions from relative to actual values.

9-4.5 St. Venant Torsion of Thin Plate Sections. A membrane under pressure, stretched over a long thin rectangular opening, is shown in Fig. 9-4.3. In the left view, the membrane is shown displacing as it actually does. By contrast the membrane on the right is shown with its narrow ends restrained perpendicular to the plate. If the hole is much longer than it is wide, then the volume under the two membranes is nearly equal; at some reasonable distance from the narrow ends, the shapes of both membranes would be almost identical. Because errors are not excessive and the calculations are much easier for the second shape, it is used in the following derivation.

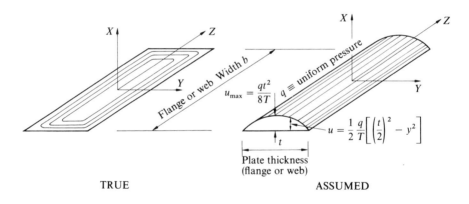

Fig. 9-4.3   Possible Membrane Displacements

A simplified surface is parabolic, and displacement is expressed as

$$u = \frac{1}{2}\frac{q}{T}\left[\left(\frac{t}{2}\right)^2 - y^2\right],$$
(9-4.28)

where $u \equiv$ vertical displacement, $q \equiv$ uniform pressure acting on membrane, $T \equiv$ uniform tension induced in membrane, $t \equiv$ narrow dimension of rectangular opening, and $y \equiv$ horizontal distance to the point where u is computed, zero at centerline.

By analogy u is related to Φ; specifying shearing stress by Eq. (9-4.9) with Φ replace by u results in

$$\tau_{xz} = -\frac{\partial \Phi}{\partial y} = -\frac{\partial}{\partial y}\left\{\frac{1}{2}\frac{q}{T}\left[\left(\frac{t}{2}\right)^2 - y^2\right]\right\} = y(q/T). \qquad (9\text{-}4.29)$$

If $2G\theta$ is substituted for q/T [by Eqs. (9-4.27)], shearing stress is represented by

$$\tau_{xz} = 2\,G\theta y. \qquad (9\text{-}4.30)$$

Concurrently, volume under a parabolic membrane is two-thirds of the base times height times length, so

$$V = \frac{2}{3}t\left(\frac{q}{T}\frac{t^2}{8}\right)b = \frac{1}{12}\frac{q}{T}bt^3 = \left(\frac{G\theta}{2}\right)\left(\frac{bt^3}{3}\right). \qquad (9\text{-}4.31)$$

But by sec. 9-4.4, M_x^t is analogous to twice the volume under the membrane (M_x^t is the torsional resisting moment), so

$$M_x^t = G\theta\left(\frac{bt^3}{3}\right) \qquad (9\text{-}4.32)$$

or

$$\theta = \frac{M_x^t}{G(bt^3/3)}. \qquad (9\text{-}4.33)$$

Substitute this value of θ into Eq. (9-4.30), and we have

$$\tau_{xz} = \frac{2M_x^t y}{bt^3/3} = \frac{2M_x^t y}{C_t}. \qquad (9\text{-}4.34)$$

where C_t is defined as the torsional constant and y is measured from the center-line of the element where the torsional stress is zero, that is, along the line where the analogous membrane would have zero slope. Maximum stress occurs along the edge of the element analogous to the maximum slope of the membrane and is

$$\tau_{max} = \frac{M_x^t t}{C_t}. \qquad (9\text{-}4.35)$$

Brief reflection on the sketches in Fig. 9-4.3 shows that the torsional constant C_t for members formed by joining *thin* plate structures (T, H, WF, etc.) is the sum of each of their respective C_t values, or

$$C_t = \Sigma\, bt^3/3, \qquad (9\text{-}4.36)$$

and

$$\tau_{max} = \frac{M_x^t}{C_t}\,t_{max}, \qquad (9\text{-}4.37)$$

where t_{max} is the thickness of the *thickest* element. The thickest element is used because the derivative of Eq. (9-4.28) with respect to y yields yq/T, maximum for maximum y. Note particularily, however, that y is measured across the *narrow* dimension.

9-4.6 General Comments. Although not required for derivation of code equations, secs. 9-5 through 9-8 are important. Section 9-5 illustrates a direct stress function solution as applied to a bar with elliptical cross section. This leads to validation of $\tau = Tp/J$ when the ellipse is degenerated to a circular cross section.

Section 9-6 provides insight into rectangular sections, verifying use of one-third as a reasonable and conservative coefficient for the torsional constant [Eq. (9-4.36)]. It also shows why one-third (actually reciprocal) is used with $\Sigma\ x^2y$ of ACI Code formula 11–16:

$$v_{tn} = \frac{3T_u}{\Phi \Sigma\ x^2y} \tag{9-4.38}$$

Section 9-7 develops a mathematical formulation which aids in visualization of warping and verifies that circular sections do not warp.

Section 9-8 considers a theoretically correct solution for an elliptical tube subject to torsion, and compares the results with a simpler formula that is applicable to tubes of many shapes.

Section 9-9 treats formulation of equations for nonuniform torsion, obtained when members are subject to constrains against warping. This provides necessary background for deriving AISC beam formulas and does not depend upon material presented in secs. 9-5 through 9-8.

9-5   DIRECT SOLUTION—ELLIPTIC AND ROUND CROSS SECTIONS

Although membrane analogy is very useful both analytically and as a visualization aid, it is possible to solve some torsional problems by assuming a stress function. As an example, consider the ellipse of Fig. 9-5.1, for which

$$\Phi = m\left(\frac{y^2}{a^2} + \frac{z^2}{b^2} - 1\right), \tag{9-5.1}$$

where $\Phi \equiv$ stress function, $m \equiv$ a constant, $a \equiv$ one-half of major axis $(b < a)$, $b \equiv$ one-half of minor axis, $y \equiv$ coordinate measured parallel to a, and $z \equiv$ coordinate measured parallel to b.

9-5.1 Evaluate the Constant m. Differentiation of Φ twice with respect to y and twice with respect to z produces

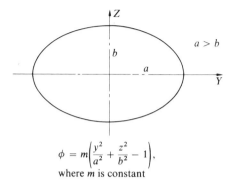

$$\phi = m\left(\frac{y^2}{a^2} + \frac{z^2}{b^2} - 1\right),$$

where m is constant

Fig. 9-5.1   Ellipse and Stress Function

$$\frac{\partial^2 \Phi}{\partial y^2} = \frac{2m}{a^2} \quad \text{and} \quad \frac{\partial^2 \Phi}{\partial z^2} = \frac{2m}{b^2}.$$

Substitution of these values in compatibility Eq. (9-4.13) shows that

$$\frac{2m}{a^2} + \frac{2m}{b^2} = -2G\theta, \tag{9-5.2}$$

where $G \equiv$ modulus of rigidity and $\theta \equiv$ angle of twist *per unit length*. Solving Eq. (9-5.2) for m:

$$m = \frac{-G\theta a^2 b^2}{a^2 + b^2} \tag{9-5.3}$$

so that

$$\Phi = \frac{-G\theta a^2 b^2}{a^2 + b^2}\left(\frac{y^2}{a^2} + \frac{z^2}{b^2} - 1\right). \tag{9-5.4}$$

9-5.2 Torsional Resisting Moment.   When the expression for Φ in Eq. (9-5.4) is used in Eq. (9-4.19),

$$M_x^t = \frac{-2G\theta a^2 b^2}{a^2 + b^2}\int\int\left(\frac{y^2}{a^2} + \frac{z^2}{b^2} - 1\right)dA, \tag{9-5.5}$$

which can be expanded to

$$M_x^t = \frac{-2G\theta a^2 b^2}{a^2 + b^2}\left(\frac{1}{a^2}\int y^2 dA + \frac{1}{b^2}\int z^2 dA - \int dA\right). \tag{9-5.6}$$

For an ellipse, $\int y^2 dA = I_z = \pi a^3 b/4$, $\int z^2 dA = I_y = \pi a b^3/4$, and $\int dA = A = \pi ab$, so Eq. (9-5.6) reduces to

$$M^t_x = \frac{-2G\theta a^2 b^2}{a^2 + b^2} \left[\frac{\pi ab}{4} + \frac{\pi ab}{4} - \pi ab \right] \qquad (9\text{-}5.7)$$

or

$$M^t_x = \frac{\pi G\theta a^3 b^3}{a^2 + b^2}. \qquad (9\text{-}5.8)$$

9-5.3 Stresses. From this latter equation, constant angle of twist per unit length is evaluated as

$$\theta = \frac{M^t_x(a^2 + b^2)}{\pi G a^3 b^3}. \qquad (9\text{-}5.9)$$

With this value of θ, Φ can be expressed, from Eq. (9-5.4), by use of only known parameters:

$$\Phi = \frac{-Ga^2 b^2}{(a^2 + b^2)} \left[\frac{M^t_x(a^2 + b^2)}{\pi G a^3 b^3} \right] \left(\frac{y^2}{a^2} + \frac{z^2}{b^2} - 1 \right) \qquad (9\text{-}5.10)$$

which reduces to

$$\Phi = \frac{-M^t_x}{\pi ab} \left(\frac{y^2}{a^2} + \frac{z^2}{b^2} - 1 \right). \qquad (9\text{-}5.11)$$

Basic definitions of the stress function [Eqs. (9-4.9)] yield

$$\tau_{xy} = \frac{\partial \Phi}{\partial z} = \frac{-2M^t_x z}{\pi ab^3} \qquad (9\text{-}5.12)$$

and

$$\tau_{xz} = \frac{-\partial \Phi}{\partial y} = \frac{+2M^t_x y}{\pi a^3 b}. \qquad (9.5\text{-}13)$$

By use of these two equations, shear at any point (y, z) can be expressed as [7]

$$\tau = \sqrt{[(-2M^t_x z)/\pi ab^3]^2 + [(2M^t_x y)/\pi a^3 b]^2}. \qquad (9\text{-}5.14)$$

Multiplying numerators and denominators within the radical by $a^4 b^4$, this equation for shear at point (y, z) may be expressed:

$$\tau = \frac{2M^t_x}{\pi a^2 b^2} \sqrt{[(a^2 z^2)/b^2] + [(b^2 y^2)/a^2]}. \qquad (9\text{-}5.15)$$

All points interior to the basic elliptical cross section can be represented by ellipses having semi-major and semi-minor axes defined as α and β, where $\alpha/a = \beta/b$ and $\alpha \leq a$. Points on these auxiliary ellipses can be represented in parametric form as $y = \alpha \cos \psi$ and $z = \beta \sin \psi$, so that Eq. (9-5.15) can be written as:

$$\tau_{(y,z)} = \frac{2M_x^t}{\pi a^2 b^2} \sqrt{a^2(\beta/b)^2 \sin^2 \psi + b^2(\alpha/a)^2 \cos^2 \psi}. \qquad (9\text{-}5.16)$$

Use the identity $\beta/b = \alpha/a$, add and subtract $b^2 \sin^2 \psi$, and factor; this equation then takes the form

$$\tau_{(y,z)} = \frac{2M_x^t}{\pi a^2 b^2} \left(\frac{\alpha}{a}\right) \sqrt{b^2 + (a^2 - b^2) \sin^2 \psi}, \qquad (9\text{-}5.17)$$

which is maximum when $\alpha = a$ and $\sin \psi = 1$ (because $a > b$ by definition):

$$\tau_{max} = \frac{2M_x^t}{\pi ab^2}. \qquad (9\text{-}5.18)$$

This equation is simply Eq. (9-5.12) with z replaced by b, proving that for an ellipse maximum shearing stress occurs on the boundary nearest the centroid of the cross section. This conclusion appears valid for most usual solid structural shapes, however, Timoshenko [8, page 267] references an article by Filon showing that exceptions occur.

9-5.4 Specialization to Circular Section. For a solid cricular cross section,

$$a = b = R \quad \text{and} \quad J = I_p = I_y + I_z = \pi R^4/2, \qquad (9\text{-}5.19)$$

where $R \equiv$ radius of cross section, and $I_p \equiv J \equiv$ polar moment of inertia.
    From Eq. (9-5.8), torsional resisting moment is

$$M_x^t = \pi G\theta(R^4/2) = G\theta J, \qquad (9\text{-}5.20)$$

so that

$$\theta = M_x^t/GJ. \qquad (9\text{-}5.21)$$

    Shearing stress from Eq. (9-5.13) validates simple theory presented in preliminary mechanics courses:

$$\tau_{xz} = \frac{2M_x^t z}{\pi R^4} = \frac{T\rho}{J} \qquad (9\text{-}5.22)$$

where $T \equiv$ torsional moment, $\rho \equiv$ distance to point, and $J \equiv$ polar moment of inertia.

9-5.5 St. Venant's Paradox. Comparison of Eqs. (9-4.33) and (9-5.9) suggests that both can be expressed as

$$\theta = \frac{M_x^t}{GC_t}, \tag{9-5.23}$$

where in the former

$$C_t = \frac{bt^3}{3}$$

and in the latter

$$C_t = \frac{\pi a^3 b^3}{a_2 + b_2}. \tag{9-5.24}$$

With area $A = \pi ab$ and polar moment of inertia $I_p = \pi ab(a^2 + b^2)/4$, Eq. (9-5.24) can be reordered as

$$C_t = \left(\frac{1}{4\pi^2}\right)\left(\frac{A^4}{I_p}\right). \tag{9-5.25}$$

This equation states that for constant area, torsional rigidity GC_t *decreases* as polar moment of inertia increases! This is St. Venant's paradox which can be extended to cover all cross sections without interior holes.

This shows that an elliptical cross section has less rigidity than a circular one of the same area because, with $R^2 = ab$, $I_{p(\text{ellipse})} > I_{p(\text{circle})}$ based on

$$\frac{\pi ab(a^2 + b^2)}{4} > \frac{\pi R^4}{2} = \frac{\pi(ab)(ab)}{2},$$

which reduces to $a^2 + b^2 > 2ab$, and this yields the starting premise that $a > b$.

In a similar manner, maximum stress on an elliptical section (higher I_p) is larger than on a circular one (lower I_p) because

$$\frac{2M_x^t}{\pi ab^2} > \frac{2M_x^t}{\pi R^3} = \frac{2M_x^t}{\pi(ab)\sqrt{ab}}$$

reduces to $1/b > 1/\sqrt{ab}$. Again, this yields proof that the original definition $a > b$ is valid.

In summary, for solid cross sections of the same area, as the polar moment of inertia increases, (a) rigidity decreases, that is, GC_t or simply C_t decreases, and (b) maximum torsional stresses increase (for the same applied torque).

9-6 RECTANGULAR CROSS SECTION

For a rectangular section, Timoshenko[8, page 276] supplies a stress function:

$$\Phi = \frac{32G\theta a^2}{\pi^3} \sum_{n=1,3,5\ldots}^{\infty} \frac{1}{n^3}(-1)^{(n-1)/2}\left[1 - \frac{\cosh(n\pi z/2a)}{\cosh(n\pi b/2a)}\right]\cos\frac{n\pi y}{2a}, \tag{9-6.1}$$

where $a \equiv$ one-half the width of rectangle A and $b \equiv$ one-half the depth of rectangle B ($b > a$); $y \equiv$ coordinate of width, and $z \equiv$ coordinate of depth, both from centroid.

It can be seen that Φ vanishes on the boundaries because (1) when $z = \pm b$, the bracketed term in Eq. (9-6.1) is zero, and (2) when $y = \pm a$, $\cos n\pi y/a$ is zero for odd n.

Proof of compatibility is left as a student exercise.

Shearing stress $\tau_{xz} = -\partial\Phi/\partial y$ [Eqs. (9-4.9)] , so

$$\tau_{xz} = \frac{16G\theta a}{\pi^2} \sum_{n=1,3,5\ldots}^{\infty} \frac{1}{n^2}(-1)^{(n-1)/2} \left[1 - \frac{\cosh{(n\pi z/2a)}}{\cosh{(n\pi b/2a)}}\right] \sin\frac{n\pi y}{2a}. \quad (9\text{-}6.2)$$

Equation (9-6.2) is maximum (for $b > a$) when $z = 0$ and $y = \pm a$, so

$$\tau_{max} = \frac{16G\theta a}{\pi^2} \sum_{n=1,3,5\ldots}^{\infty} \frac{1}{n^2}\left[1 - \frac{1}{\cosh{(n\pi b/2a)}}\right]. \quad (9\text{-}6.3)$$

The term $\sin n\pi y/2a$ of Eq. (9-6.2) for $y = \pm a$ becomes $\sin n\pi/2$, which when evaluated for $n = 1, 3, 5 \ldots$ effectively counters $(-1)^{(n-1)/2}$. Therefore the term for alternating sign and for sin both disappeared in Eq. (9-6.3). However summation over the odd factors $n = 1, 3, 5\ldots$ for $(1/n^2) = \pi^2/8$ yields a simpler form:

$$\tau_{max} = 2G\theta a \left[1 - \frac{8}{\pi^2}\sum_{n=1,3,5\ldots}^{\infty}\frac{1}{n^2\cosh{(n\pi b/2a)}}\right] = 2G\theta ak_V, \quad (9\text{-}6.4)$$

where k_V represents the braketed term.

In a similar manner, torsional resisting moment is found by integration of

$$M_x^t = 2\int_{-a}^{a}\int_{-b}^{b}\Phi\,dx\,dy,$$

which, if $1/1^4 + 1/3^4 + 1/5^4\ldots$ is replaced by $\pi^4/96$, results in

$$M_x^t = G\theta A^3 B\left[\frac{1}{3}\left(1 - \frac{192}{\pi^5}\frac{A}{B}\sum_{n=1,3,5\ldots}^{\infty}\frac{1}{n^5}\tanh\frac{n\pi B}{2A}\right)\right] = G\theta A^3 Bk_m, \quad (9\text{-}6.5)$$

where $A = 2a$, $B = 2b$, $B > A$, and k_m represents the bracketed term.

The bracketed terms of both Eqs. (9-6.4) and (9-6.5), k_V and k_m, respectively, are dependent upon only aspect ratio B/A (or b/a). Using these representations, Eq. (9-6.5) can be rearranged as

$$\theta = M_x^t/GA^3 Bk_m. \quad (9\text{-}6.6)$$

In general, applied torque is known and stress is desired, so Eq. (9-6.6) is substituted in Eq. (9-6.4) to obtain

$$\tau = 2G\left(\frac{M_x^t}{GA^3 Bk_m}\right)ak_V = \frac{2M_x^t y}{A^3 B}\left(\frac{1}{K_\varrho}\right), \quad (9\text{-}6.7)$$

where y, the variable distance from plate centroid, replaces the constant one-half thickness a, and $K_\varrho = k_m/k_V$.

Values of these coefficients are tabulated by Timoshenko[8, page 276]. The value of K_Q is 0.208 when $B/A = 1$ and increases asymptotically to one-third for $B/A = \infty$, plotted in Fig. 9-6.1. Also shown are curves for k_V and k_m.

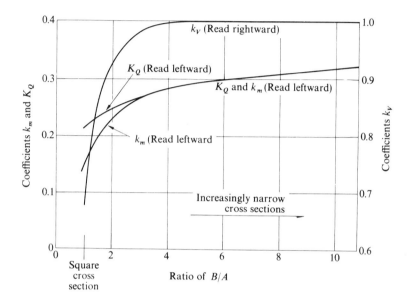

Fig. 9-6.1 Torsional Coefficients for a Rectangular Bar

If A is replaced by element plate thickness t, B by plate width b, and $k_m = K_Q = 1/3$, previous solutions for thin plate sections are validated:

1. Equation (9-6.5) becomes Eq. (9-4.32).
2. Equation (9-6.6) becomes Eq. (9-4.33).
3. Equation (9-6.7) becomes Eq. (9-4.34).

If Eq. (9-6.7) is rewritten with product AB replaced by area, and $2y$ replaced by A (to obtain maximum stress), then

$$\tau_{max} = \frac{M_x^t}{(\text{thinner dimension})(\text{cross-sectional area})K_Q}. \qquad (9\text{-}6.8)$$

From Eq. (9-6.8) it can be seen that for a given torque and a constant cross-sectional area, maximum stress is inversely proportional to the thinner dimension, if K_Q is neglected. Because K_Q, although nonlinear, always has a flat slope, τ_{max} is always greater for thinner sections, torque and area held constant. Thus one-third is a conservative value of K_Q for any rectangular section loaded in the elastic range.

9-7 WARPING OF ELLIPTICAL CROSS SECTION

An alternate and consistent formulation can be used instead of the assumption of sec. 9-4.1. It is based upon

$$v = -\theta xz; \qquad w = \theta xy; \qquad u = \theta\phi(y,z), \qquad (9\text{-}7.1)$$

where $v \equiv$ displacement in Y direction, $\theta \equiv$ angle of twist *per unit length*, $x \equiv$ distance along the X axis, $y \equiv$ distance from center of twist at a particular x point, $z \equiv$ distance from center of twist at a particular x point, and $\phi \equiv$ function of y and z only.

From strain-displacement equations (Tables B-6.1 and B-6.2), then

$$\epsilon_x = \epsilon_y = \epsilon_z = 0,$$

$$\gamma_{xy} = \frac{\partial u}{\partial y} + \frac{\partial v}{\partial x} = \theta\left(\frac{\partial\phi}{\partial y} - z\right), \qquad (9\text{-}7.2)$$

$$\gamma_{yz} = \frac{\partial v}{\partial z} + \frac{\partial w}{\partial y} = -\theta x + \theta x = 0,$$

and

$$\gamma_{zx} = \frac{\partial w}{\partial x} + \frac{\partial u}{\partial z} = \theta\left(y + \frac{\partial\phi}{\partial z}\right).$$

Then by stress-strain equations:

$$\tau_{xy} = G\gamma_{xy} = G\theta\left(\frac{\partial\phi}{\partial y} - z\right) \quad \text{and} \quad \tau_{xz} = G\gamma_{xz} = G\theta\left(y + \frac{\partial\phi}{\partial z}\right). \qquad (9\text{-}7.3)$$

The first of Eqs. (9-7.3) can be expanded by (a) replacing τ_{xy} by its representation in Eq. (9-5.12), (b) θ in $-G\theta z$ by the expression of Eq. (9-5.9), and (c) dividing by G:

$$\frac{-2M_x^t z}{G\pi ab^3} = \frac{G\theta}{G}\frac{\partial\phi}{\partial y} - \frac{G}{G}\left[\frac{M_x^t(a^2 + b^2)}{\pi Ga^3b^3} z\right]. \qquad (9\text{-}7.4)$$

By collecting terms and rearranging, an expression can be obtained for $\partial\phi/\partial y$. Subsequent integration with respect to y yields ϕ. This expression for ϕ is then used in the basic definition of u [Eq. (9-7.1)] to obtain an expression for warping displacement as:

$$u = \theta\phi = \frac{M_x^t(b^2 - a^2)}{\pi Ga^3b^3}zy \qquad . \qquad (9\text{-}7.5)$$

Equation (9-7.5) is (1) zero along either the Y or the Z axis, (2) increases as y and/or z increases, (3) is positive in the first and third quadrants, and (4) is negative in the second and fourth quadrants, with M_x^t defined as *resisting* torsional moment.

When $a = b$, an ellipse becomes a circle and u is zero; a circular cross section subject to torsion does not warp.

9-8   HOLLOW TUBES

Figure 9-8.1 shows the plan view of a soap film, subject to uniform pressure, spanning an elliptical hole. Auxiliary lines represent contour lines along the surface of the soap film. Arc length s is measured along a contour, so $\partial x/\partial s = 0$ because the height above the base is constant; that is, a contour has no slope with respect to its longitudinal (vertical) X axis.

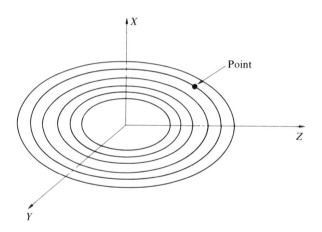

Fig. 9-8.1   Point on Elliptical Contour

The point on one of these contours is shown magnified in Fig. 9-8.2. Components of an assumed τ_{xz} and τ_{xy}, parallel to a line drawn normal to the contour at that point, are $\tau_{xz} \cos \theta = \tau_{xz}(dy/ds)$ and $\tau_{xy} \cos \alpha = \tau_{xy}(dz/ds)$, respectively. Using membrane analogy, $\partial x/\partial s$ is equivalent to $\partial \Phi/\partial s$, and expansion of the latter by the chain rule yields

$$\frac{\partial \Phi}{\partial s} = \frac{\partial \Phi}{\partial z}\left(\frac{dz}{ds}\right) + \frac{\partial \Phi}{\partial y}\left(\frac{dy}{ds}\right), \tag{9-8.1}$$

which results [(by Eqs. (9-4.9)] in

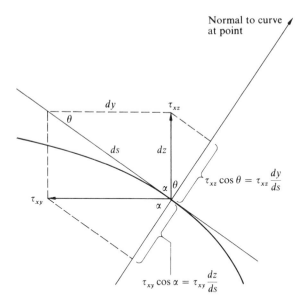

Fig. 9-8.2 Magnified Point on Elliptical Contour

$$0 = \tau_{xy} \frac{dz}{ds} - \tau_{xz} \frac{dy}{ds}. \tag{9-8.2}$$

Interestingly, this equation states mathematically that *no* stresses act across a contour line! That is, shearing stresses act tangentially to contour lines.

From this conclusion, Eq. (9-5.1) can be altered to represent a contour line if a and b are scaled by the same constant K. Thus for $K < 1$

$$\frac{y^2}{(aK)^2} + \frac{z^2}{(bK)^2} - 1 = 0 \tag{9-8.3}$$

represents a general equation for a contour line for an elliptical cross section subject to torsion.

Further, because no stresses cross contour lines, removal of all interior material bounded by a contour line (a) does not change stresses and (b) satisfies boundary conditions. More precisely, for a given angle of twist per unit length, stresses in a hollow and solid elliptical shaft are the same.

To find angle of twist per unit length of a hollow ellipse θ_{hollow}, it is first necessary to find the relationship between M_x^S for a full size ellipse and

M_x^K for an ellipse with semiaxes multiplied by $K < 1$, both twisted through the same angle θ. Then $\theta_{\text{full size}} = \theta_{\text{reduced size}}$ implies that

$$\frac{M_x^S(a^2 + b^2)}{\pi G a^3 b^3} = \frac{M_x^K[(Ka)^2 + (Kb)^2]}{\pi G(Ka)^3(Kb)^3}. \tag{9-8.4}$$

Thus, the moment required to twist the smaller ellipse through the same angle as the larger ellipse is K^4 times the moment required for the larger ellipse. If the hole in the larger ellipse is represented by the smaller ellipse, then the moment required to produce the same rotation in the hollow ellipse is $M_x^H = M_x^S - M_x^K = M_x^S(1 - K^4)$. Therefore for the same torsional moment in both the full size and the hollow ellipses, $\theta_{\text{hollow}} = \theta_{\text{solid}}/(1 - K^4)$, or

$$\theta_{\text{hollow}} = \frac{1}{1 - K^4}\left(\frac{M_x^H(a^2 + b^2)}{\pi G a^3 b^3}\right). \tag{9-8.5}$$

By the use of this value of θ in Eq. (9-5.4),

$$\tau_{xy} = \frac{\partial \Phi}{\partial z} = \frac{-2M_x^t z}{\pi a b^3}\left(\frac{1}{1 - K^4}\right), \tag{9-8.6}$$

where $b \equiv$ minor semi-axis; $a \equiv$ major semi-axis ($b < a$); $z \equiv$ coordinate measured parallel to b; $K \equiv$ a constant such that the semi-minor axis of hole $= bK$ and the semi-major axis of hole $= aK$; and the thickness of the ring *varies* between $a(1 - K)$ and $b(1 - K)$.

Any generalized open tube may be represented by membrane analogy if the "hole" is replaced by a flat plate of identical configuration. Assume a very thin tube, as in Fig. 9-8.3; the average membrane slope is its height above the base divided by tube thickness t. So $h = t\tau$. But, twice the volume under the membrane is analogous to M_x^t, so

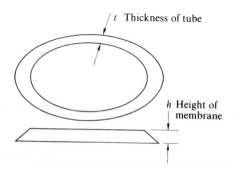

Fig. 9-8.3 Membrane Representing a Tube

$$M_x^t = 2Ah = 2At\tau \tag{9-8.7}$$

or

$$\tau = M_x^t/2At, \tag{9-8.8}$$

where $\tau \equiv$ average shearing stress due to torque, $M_x^t \equiv$ torque, $A \equiv$ average area = (outside + inside areas)/2, and $t \equiv$ tube thickness.

Adequacy of such a formulation can be most easily demonstrated by comparing Eqs. (9-8.6) and (9-8.8) for a circular tube of constant thickness. Outside radius R replaces a, b, and z. Tube thickness is $R(1 - K)$ and average tube radius is $R(1 + K)/2$. Thus Eq. (9-8.6) equated to (9-8.8) yields (note average areas is $\pi[R(1 + K)/2]^2$)

$$\frac{2M_x^t R}{\pi R^4}\left(\frac{1}{1 - K^4}\right) \doteq \frac{M_x^t}{\underset{4}{2\pi(R^2)(1 + K)^2(R)(1 - K)}}, \tag{9-8.9}$$

which reduces to

$$\frac{1}{1 - K^4} \doteq \frac{1}{(1 + 2K + K^2)(1 - K)}$$

Comparison for several values of K is shown in Table 9-8.1, which shows that the approximate method gives very good results when the tube has relatively thin walls.

Table 9-8.1   Comparison of Tube Formulations

K	$\dfrac{1}{1 - K^4}$	$\dfrac{1}{(1 + 2K + K^2)(1 - K)}$	Difference	Percent Difference
0.99	25.378	25.252	0.126	0.497
0.98	12.881	12.754	0.127	0.990
0.95	5.391	5.260	0.131	2.44
0.90	2.908	2.770	0.138	4.74
0.85	2.092	1.948	0.141	6.89
0.80	1.694	1.543	0.151	8.89
0.75	1.463	1.306	0.157	10.71

9-9   NONUNIFORM TORSION

When an unrestrained bar is subject to a torque, M_x^t equals $GC_t\theta$ by Eq. (9-4.32). But when torsional restraint is provided at one or more points, the rate of twist θ varies between zero and its maximum value. Thus the constant angle of twist θ per unit length employed in analysis of St. Venant torsion must be replaced in the presence of torsional restraint by $d\beta/dx$ (rate of change of twist at one point), where β is a measure of actual twist at any point along the longitudinal X axis, and $M_x^t = GC_t\, d\beta/dx$. To aid in visualization of an analytical approach, Fig. 9-9.1 shows a cantilever beam loaded by torque M^T. To provide a complete physical picture, a typical cross section and a view of the bottom flange are also shown. The differential equation (9-9.8) for flange moment and an equation (9-9.12) for β are generalized from this cantilever solution to any case by use of appropriate end conditions.

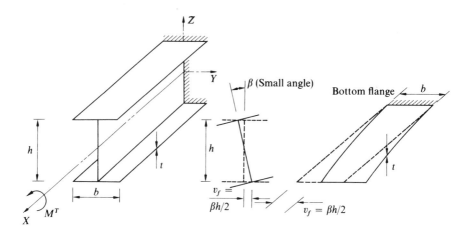

Fig. 9-9.1   Torque on WF Beam

9-9.1 Flange Moment. The bottom flange is displaced laterally a distance v_f, equal to angle of twist β times half the beam depth. Based on $EIy'' = M$ (sec. A-3), bending moment in the bottom flange due to bending *parallel* to the flange (perpendicular to the beam axis), is

$$M_f = -EI_f\frac{d^2v_f}{dx^2},\qquad (9\text{-}9.1)$$

where I_f is the moment inertia of the flange $tb^3/12$. By differentiation, flange shear due to this displacement is

$$V_f = -EI_f\frac{d^3v_f}{dx^3}.\qquad (9\text{-}9.2)$$

Concurrently, torsional moment due to flange forces can be expressed as a couple computed as flange shear times distance between flanges:

$$M_x^w = V_f h \tag{9-9.3}$$

so that

$$M_x^w = -EI_f \frac{d^3 v_f}{dx^3} h. \tag{9-9.4}$$

But, because $v_f = \beta h/2$ (by geometry)

$$\frac{d^3 v_f}{dx^3} = \frac{d^3(\beta h/2)}{dx^3} = \frac{h}{2}\left(\frac{d^3\beta}{dx^3}\right). \tag{9-9.5}$$

so that Eq. (9-9.4) can be restated as

$$M_x^w = -EI_f \frac{h^2}{2}\frac{d^3\beta}{dx^3}. \tag{9-9.6}$$

Defining an additional term, called the *warping constant*, as

$$I_w = \frac{h^2}{2} I_f \tag{9-9.7}$$

(note h is distance between centroid of flanges), the final expression for transverse *resisting* moment in the flange due to nonuniform torsion is obtained:

$$M_x^w = -EI_w \frac{d^3\beta}{dx^3}. \tag{9-9.8}$$

For sections which meet at a common point, angles, tees, etc., $M_x^w = 0$, meaning that no warping moment occurs.

9-9.2 Torsional Stresses. To determine axial and shearing stresses produced in the flange by nonuniform torsion, it is necessary to find a correct expression for β. Because both St. Venant and nonuniform torsional moments are functions of β, both expressions are combined to form a single linear differential equation.

Applied torsional moment M^T is equal to St. Venant M_x^t added to non-uniform torsional resisting moments M_x^w, and is expressed by

$$M^T = M_x^t + M_x^w = GC_t \frac{d\beta}{dx} - EI_w \frac{d^3\beta}{dx^3}. \tag{9-9.9}$$

To write Eq. (9-9.9) in operator form, it is helpful to define a constant

$$\lambda = \sqrt{GC_t/EI_w} \tag{9-9.10}$$

so

$$(\mathscr{D}^3 - \lambda^2 \mathscr{D})\beta = -M^T/EI_w. \tag{9-9.11}$$

A solution to Eq. (9-9.11) is

$$\beta = A \sinh \lambda x + B \cos \lambda x + D + M^T x / GC_t, \qquad (9\text{-}9.12)$$

where A, B, and D are arbitrary constants of integration evaluated for end conditions corresponding to the particular type of beam being analyzed.

Once β is evaluated, it is possible to find normal stresses and shearing stresses produced by bending of the flanges, as well as St. Venant shearing stresses. In summary, maximum stresses using Eq. (9-9.1) for M_f and (9-9.2) for V_f are

$$\sigma_x = \frac{M_f(b/2)}{I_f} = -EI_f \frac{h}{2} \frac{d^2\beta}{dx^2} \frac{b/2}{I_f} = -E \frac{bh}{4} \beta'', \qquad (9\text{-}9.13)$$

$$\tau_T = \frac{V_f Q_f}{I_f t_f} = \frac{[-EI_f(h/2)(d^3\beta/dx^3)] \{(bt_f/2)[(b/2)/2]\}}{I_f t_f} = -E \frac{b^2 h}{16} \beta''', \qquad (9\text{-}9.14)$$

and the St. Venant shearing stress,

$$\tau = -\frac{M^T t}{C_t}, \qquad (9\text{-}9.15)$$

where Eq. (9-9.15) is obtained from Eq. (9-4.34) with $2y$ replaced by flange or web thickness t and $C_t = \sum (bt^3/3)$ for web and flanges. Equations (9-9.13) and (9-9.14) apply *only* to symmetrical W shapes and σ_x and τ_T occur only in flanges; σ_x is for maximum normal stress occurring at free end of flanges and τ_T is for maximing shearing stress occurring at flange centerline. Extension of these equations is discussed briefly in sec. 9-10.4.

In *Torsion Analysis* [4], $bh/4$ is replaced by W_{no} and $b^2h/16$ by S_{wl}/t, the later values being computed for typical rolled sections to account for flange tapering, etc.

9-10   ANALYSIS WITH COMBINED BENDING AND TORSION

For development of code equations, formulations of secs. 9-1 throughout 9-4 and 9-9 are sufficient. However, it is instructive to consider application of Eqs. (9-9.12) through (9-9.15) to analysis of actual structures. For efficient solutions, *Torsion Analysis*[4] is recommended. Section 9-10.1 illustrates the development of a typical equation by use of reference 4; discussions relate the development and corresponding plots of β, β', β'', and β'''. An alternate approach is discussed in sec. 9-10.2. Finally, in sec. 9-10.3 distributed torsional loads are briefly treated.

9.10-1. The β Equations. Equation (9-9.12) applies to beams with a variety of loading and support conditions once constants have been appropriately evaluated. Figure 9-10.1 tabulates end conditions for four typical support conditions. In the table, *yes* signifies that β or its corresponding derivative is other than zero. The symbol $\sim$ indicates that two events occur *simultaneously*; that is, they are functionally related to each other.

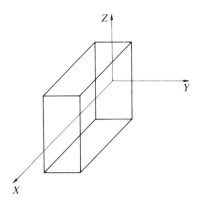

	β and its derivatives	End conditions			
		Free	Simple	Fixed	Guided
Angle of twist per unit length	β	yes	0	0	yes
Rate of change of angle of twist, per unit length $\sim$ to warping displacement u	β'	yes	yes	0	0
Flange bending [Eq. (9-9.13)] $\sim$ to torsional moment	β''	0	0	yes	yes
Flange shear [Eq. (9-9.14)] $\sim$ to first derivative of torsional moment	β'''	0	yes	yes	0

Fig. 9-10.1   Boundary Conditions, Torsion

Rate of change of angle of twist β'' is related to warping displacement u by Eq. (9-9.13):

$$\varepsilon_x = \sigma_x/E = bh\beta''/4 = bh(d^2\beta/dx^2)/4 = du/dx$$

By integration, then, $\beta' \sim u$.

Simple ends prevent rotation about *longitudinal X* axis but not about *horizontal Y* axis. Warping is permitted; that is, u displacement parallel to the X axis occurs. (Specific details may change these relationships.)

Guided ends prevent rotation about *horizontal Y* axis but not about *longitudinal X* axis. Warping is prohibited; that is, no *u* displacement occurs parallel to the *X* axis.

In addition to support conditions, at a point where concentrated torsional moment is applied:

$$\beta_R = \beta_L \qquad \beta_R' = \beta_L' \qquad \text{and} \qquad \beta_R'' = \beta_R'',$$

where subscripts *R* and *L* refer to *right* and *left* of load application point.

Figure 9-10.2 shows a simply supported beam of length *L* subject to a concentrated torsional moment located at distance *a* from the left support. Beam is assumed weightless and *EI* is constant. Equation (9-9.12) is written twice because of discontinuity at the point of load, and produces six unknown constants in combination with two unknown components of torsional moment.

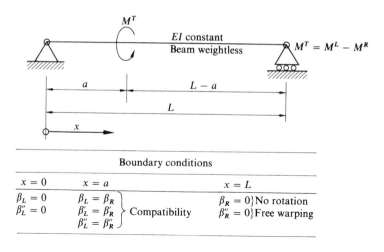

Boundary conditions

$x = 0$	$x = a$		$x = L$	
$\beta_L = 0$	$\beta_L = \beta_R$		$\beta_R = 0$	No rotation
$\beta_L'' = 0$	$\beta_L' = \beta_R'$	Compatibility	$\beta_R'' = 0$	Free warping
	$\beta_L'' = \beta_R''$			

Fig. 9-10.2   Torsional Moment on Simply Supported Beam

$$\beta_L = A_L \sinh \lambda x + B_L \cosh \lambda x + D_L + M^L x / GC_t$$

and

$$\beta_R = A_R \sinh \lambda x + B_R \cosh \lambda x + D_R + M^R x / GC_t. \qquad (9\text{-}10.1)$$

The six constants $A_L, B_L, D_L, A_R, B_R, D_R$ and two unknown torsional moments M^L and M^R (in terms of M_T) are obtained from eight equations as follows:

At $x = 0$ $\beta_L = 0$ (no rotation) $\left[\text{Step 1: } B_L = -D_L\right]$ (9-10.2)
              $\beta'' = 0$ (free warping) $\left[\text{Step 2: } B_L = D_L = 0\right]$ (9-10.3)
                                                  $\left[\text{Step 8: form for } A_R\right]$

at $x = a$ $\left.\begin{array}{l}\beta_L = \beta_R \\ \beta'_L = \beta'_R \\ \beta''_L = \beta''_R\end{array}\right\}$ (compatibility) $\begin{array}{l}\left[\text{Step 6}\right] \\ \left[\text{Step 8}\right]\end{array}$ $\begin{array}{l}(9\text{-}10.4) \\ (9\text{-}10.5)\end{array}$

$$\left[\text{Step 5: } A_L = A_R\left(1 - \frac{\tanh \lambda L}{\tanh \lambda a}\right)\right]$$
(9.10.6)

at $x = L$ $\beta_R = 0$ (no rotation) $\left[\text{Step 4: } D_R = -M^R L/GC_t\right]$
(9-10.7)
              $\beta''_R = 0$ (free warping) $\left[\text{Step 3: } B_R = -A_R \tanh \lambda L\right]$
(9-10.8)

$$M^T = M^L - M^R$$ $\left[\text{Step 7: Eq. (9-10.21)}\right]$.   (9-10.9)

Signs of Eq. (9-10.9) are necessary to conform to

$$|M^T| = |M^L| + |M^R|,$$

with M^L applied to the right-hand end of the left part, while M^R is applied to the left-hand end of the right part.

As an initial step, first and second derivatives of β are obtained as

$$\beta = A \sinh \lambda x + B \cosh \lambda x + D + Mx/GC_t,$$ $[9\text{-}9.12]$

$$d\beta/dx = \beta' = A\lambda \cosh \lambda x + B\lambda \sinh \lambda x + M/GC_t$$ (9-10.10)

$$d^2\beta/dx^2 = \beta'' = A\lambda^2 \sinh \lambda x + B\lambda^2 \cosh \lambda x.$$ (9-10.11)

Step 1. From Eq. (9-10.1), with $\beta_L = 0$ at $x = 0$ by Eq. (9-10.2), $0 = 0 + B_L + D_L + 0$, so $B_L = -D_L$.

Step 2. Use of boundary conditions of Eq. (9-10.3) in Eq. (9-10.11) gives $B_L = 0$. Therefore

$$B_L = D_L = 0,$$ (9-10.12)

and two constants vanish.

Step 3. Invoke the boundary condition established by Eq. (9-10.8) into (9-10.11) to get

$$B_R = -A_R \tanh \lambda L.$$ (9-10.13)

Step 4. Substitution of (9-10.13) into Eq. (9-9.12) by use of the boundary conditions of Eq. (9-10.7) yields

$$0 = A_R \sinh \lambda L + (-A_R \tanh \lambda L) \cosh \lambda L + D_R + M^R L/GC_t$$ (9-10.14)

or

$$D_R = -M^R L/GC_t. \qquad (9.10\text{-}15)$$

Step 5. Equation (9-10.6), with boundary conditions at $x = a$, $B_L'' = B_R''$ produces [with $B_L = 0$ and B_R by Eq. (9-10.13)]

$$A_L \lambda^2 \sinh \lambda a + 0 = A_R \lambda^2 \sinh \lambda a + (-A_R \tanh \lambda L) \lambda^2 \cosh \lambda a \qquad (9\text{-}10.16)$$

which, when divided throughout by $\lambda^2 \sinh \lambda a$, upon collecting terms yields

$$A_L = A_R (1 - \tanh \lambda L/\tanh \lambda a). \qquad (9\text{-}10.17)$$

Step 6. A more complicated expression is obtained from Eq. (9-10.4) [with $B_L = D_L = 0$, B_R by Eq. (9-10.13), and D_R by Eq. (9-10.15)]:

$$A_L \sinh \lambda a + M^L a/GC_t = A_R \sinh \lambda a + (-A_R \tanh \lambda L) \cosh \lambda a$$
$$- M^R L/GC_t + M^R a/GC_t.$$

This can be rearranged to

$$A_L = A_R \left[1 - \frac{\tanh \lambda L}{\tanh \lambda a} \right] - \frac{1}{GC_t \sinh \lambda a} (-M^L a - M^R L + M^R a). \qquad (9\text{-}10.18)$$

Comparison of Eqs. (9-10.18) and (9-10.17) show that

$$-M^L a - M^R L + M^R a = 0, \qquad (9\text{-}10.19)$$

so that

$$-M^R (L - a)/a = M^L. \qquad (9\text{-}10.20)$$

Step 7. Substitution of Eq. (9-10.20) into Eq. (9-10.9) produces

$$M^R = -M^T a/L \qquad (9\text{-}10.21)$$

and, likewise,

$$M^L = M^T (L - a)/L. \qquad (9\text{-}10.22)$$

Step 8. Conditions of Eq. (9-10.5) used with (9-10.10), where on the left $B_L = 0$, $M \equiv M^L = M^T(L - a)/L$ and on the right B_R replaced by (9-10.13) and $M \equiv M^R = -M^T a/L$) results in:

$$A_L \lambda \cosh \lambda a + \frac{M^T(L - a)}{LGC_t} = A_R \lambda \cosh \lambda a + (-A_R \tanh \lambda L) \lambda \sinh \lambda a$$

$$+ \left(\frac{-M^T a}{LGC_t} \right) \qquad (9\text{-}10.23)$$

To reduce, divide throughout by $\lambda \cosh \lambda a$ and replace A_R by $\tanh \lambda a/(\tanh \lambda a - \tanh \lambda L)$ from Eq. (9-10.17), resulting in

$$\frac{M^T}{GC_t\lambda \cosh \lambda a} = A_L\left[\frac{\tanh \lambda a}{\tanh \lambda a - \tanh \lambda L}(1 - \tanh \lambda L \tanh \lambda a) - 1\right].$$

When reduced and rearranged this would appear as

$$\frac{M^T(\tanh \lambda a - \tanh \lambda L)}{GC_t\lambda} = A_L \cosh \lambda a \,[\tanh \lambda a - \tanh^2 \lambda a \tanh \lambda L \\ - \tanh \lambda a + \tanh \lambda L].$$

After cancelling, factoring, and recognizing $1 - \tanh^2 \lambda a$ as $\text{sech}^2 \lambda a$, we may simplify this to

$$\frac{M^T(\tanh \lambda a - \tanh \lambda L)}{GC_t\lambda} = A_L \cosh \lambda a \,\text{sech}^2 \lambda a \tanh \lambda L, \qquad (9\text{-}10.24)$$

from which A_L is determined as

$$A_L = \frac{M^T}{GC_t\lambda}(\tanh \lambda a - \tanh \lambda L)\frac{\cosh \lambda a}{\tanh \lambda L}. \qquad (9\text{-}10.25)$$

Equating Eqs. (9-10.25) and (9-10.17):

$$A_R = \frac{(M^T/GC_t\lambda)(\tanh \lambda a - \tanh \lambda L)(\cosh \lambda a/\tanh \lambda L)}{[(\tanh \lambda a - \tanh \lambda L)/\tanh \lambda a]} \qquad (9\text{-}10.26)$$

which reduces to

$$A_R = \frac{M^T}{GC_t\lambda}\frac{\sinh \lambda a}{\tanh \lambda L}. \qquad (9\text{-}10.27)$$

Finally, by use of Eq. (9-10.13), the last constant is evaluated as

$$B_R = \frac{-M^T}{GC_t\lambda}\sinh \lambda a. \qquad (9\text{-}10.28)$$

Substitution of A_L of Eq. (9-10.25) (slightly rearranged) and M^L from (9-10.22) into the first of Eq. (9-10.1) ($B_L = D_L = 0$) results in an expression for angle of twist per unit length for points to the left of the concentrated torsional moment on the beam:

$$\beta_L = \frac{M^T L}{GC_t}\left[\frac{(L-a)x}{L^2} + \frac{1}{\lambda L}\left(\frac{\sinh \lambda a}{\tanh \lambda L} - \cosh \lambda a\right)\sinh \lambda x\right]. \quad (9\text{-}10.29)$$

Likewise, substitution of values of constants A_R from Eq. (9-10.27), B_R from Eq. (9-10.28), D_R from Eq. (9-10.15), and M^R from Eq. (9-10.21) into the second of Eqs. (9-10.1) yields

$$\beta_R = \frac{M^T L}{GC_t}\left[\frac{a}{L}\left(1 - \frac{x}{L}\right) + \frac{1}{\lambda L}\left(\frac{\sinh \lambda a}{\tanh \lambda L}\sinh \lambda x - \sinh \lambda a \cosh \lambda x\right)\right].$$
$$(9\text{-}10.30)$$

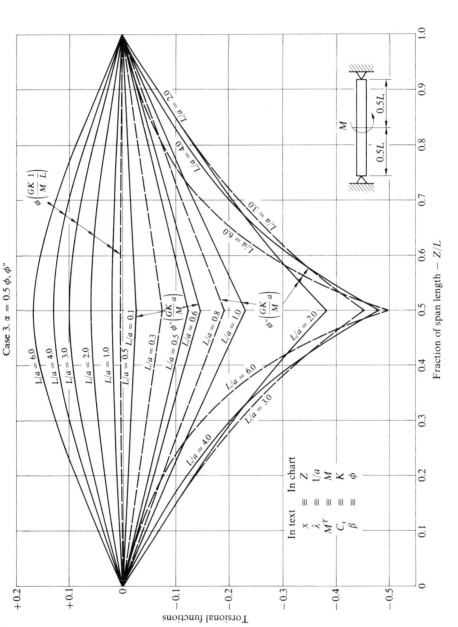

Fig. 9-10.3 β and β'' Curves [4, page 28]

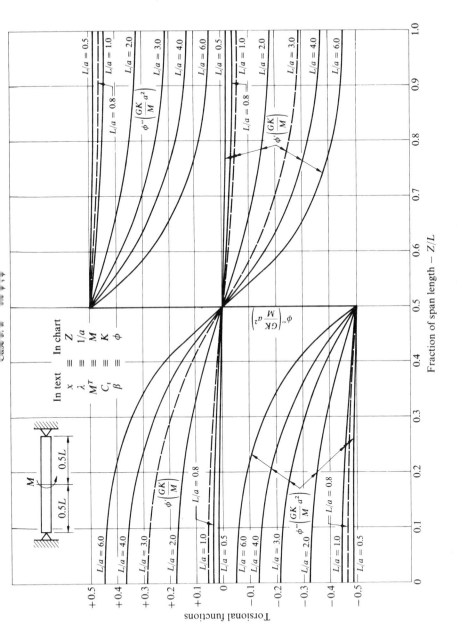

Fig. 9-10.4 β' and β'' Curves [4, page 29]

Equations (9-10.29) and (9-10.30) correspond to Case 3 equations of *Torsion Analysis*[4, page 76] if a/L is replaced by α, M^T by M, C_t by K, β by ϕ, x by Z, and λ by $1/a$. Other equations have been derived in a similar manner.

Figures 9-10.3 and 9-10.4 are typical curves from *Torsion Analysis*[4, pages 28–29] which show numerical values for $\beta(\phi)$ and its derivatives at all points for a torsional moment applied at mid-span to a simply supported beam. Member rotation is greatest at the center and zero at the ends $[\phi(GK/ML)$ curve]. Warping, on the other hand, is zero at mid-span and maximum at the ends $[\phi'(GK/M)$ curve]. This reference work has 50 pages of such curves corresponding to 12 different loading cases.

9-10.2 Alternate Approach. For symmetrical spans with the torsional moment applied at mid-span, a simpler approach is possible. Because no warping occurs at mid-span, computations are shortened if the origin is chosen to be at mid-span (Fig. 9-10.5). With an assumed span length of $2L$ and the "vertical" axis rotated to the angle of maximum twist, the boundary conditions are:

$$\text{At } x = 0 \qquad \beta = 0 \text{ (no twist)}$$
$$\beta' = 0 \text{ (no warping)} \qquad (9\text{-}10.31)$$
$$\text{at } x = L \qquad \beta'' = 0 \text{ (free warping).}$$

Use of these boundary conditions with β from Eq. (9-9.12) and its derivatives from Eqs. (9-10.10) and (9-10.11) gives

$$\beta = \frac{M^T}{GC_t\lambda}\left[\tanh \lambda L \left(\cosh \lambda x - 1\right) - \sinh \lambda x + \lambda x\right], \quad (9\text{-}10.32)$$

$$\beta' = \frac{M^T}{GC_t}\left[1 - \frac{\cosh \lambda (L - x)}{\cosh \lambda L}\right], \quad (9\text{-}10.33)$$

$$\beta'' = \frac{M^T\lambda}{GC_t}\frac{\sinh \lambda (L - x)}{\cosh \lambda L}, \quad (9\text{-}10.34)$$

$$\beta''' = -\frac{M^T\lambda^2}{GC_t}\frac{\cosh \lambda (L - x)}{\cosh \lambda L}. \quad (9\text{-}10.35)$$

9-10.3 Distributed Loads. Equations for distributed torsional loads are obtained in two steps. First, differentiate Eq. (9-9.9) so that

$$\frac{dM}{dx} = GC_t\frac{d^2\beta}{dx^2} - EI_w\frac{d^4\beta}{dx^4}. \quad (9\text{-}10.36)$$

An expression for dM/dx is then found from summing moments for an elemental section as shown in Fig. 9-10.6. Thus for an nth order torsional moment originating at $x = 0$,

$$\Sigma M = 0 \Rightarrow -M + m\left(\frac{x}{L}\right)^n dx + M + dM = 0 \quad (9\text{-}10.37)$$

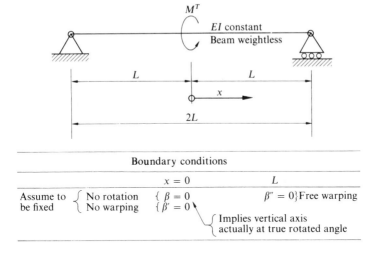

Boundary conditions		
	$x = 0$	L
Assume to be fixed { No rotation { $\beta = 0$	No warping { $\beta' = 0$	$\beta'' = 0$} Free warping

{ Implies vertical axis actually at true rotated angle

Fig. 9-10.5 Torsional Moment on Symmetrical Beam

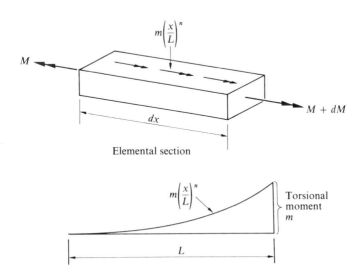

Elemental section

Distribution of torsional moment along span

Fig. 9-10.6 Distributed Torsional Moment

from which

$$\frac{dM}{dx} = -m\left(\frac{x}{L}\right)^n.$$ (9-10.38)

Equating Eqs. (9-10.36) and (9-10.38) and using operator notation, we get

$$(\mathscr{D}^4 - \lambda^2 \mathscr{D}^2)\beta = -\frac{mx^n}{EI_w L^n}$$ (9-10.39)

which has the general solution

$$\beta = A + Bx + D \sinh \lambda x + E \cosh \lambda x - \frac{mx^{(n+2)}}{GC_t L^n (n+1)(n+2)}$$

for $n \leq 2$ (0, uniform; 1, triangular; and 2, parabolic load distribution), where $A, B, D,$ and $E \equiv$ constants of integration (values to be established by boundary conditions); $m \equiv$ torsional moment per unit length at maximum point, that is, at end of beam; $L \equiv$ span length; $\lambda \equiv \sqrt{GC_t/EI_w}$; $G \equiv$ modulus of rigidity; $E \equiv$ modulus of elasticity; $I_w \equiv$ warping constant and $C_t \equiv$ torsional constant (see sec. 9-10.4).

9-10.4 Constants.   To this point, λ has been defined for a wide-flange beam as

$$\lambda = \sqrt{GC_t/EI_w}.$$ [9-9.10]

However, if the member is a channel, the basic differential equation includes a term for bending of the web as well as of flanges. Timoshenko [6, page 292] has shown that all equations apply to channels if a modification is made for λ, so that

$$\lambda = \sqrt{(GC_t/EI_w)/[1 + (t_w h^3)/4I_x]},$$ (9-10.40)

where $t_w \equiv$ thickness of web and $I_x \equiv$ total moment of inertia of section about X axis; and, as before, $G \equiv$ modulus of rigidity; $h \equiv$ section depth (distance between centroid of flanges) $E \equiv$ modulus of elasticity; $I_w \equiv I_f h^2/2$ for WF sections $\equiv$ warping constant, where $I_f \equiv$ maximum principal moment of inertia of one flange and $C_t \equiv \Sigma(1/3)bt^3 \equiv$ torsional constant when $b \equiv$ flange or web width and $t \equiv$ corresponding flange or web thickness.

An alternate approach is to redefine I_w so that Eq. (9-9.10) applies to channels as well as wide-flange sections. This is the approach used in *Steel Construction* and *Torsion Analysis*, and makes it possible to *always* evaluate λ as $\sqrt{CG_t/EI_w}$, even for members with sloping flanges and taking fillets into account.

The torsional constant C_t is valid for most thin open sections. However, a more precise value can be obtained for special shapes, particularly when fillets are considered [4, page 75].

Table 9-10.1 Comparison of Torsional Constants for C15 × 50

$I_x = 404$ in.^{4}

(avg.) $t_f = 0.650$ in. $\quad b = 3.716$ in. $\quad h = 15.00 - 0.65 =$ say, 14.35 in. $\quad t_w = 0.716$ in.

	Calculated Values	Torsion Analysis Values	Steel Construction Values
$I_f = t_f b^3/12 = 0.650(3.716)^3/12$	$= 2.78$ in.^{4}		
$I_w = I_f h^2/2 = 2.78(14.35)^2/2$	$= 286$ in.^{6}	490 in.^{6}	$C_w = 492$ in.^{6}
$C_t = \sum (1/3)bt^3 = \dfrac{1}{3}(3.716)(0.650)^3 \times 2$	$= .68$		
$\qquad + \dfrac{1}{3}(13.70)(0.716)^3$	$= \underline{1.68}$		
	2.36 in.^{4}	2.66 in.^{4}	$J = 2.66$ in.^{4}
$e = \dfrac{tb^2h^2}{4I_x} = \dfrac{(0.650)(3.716)^2(14.35)^2}{4(404)}$	$= 1.14$ in.	0.94 in.	$E_o = 0.941$ in.

$$\lambda = \sqrt{(GC_t/EI_w)/\left[1 + (t_w h^3)/4I_x\right]}$$

$$= \sqrt{[(11.2 \times 10^6)2.36]/[(29 \times 10^6)286]/\underbrace{\{1 + [(0.716)(14.35)^3/4(404)]\}}_{2.309}}$$

$$= \sqrt{0.00138} = 0.0371 \text{ in.}^{-1}$$

$$\lambda = \sqrt{\frac{GC_t}{EI_w}} = \sqrt{\frac{11.2(10^6)\,2.66}{29(10^6)\,492}}$$

$$= 0.0458 \text{ in.}^{-1}$$

or from table,

$$\lambda = 1/a = 1/21.8 = 0.0458 \text{ in.}^{-1}$$

Steel Construction tabulates values of torsional constant, warping constant, and shear center location, as appropriate, for standard rolled sections (respectively called C_t, I_w, and e in this text, but J, C_w, and E_o in *Steel Construction*).

Section properties for a C15 x 50 computed according to approximate formulas developed in this section are compared in Table 9-10.1 to the more exact values presented in *Steel Construction* and *Torsion Analysis*. Note that the large difference in I_w includes provisions for tapered flanges and for fillets and makes it possible to use the simplified quotient

$$\lambda = \sqrt{GC_t/EI_w}.$$

Equations (9-10.32) through (9-10.35) can be used to verify curves of Figs. 9-10.3 and 9-10.4. Because of the difference in derivations, however, L, as used in the equations is $\frac{1}{2}L$ in the curves and M^T is one-half the moments used in the figures. By rearrangement of terms, different notations equated, and inclusion of factors to account for differences in M and L, Eqs. (9-10.32) through (9-10.35) can be rewritten as:

$$\phi\left(\frac{GK}{M}\frac{1}{L}\right) = \beta\left(\frac{GC_t}{M^T}\frac{1}{L}\right)\frac{1}{4}$$
$$= \frac{1}{4\lambda L}[\tanh \lambda L\,(\cosh \lambda x - 1) - \sinh \lambda x + \lambda x], \quad (9\text{-}10.41)$$

$$\phi'\left(\frac{GK}{M}\right) = \beta'\left(\frac{GC_t}{M^t}\right)\frac{1}{2}$$
$$= \frac{1}{2}[1 - (\cosh \lambda(L - x)/\cosh \lambda L)], \quad (9\text{-}10.42)$$

$$\phi''\left(\frac{GK}{M}a\right) = \beta''\left(\frac{GC_t}{M^T\lambda}\right)\frac{1}{2}$$
$$= \frac{1}{2}[\sinh \lambda(L - x)/\cosh \lambda L], \quad (9\text{-}10.43)$$

$$\phi'''\left(\frac{GK}{M}a^2\right) = \beta'''\left(\frac{GC_t}{M^T\lambda^2}\right)\frac{1}{2}$$
$$= \frac{1}{2}[\cosh \lambda(L - x)/\cosh \lambda L]. \quad (9\text{-}10.44)$$

Several combinations are shown evaluated in Table 9-10.2 to clarify the relation between λL and x as differentiated by the two derivations used. For all derivative terms, real $x = (L \text{ used} - x \text{ used})/2$. Because the axis was rotated to coincide with maximum actual rotation, use of $x = L$ gives true maximum

Table 9-10.2 Verification of Curves

	Values used in equation for:			Result applies in figures to:	
Term	λL	x	Results	λL	x
$\phi\left(\dfrac{GK}{M}\dfrac{1}{L}\right)$	3	L	$0.167-0.167 = 0$	6	0
	3	$4L/5$	$0.167-0.122 = 0.045$	6	$0.1L$
	3	$3L/5$	$0.167-0.080 = 0.087\dagger$	6	$0.2L$
	2	L	0.130	4	$L/2$
	1	L	0.060	2	$L/2$
	5.484	L	0.208	10.968	$L/2$
$\phi'\left(\dfrac{GK}{M}\right)$	3	L	0.450	6	0
	2	L	0.367	4	0
	1	L	$0.176\dagger\dagger$	2	0
	5.484	0	0	10.968	$L/2$
$\phi''\left[\dfrac{GK(a)}{M}\right]$	3	L	0	6	0
	3	$4L/5$	0.032	6	$0.1L$
	3	$3L/5$	$0.075\dagger\dagger\dagger$	6	$0.2L$
	5.484	0	$\doteq 0.500$	10.968	$L/2$
$\phi'''\left[\dfrac{GK(a^2)}{M}\right]$	3	L	0.050	6	0
	3	$4L/5$	0.059	6	$0.1L$
	3	$3L/5$	0.090	6	$0.2L$
	5.484	0	0.500	10.968	$L/2$
	5.484	L	0.00415	10.968	0

† This computation is in two parts. First, it is necessary to figure the mid-span rotation because in this formulation zero rotation is assumed to occur there. Replace x by L in Eq. (9-10.41):

$$\frac{1}{4\lambda L}\left[\tanh \lambda L(\cosh \lambda L - 1) - \sinh \lambda L + \lambda L\right]$$

Replace λL by 3:

$$\frac{1}{12}\left[\sinh 3 - \tanh 3 - \sinh 3 + 3\right] = 0.167.$$

Then figure the change from mid-span rotations. Replace x by $3L/5$ in Eq. (9-10.41):

$$\frac{1}{4\lambda L}\left[\tanh \lambda L(\cosh 0.6\,\lambda L - 1) - \sinh 0.6\,\lambda L + 0.6\,\lambda L\right]$$

Replace λL by 3:

$$\frac{1}{12}\left[\tanh 3(\cosh 1.8 - 1) - \sinh 1.8 + 1.8\right] = 0.080.$$

†† Replace x by L in Eq. (9-10.42):

$$\frac{1}{2}\left[1 - \cosh \lambda(L - L)/\cosh \lambda L\right]$$

Replace λL by 1:

$$\frac{1}{2}\left[1 - 1/\cosh 1\right] = 0.176.$$

††† Replace x by $3L/5$ in Eq. (9-10.43):

$$\frac{1}{2}\left[\sinh \lambda(L - 3L/5)/\cosh \lambda L\right]$$

Replace λL by 3:

$$\frac{1}{2}\left[\sinh 1.2/\cosh 3\right] = 0.075.$$

at mid-span; all intermediate values of rotation must therefore be subtracted from this maximum to give true values at real x locations.

Torsion Analysis gives additional torsional properties useful for more precise analysis: Using these relations, Eqs. (9-9.13) and (9-9.14) can be altered to:

$$\sigma_x = -E\frac{bh}{4}\beta'' = -EW_{ns}\beta'' \tag{9-10.45}$$

and

$$\tau_T = -E\frac{b^2 h}{16}\beta''' = -E\left(\frac{S_{ws}}{t}\right)\beta'''. \tag{9-10.46}$$

Parameters W_{ns} and S_{ws} apply to specific points on a cross section. For a symmetrical W shape, with nontapered flanges $W_{n0} = bh/4$ (point 0 is at flange ends where maximum warping normal stress occurs) and $S_{w1} = b^2 ht/16$ (point 1 is located at center of flange where maximum warping shear stress occurs).

A comparison for two sections, given in Table 9-10.3, suggests that errors in stresses computed for channels, assuming nontapering flanges, can be upward of 40%.

Table 9-10.3   Comparison of Torsional Properties

	W 14 X 176	C15 X 50
$(b_f)(d - t_f)/4$	$(15.640)(15.25{-}1.313)/4$	$(3.716)(15.0{-}0.650)/4$
$= bh/4$ (W_{ns})	$= 54.5$ (54.5)	$= 13.3$ (17.3)
$b_f^2(d - t_f)/16$	$(15.640)^2(15.25{-}1.313)/16$	$(3.716)^2(15.0{-}0.650)/16$
$= bh^2/16$	$= 213.1$	$= 12.4$
S_{ws}/t_t	$280/1.313$	$13.6/0.650$
$= S_{ws}/t$	$= 213.3$	$= 20.9$

Note. Values of W_{ns} and S_{ws} taken from *Torsion Analysis*; all other values from *Steel Construction*.

Torsion Analysis gives coefficients for W_{ns} and S_{ws} for critical points on various members. For a channel section, for example, S_{w1} is for maximum warping shear stress in flange, S_{w2} at junction of centerlines of web and flange, S_{w3} at mid-depth of web, W_{n0} at free end of flanges, and W_{n2} at junction of centerlines of web and flange. Adequate design and analysis of channel or zee sections is very difficult without these coefficients.

9-11   ANALYSIS AND DESIGN

In general, analysis proofing for steel beams subject to torsion is quite necessary because simplifications used in design may lead to serious error.

9-11.1 Steel Design. Design of beams subject to torsion is essentially by trial and error with publications such as *Torsion Analysis* used to decrease time and effort required. However, some simplified techniques also are very useful.

One typical simplification which leads to reasonable results for WF sections subject to torsion producing loads is to resolve them into vertical and horizontal components, adding flexural stresses produced about both principal axes. As an example, the beam of Fig. 9-11.1 can be designed by the method of sec. 9-2 if $2Z_{y'R}$ is plotted rather than $Z_{y'R}$. Rationale for such a procedure is obtained when lateral flange stiffness is compared to that of the web; it is difficult to visualize the web transmitting enough force to the lower flange to cause the latter to bend (Fig. 9-11.2). Likewise, it is reasonable to use Z_y rather than flange section modulus because of minor web contribution.

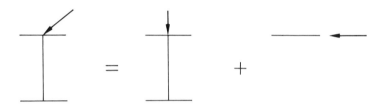

Fig. 9-11.1   Simplified Analysis Resolution

For torsion caused by an eccentrically applied load, resolution according to Fig. 9-11.3 is appropriate; design may then proceed as for the beam of Fig. 9-11.1.

It is left as a student exercise to obtain typical errors that result from these simplifications when compared to more exact analysis. Remember that warping normal stresses vary from zero at flange centerline to maximum at free ends whereas warping shearing stresses are zero at free ends and maximum at flange center. No warping stresses occur in webs of symmetrical W shapes. St. Venant torsion is maximum on element edges, maximum where elements are the thickest.

9-11.2 Torsion of Structural Concrete. Theory applied to reinforced concrete is adapted from that developed earlier in this chapter for a perfectly elastic material. The basic equations evolve from Eq. (9-4.34) with C_t replaced by a summation, repeated here as

$$\tau_{xy} = \frac{2M_x^t\, y}{\sum bt^3/3}. \qquad [9\text{-}4.34]$$

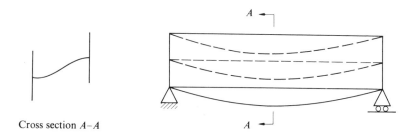

Cross section A–A

Fig. 9-11.2   Load on One Flange

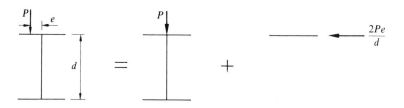

Fig. 9-11.3   Simplification—Load Eccentrically Applied

With τ_{xy} replaced by v_{tu}, M'_x replaced by T_u, b replaced by y (larger dimension), t replaced by x (small dimension), $2y$ equated to t, and reduction of x from a cubic to a squared term, Eq. (9-4.34) results in ACI formula 11-16:

$$v_{tu} = \frac{3T_u}{\phi \Sigma x^2 y},\qquad(9\text{-}11.1)$$

where the capacity reduction factor ϕ has been added.

As illustrated in Fig. 9-11.4, restrictions are also placed on x and y such that:

1. Usable overhanging flange width is limited to three times flange depth.
2. Box sections with wall thicknesses $h \geq x/4$ can be considered as solid sections.
3. Box sections with thinner wall thicknesses, but $h \geq x/10$ require $\Sigma x^2 y$ to be multiplied by $4h/x$.
4. Box sections with thinner walls require more accurate analysis to include wall stiffnesses.

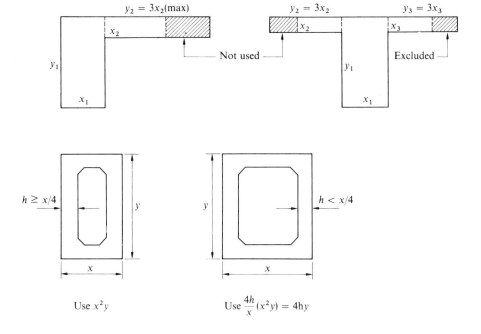

Fig. 9-11.4 Modified Torsional Constant for Reinforced Concrete

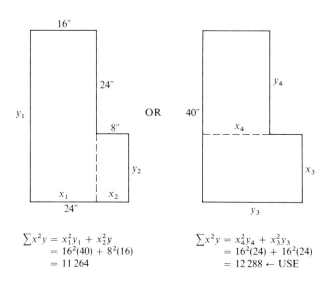

$$\sum x^2 y = x_1^2 y_1 + x_2^2 y \qquad\qquad \sum x^2 y = x_4^2 y_4 + x_3^2 y_3$$
$$= 16^2(40) + 8^2(16) \qquad\qquad = 16^2(24) + 16^2(24)$$
$$= 11\,264 \qquad\qquad\qquad = 12\,288 \leftarrow \text{USE}$$

Fig. 9-11.5 Using Largest $\Sigma x^2 y$

The discussion of secs. 9-6 and 9-8 adds validity to these procedures. When subdividing sections, always use the largest $\sum x^2 y$ as illustrated in Fig. 9-11.5.

For combined shear and torsion, allowable shear stress without torsion is set at $2\sqrt{f_c'}$ and allowable torsional stress without shear at $2.4\sqrt{f_c'}$. These are combined by an elliptical equation, and total stresses (obtained by use of stirrups) are considered proportional to concrete stresses. That is,

$$\frac{v_u}{v_{tu}} = \frac{v_c}{v_{tc}}, \tag{9-11.2}$$

where $v_u \equiv$ nominal total design stress due to shear, $v_{tu} \equiv$ nominal total design stress due to torsion, $v_c \equiv$ nominal permissible shear stress carried by concrete, and $v_{tc} \equiv$ nominal permissible torsional stress carried by concrete.

These relationships are pictured in Fig. 9-11.6 which yields from the standard elliptical equation:

$$\left(\frac{v_{tc}}{2.4\sqrt{f_c'}}\right)^2 + \left(\frac{v_c}{2\sqrt{f_c'}}\right)^2 = 1. \tag{9-11.3}$$

Multiplication by $(2.4\sqrt{f_c'})^2$ and factoring out $(v_{tc})^2$ yields

$$(v_{tc})^2 \left[1 + \left(\frac{1.2v_c}{v_{tc}}\right)^2\right] = (2.4\sqrt{f_c'})^2. \tag{9-11.4}$$

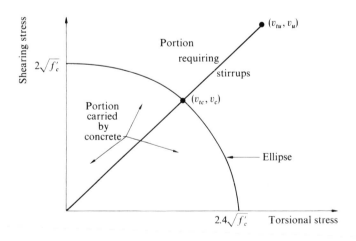

Fig. 9-11.6   Relationship Between Shear and Torsion

With substitution from Eq. (9-11.2) and rearrangement, ACI formula 11-17 is obtained:

$$v_{tc} = \frac{2.4\sqrt{f_c'}}{\sqrt{1 + (1.2v_u/v_{tu})^2}}$$ (9-11.5)

When torsional stress exceeds this amount, torsional reinforcement must be used. However, v_{tu} cannot exceed $5v_{tc}$. Torsional reinforcement, consisting of closed stirrups and longitudinal bars *in excess* of those required for flexure shear and bending, is determined as follows: (ACI formula 11-19)

$$A_t = \frac{(v_{tu} - v_{tc})s\sum x^2 y}{3\alpha_t x_1 y_1 f_y},$$ (9-11.6)

where

$A_t \equiv$ area of *one* leg of closed stirrup within s;

$s \equiv$ stirrup spacing $\leq (x_1 + y_1)/4 \leq 12$ in.;

$\alpha_t \equiv 0.66 + 0.33\,(y_1/x_1) \leq 1.5$;

$x_1 \equiv$ shorter dimension of closed stirrup;

$y_1 \equiv$ longer dimension of closed stirrup.

Amount of *extra* longitudinal steel A_l is the maximum of (ACI formulas 11-20 and 11-21, respectively)

$$A_l \geq 2A_t(x_1 + y_1)/s$$ (9-11.7)

or

$$A_l \geq \left[\frac{400\,xs}{f_y}\left(\frac{v_{tu}}{v_{tu} + v_u}\right) - 2A_t\right](x_1 + y_1)/s.$$ (9-11.8)

But A_l computed by Eq. (9-11.8) need not exceed the computed value if $2A_t$ is replaced by $50\,b_w s/f_y$.

Longitudinal bars must be

1. #3 or greater,

2. spaced less than or equal to 12 in. around the perimeter of stirrups,

3. occupy each stirrup corner.

Both longitudinal and stirrup reinforcement for torsion must extend $d + b$ beyond the point where no longer theoretically required.

PROBLEMS

9-1 given: L 8 x 6 x _____ $(1, \frac{3}{4}, \frac{5}{8}, \frac{9}{16}, \frac{1}{2}, $ or $\frac{7}{16})$,
        $F_y =$ _____ (36 or 50 ksi),
        $M_y =$ _____ (20, 30, 40, or 50%) of M_x.

   find: Maximum M_x using (a) principal axes solution and (b) direct method.

9-2 given: _____ (20, 22, or 24 ft) simple span. Mid-span concentrated load of _____ (20, 21, or 22 kips). Beam's own dead load. Mid-span lateral concentrated load _____ (40, 42, or 44%) of vertical concentrated load.

   design: Most economical rolled A36 steel section using method of sec. 9-2.

   present: Analysis proof.

9-3 given: Sketch _____ [(a), (b), (c), or (d) of Fig. 9-P.1] with $t =$
        _____ $(\frac{1}{4}, \frac{3}{8}, \frac{1}{2}, \frac{3}{4},$ or 1 in.), $h =$ _____ (10, 12, or 14 in.),
        $a =$ _____ (0.3, 0.4, 0.5, or 0.6), $b =$ _____ (0.7, 0.8, or 0.9),
        $\theta =$ _____ (30, 33.6, 41.7, or 53°).

   find: (a) Location of shear center and (b) torsional constant C_t.

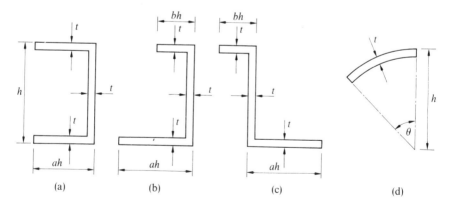

Fig. 9-P.1   Cross Sections for Shear Center Computations

9-4 given: W14 x _____ (730, 550, 314, 228, 142, 87, 53, or 22).

   verify: (a)Torsional constant and (b) warping constant as published in *Steel Construction*.

9-5 given: Tubes of Fig. 9-P.2 with $t =$ _____ ($\frac{1}{2}, \frac{3}{4}$, or 1 in.), $d =$ _____ (10, 15, or 20 in.), $b =$ _____ (0.5, 0.6, or 0.7 of d). A36 steel.

find: (a) Maximum torsional moment each configuration can carry. (b) use the smallest maximum torsional moment as a load to determine (1) maximum shear and (2) θ for each configuration.

9-6 find: Equation for β with constants evaluated for _____ (uniform *or* triangular) distributed load for _____ (simple span *or* fixed-end beam).

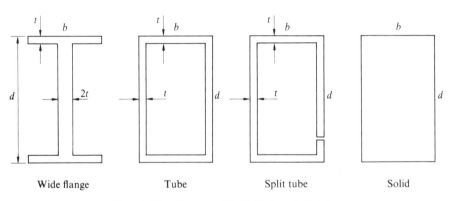

Wide flange   Tube   Split tube   Solid

Fig. 9-P.2 Open and Split Tubes, Solid Bar

9-7 given: Beam of Fig. 9-P.3 with $L =$ _____ (30, 34, 38, or 40 ft), $P =$ _____ (10, 12, or 14 kips), $Q =$ _____ (0.1, 0.2, 0.3, or 0.4) times P.

design: Most economical A36 rolled section by method of sec. 9-2.

present: Analysis proof *including* effects of torsion.

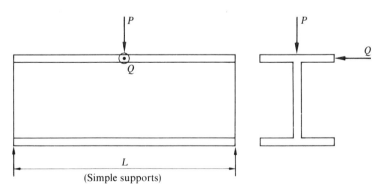

(Simple supports)

Fig. 9-P.3 Simple Beam—Biaxial Load—Torsion

9-8 given: Beam of Fig. 9-P.3 with $P = 0$, beam weightless, and $L =$ _____ (20, 24, or 28 ft), $Q =$ _____ (4, 6, or 7 kips).

design: Most economical A36 rolled section assuming equivalent section modulus is $\frac{1}{2}S$ about axis Y–Y.

present: Analysis proof *including* effects of torsion, that is, do not use simplifying assumption of design.

9-9 given: Spandrel beam of Fig. 9-P.4 with dimensions set by architectural requirements as $b =$ _____ (13, 15, or 17 in.), $h =$ _____ (29, 30, or 31 in.), $t =$ _____ (4, $4\frac{1}{2}$, or 5 in.), $L =$ _____ (12, 13, or 14 ft). Loads *including* load factors induce $M_A =$ _____ ($+180$, $+190$, $+200$, or $+210$ ft-kips), $M_B =$ _____ (-180, -185, -210, or -215 ft-kips), $V_A =$ _____ (26, 30, 32, or 34 kips), $V_B =$ _____ (28, 33, 35, or 37 kips), $T_A =$ _____ (60, 65, or 70 ft-kips). $T_B =$ _____ (70, 75, or 80 ft-kips).

design: Beam steel including stirrups and ties.

present: Analysis proof and/or stress sheet. Include spacing and cutoff points.

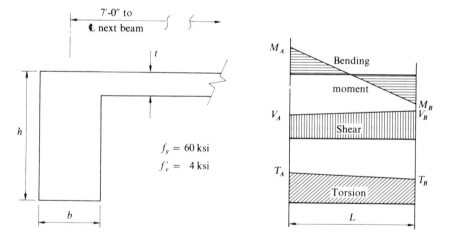

Fig. 9-P.4 Spandrel Beam

9-10 given: Ledger beam of Fig. 9-P.5 with dimensions set by architectural requirements as $b =$ _____ (15, 16, or 17 in.), $h =$ _____

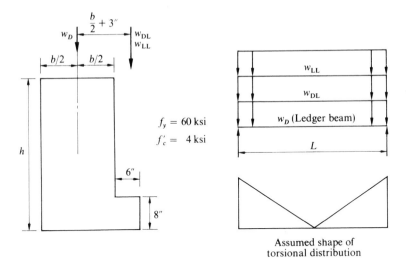

$f_y = 60 \text{ ksi}$

$f'_c = 4 \text{ ksi}$

Fig. 9-P.5   Ledger Beam

(30, 31, or 32 in.), $L =$ _____ (30, 32, or 34 ft). Service loads *not* including load factors are $w_{DL} =$ _____ (1.2, 1.3, or 1.4k/ft), $w_{LL} =$ _____ (1.1, 1.15, or 1.2 k/ft).

design: Beam steel including stirrups and ties.

present: Analysis proof and/or stress sheet.

REFERENCES

1. Cross, Hardy, "The Column Analogy," *Bulletin 215*, Engineering Experimental Station, University of Illinois (1930).

2. Seely, Fred B. and James O. Smith, *Advanced Mechanics of Materials*, Wiley, New York (1952).

3. Popov, Egor P., *Introduction to Mechanics of Solids*, Prentice-Hall, New Jersey (1968).

4. Heins, C. P., Jr. and P. A. Seaburg, *Torsion Analysis of Rolled Steel Sections*, Bethlehem Steel Corporation, AIA File No. 13-A-1, Bethlehem, Pennsylvania.

5. St. Venant, *Mém des Savans Étrangers*, V. 14 (1855).

6. Timoshenko, S., *Strength of Materials, Part II*, 2nd ed., D. Van Nostrand, New York (December 1954).

7. Wang, Chi-Teh, *Applied Elasticity*, McGraw-Hill, New York (1953).

8. Timoshenko, S. and J. N. Goodier, *Theory of Elasticity*, McGraw-Hill, New York (1951).

CHAPTER 10

COMBINED BENDING AND COMPRESSION

Formulations for the design and analysis of members subject to combined bending and compression are intricate, long, and contain parameters which may appear mysterious to the novice. Yet they represent the cumulative work of many of the outstanding minds in engineering. Historically, progress has been by "fits and starts" and by discarding, melding, and refining the work of still earlier investigators. Therefore, the approach of this chapter is to present the formulations in a manner which, hopefully, will *lead* toward full understanding of the current state of the art, without attention to either historical or mathematical chronology.

As alluded to in Chap. 3, real beams and columns do not often occur in "pure" form. Even a simple beam with one end on rollers resists a small axial force when temperature changes because supports are not free of friction. In a significant number of cases, however, design is (or was), according to dominant load-carrying character, made adequate by built-in safety factors. But, with the advent of electronic computers, optimizing techniques, better understanding of structural action, introduction of statistical techniques, and the capacity for better field control, factors of safety have been, and will probably continue to be lowered. As these changes occur, the subject of beam-columns becomes a more predominant consideration.

Members of simple trusses, generally considered to be in pure tension or compression, normally are subjected to bending by: (a) eccentricity of their connections; (b) loads which act transverse to the members between joints, typical of roof purlins; and (c) secondary stresses arising from rigid connections which, in analysis, were considered as perfect hinges.

Rigid frames are almost certain to be constructed of a majority of beam-columns. Brief reflection suggests that both horizontal and vertical members of a simple portal frame are beam-columns, even under gravity loads! As a structure is subjected to wind and seismic loads, beam-column characteristics become even more pronounced. For these reasons, design and analysis of beam-columns is an increasing activity in design offices.

Although this text is primarily directed toward design and analysis of single members, beam-columns are considered with respect to frame action—interaction between members—at least as related to magnification factor, slenderness, and lateral support.

Section 10–1 illustrates the development of an *interaction curve* for a particular reinforced concrete member as the ultimate loading configuration changes from pure axial, to combined axial and bending, and finally to pure bending. Section 10–2 describes the theoretical approach used to derive interaction curves for steel members. Section 10–3 uses previous results to formulate interaction equations which are based on strength for relatively stocky steel members. In sec. 10–4 interaction equations, to preclude instability failures due to insufficient stiffness, are developed for both steel and reinforced concrete. Sections 10–5 and 10–6 illustrate practical design and analysis of steel and reinforced concrete beam columns, respectively. Section 10–7 shows how column equations for reinforced concrete are derived; and sec. 10–8 discusses the historically important secant formula.

10-1   AN INTERACTION CURVE

A most interesting and easily visualized interaction curve can be constructed for a given reinforced concrete member, neglecting slenderness considerations. In this section, such a curve is built for a column as the load changes from pure compression through various combinations of axial load and moment to pure bending. The section is composed of concrete with $f'_c = 3000$ psi and reinforcing $f_y = 36$ ksi. Dimensions are width $b = 12$ in., and overall depth $t = 24$ in. In the narrow faces, 3-#9 bars are placed, furnishing (1) an area of tensile reinforcement $A_s = 3$ in.², (2) an area of compressive reinforcement $A'_s = 3$ in.², for (3) a total area of longitudinal reinforcing $A_{st} = 6$ in.². Cover is $1\frac{1}{2}$ in., ties are #3 bars, and therefore $d' = 1\frac{1}{2} + \frac{3}{8} + \frac{9}{16} = 2\frac{7}{16} = 2.44$ in. From this, $d = 24 - 2.44 = 21.56$ in. Bending is about the axis having the largest moment of inertia.

The development is made with formulas from ACI-63-1902, derived later in this chapter. The results are shown, with appropriate equations and eccentricities, in Fig. 10-1.1. Computation of each necessary value is discussed before considering the interaction curve as a whole in sec. 10-1.5.

10-1.1 No Eccentricity. ACI-63 formula 19-7 was discussed in sec. 3-5.3. This equation

$$P_o = \phi\left[0.85 f'_c (A_g - A_{st}) + A_{st} f_y\right]$$

$$\begin{array}{r}(10\text{-}1.1)\\ [3\text{-}5.3]\end{array}$$

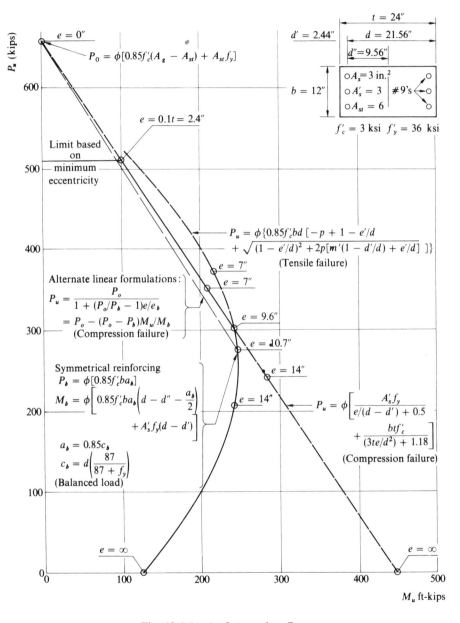

Fig. 10-1.1   An Interaction Curve

indicates the total axial force which could be carried if no eccentricity were present. This is *never* used as a design equation, because a minimum eccentricity is always tacitly assumed. But when Eq. (10-1.1) is evaluated,

$$P_o = 0.7[0.85(3)(12 \times 24 - 6) + 6(36)] = 655 \text{ kips},$$

supplies an upper point on the *linear* compression failure curve, as plotted for $e = 0$ in. in Fig. 10-1.1.

10-1.2 P_u, Compression Failure. To compute an allowable P_u for a compression failure, ACI-63 formula 19–10 is used

$$P_u = \phi \left[\frac{A'_s f_y}{[e/(d - d')] + 0.5} + \frac{btf'_c}{(3te/d^2) + 1.18} \right]. \qquad \begin{matrix} (10\text{-}1.2) \\ [3\text{-}5.6] \end{matrix}$$

For this example $\phi = 0.7$, $A'_s = 3.00$ in.^{2}, $f_y = 36$ ksi, e varies, $d = 21.56$ in., $d' = 2.44$ in., $b = 12$ in., $t = 24$ in., and $f'_c = 3$ ksi.

The eccentricity e is measured from the "plastic centroid," determined by the position of the resultant force produced by stress blocks assuming all concrete at $0.85f'_c$ and all steel at f_y, both in compression. For a symmetrical section, therefore, the plastic centroid is simply the centroid of the cross section. A nonsymmetrical section is considered in sec. 10-1.3 and illustrated in Fig. 10-1.2. If $e = 0$,

$$P_u = 0.7[(3)(36)/0.5 + (12)(24)(3)/1.18] = 664 \text{ kips}.$$

This differs from the 655 kips previously obtained because gross concrete area is used in Eq. (10-1.2) and net area in Eq. (10-1.1). (The diligent student will find that an exact comparison can be made only if $1/1.18$ is replaced by its parent value 0.85.) If $e = 7$ in.,

$$P_u = 0.7(125 + 382) = 354 \text{ kips}.$$

The corresponding M_u, equal to P_u times the eccentricity e, is therefore $354(7)/12 = 207$ ft-kips. Equivalent computations for eccentricities of 2.4 in. and 14 in., respectively, are

$$P_u = 0.7(173 + 557) = 511 \text{ kips}$$

and

$$P_u = 0.7(88 + 258) = 242 \text{ kips}.$$

As e approaches ∞, P_u approaches zero and M_u approaches 446 ft-kips. This can be verified by using L'Hôpital's rule for

$$\lim_{e \to \infty} M_u = \lim_{e \to \infty} P_u e = \lim_{e \to \infty} 0.7 \left[\frac{A'_s f_y e}{[e/(d - d')] + 0.5} + \frac{btf'_c e}{[(3te)/(d^2) + 1.18]} \right],$$

so that, taking the derivative of both the numerator and denominator with respect to e

$$\lim_{e \to \infty} 0.7 \left[\frac{A'_s f_y}{[1/(d - d')]} + \frac{btf'_c}{[(3t)/(d^2)]} \right] =$$

$$\lim_{e \to \infty} M_u = 0.7 \left[A'_s f_y (d - d') + bd^2 f'_c /3 \right]$$

$$= 0.7 \left[(3)(36)(19.12) + 12(21.56)^2(3)/3 \right] /12$$

$$= 446 \text{ ft-kips.}$$

All compressive failures plot on the straight line connecting $P_u = 664$ kips, $M_u = 0$, and $P_u = 0$, $M_u = 446$ ft-kips, implying that the compressive failure relationship between P_u and M_u is linear. Because of this, two alternate linear formulations given in ACI-63 are shown plotted on Fig. 10-1.1. The two lines are not identical because of different definitions for P_b and M_b (this is discussed further in secs. 10-1.4 and 10-1.5).

10-1.3 Tensile Failure. To compute the allowable axial load for a tensile failure, ACI-63 formula 19-5 is used:

$$P_u = \phi \left\{ 0.85 f'_c bd \left[-p + 1 - e'/d + \sqrt{(1 - e'/d)^2 + 2p[m'(1 - d'/d) + e'/d]} \right] \right\}$$
$$(10\text{-}1.3)$$

where reinforcing is symmetrical and only in two faces, and e' is defined as the eccentricity *measured from the centroid of the tensile reinforcement*, and where $p \equiv A_s/bd$ ($A_s \equiv$ area of *tensile* steel only), $m \equiv f_y/0.85f'_c$, and $m' \equiv m - 1$.

Before proceeding, it is necessary to distinguish clearly between e and e'. The value of e', as illustrated in Fig. 10-1.2, depends upon the value of e, and for nonsymmetrical sections the plastic centroid must be located before e can be determined. In this example, plastic centroid is located by the familiar (sec. A-2) tabular computation, where V represents the volume of the stress block at $0.85 f'_c$ for the concrete and f_y for the steel. Typically, the 2-#7 bars have an area of 1.2 in? at a stress of 40 ksi for a volume of 48 kips, acting 2 in. above the base axis for a first moment of 96 in-kips. In general, M_u and P_u are known, so that $e = 12.32$ in. is obtained as their quotient. Then, $e' = e + d'' = e +$ *distance between base axis and centroid of tensile steel* $- y = 12.32 + 6 - 0.88 = 17.44$ in. When reinforcing is symmetrical, e' is computed simply as $e + t/2 - d'$. Remember: e is measured from plastic centroid and e' from centroid of tensile steel.

For the given column, $d' = 2.44$ in., $d = 21.56$ in., $p = 3/(12)(21.56) = 0.0116$, and $m' = 13.12$. For an eccentricity of 7 in, $e' = 7 + 24/2 - 2.44 = 16.56$ in. and

$$P_u = 0.7 \left\{ 0.85(3)(12)(21.56) \left[0.220 + \sqrt{0.341} \right] \right\} = 372 \text{ kips}$$

so

$$M_u = 372(7/12) = 217 \text{ ft-kips.}$$

With an eccentricity of 14 in., $e' = 23.56$ in. and

$$P_u = 0.7\left\{0.85(3)(12)(21.56)\left[-0.104 + \sqrt{0.304}\right]\right\} = 206 \text{ kips.}$$

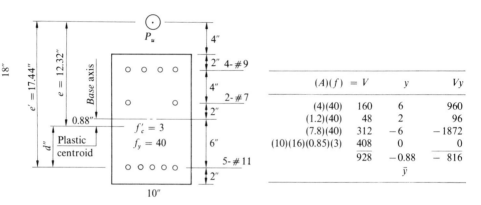

Fig. 10-1.2   Plastic Centroid; Values of e versus e' Plotted

Although the formula looks complicated, the same results can be obtained quite readily by statics. (The same is *not* possible for a compression-type failure because the stress in the *tensile* steel is unknown.) At the time of a tensile failure, both compression and tension steel are at yield while the concrete in compression can be represented by Whitney's stress block, with concrete at $0.85 f_c'$ (possibly modified for high strength concrete).

In Fig. 10-1.3, resultants of the stress blocks are shown, as well as P_u which is located 2 in. beyond the compressive face. Static summation of forces shows that equilibrium is maintained if

$$-P_u + C_s + C_c - T = 0,$$

where $P_u \equiv$ ultimate load, $C_s \equiv$ volume of stress block for compressive reinforcing steel, $C_c \equiv$ volume of stress block for compressive concrete, and $T \equiv$ volume of stress block for tensile reinforcing steel. Or, because $C_s = T$ (both have the same area of steel and both at yield point),

$$P_u = C_c = 0.85 f_c' ab$$

from which $a = P_u/30.6.$

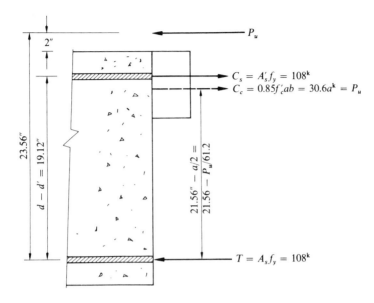

Fig. 10-1.3 Tension Failure by Statics

When moments about the tensile steel are summed, $-P_u(23.56) +$ 108(19.12) + $C_c(21.56 - a/2) = 0$ (note $C_c = P_u$), which is equivalent to setting the sum of two moment couples equal to zero: couple of steel, 108(19.12) − couple of load and concrete in compression, $P_u(2 + a/2) = 0$. In either case, replacing a by $P_u/30.6$, we have

$$108(19.12) - 2P_u - P_u^2/61.2 = 0$$

so that $P_u = 300$ kips. However, ϕ has not been included so that the allowable P_u is 0.7(300) = 210 kips, compared to 206 kips by Eq. (10-1.3). Note that eccentricity is $2 + t/2 = 14$ in.

As e approaches infinity, P_u approaches zero and M_u approaches 121 ft-kips. This is illustrated by the following table (note that 19,000 in. is 0.3 mile):

e (in.)	P_u (kips)	M_u (ft-kips)
170	9.205	130.4
1,800	0.816	122.4
19,000	0.0762	120.6

As a comparison, the allowable bending moment for this column analyzed as a doubly reinforced beam is

$$M_u = \phi A_s f_y(d - d')/12 = \phi(3)(36)(19.12) = 172\phi \text{ ft-kips.}$$

In accord with column theory, if $\phi = 0.7$ rather than 0.9,

$$M_u = 0.7(172) = 120.4 \text{ ft-kips.}$$

This agrees with results obtained using Eq. (10-1.3) with eccentricity approaching infinity.

10-1.4 Balanced Load. A column usually fails by *either* compression or tension. The point at which the failure may be of either type is called the *balanced load condition*, represented by the special symbols P_b and M_b.

The general equations for P_b and M_b are ACI-63 formulas 19-1 (with *a* replaced by a_b and f_s by f_y) and 19-3:

$$P_b = \phi [0.85 f_c' b a_b + A_s' f_y - A_s f_y] \tag{10-1.4}$$

and

$$M_b = \phi [0.85 f_c' b a_b (d - d'' - a_b/2) + A_s' f_y (d - d' - d'') + A_s f_y d''], \tag{10-1.5}$$

where $d'' \equiv$ the distance from the plastic centroid to the centroid of tensile steel and $a_b \equiv 0.85 c_b = 0.85(87d)/(87 + f_y)$, that is, depth of the equivalent rectangular stress block for balanced conditions.

With $A_s' = A_s$, the resulting computations, shown plotted on Fig. 10-1.1, are

$$P_b = 0.7[0.85(3)(12)0.85(87)(21.56)/(87 + 36)] = 277 \text{ kips}$$

and

$$M_b = 248 \text{ ft-kips, with } a_b = 12.96 \text{ and } d'' = 9.56.$$

Eccentricity e_b is $12(248)/277 = 10.7$ in. Therefore, according to code criteria, if the eccentricity is less than e_b Eq. (10-1.2) for compression failure is to be used; otherwise Eq. (10-1.3) for tensile failure is used.

The ACI-63 Commentary suggests that e_b can be determined by

$$e_b = (0.20 + 0.77 \, p_t m) t, \tag{10-1.6}$$

where $p_t \equiv A_{st}/bt$ and $m \equiv f_y/0.85 f_c'$. In this case, then

$$e_b = \left[0.20 + \frac{0.77(6)(36)}{(12)(24)(0.85)(3)} \right] 24 = 10.3 \text{ in.}$$

Therefore, two options are available for direct computations: (1) Compute approximate e_b to determine which equation to use, or (2) calculate with both equations and use the smaller P_u obtained. The second procedure leads to a slightly different defining point for P_b, M_b, located at the intersection of curves plotted from Eqs. (10-1.2) and (10-1.3).

10-1.5 The Interaction Curve. To be most useful, an interaction curve must be constructed in nondimensional form. This may be accomplished by dividing P_u by $\phi f'_c bt = (0.7)(3)(12)(24) = 604.8$ and dividing M_u by $\phi f'_c bt^2/12 = (0.7)(3)(12)(24)^2/12 = 1209.6$, where the denominator of 12 converts the answer to inch-pounds from foot-pounds.

The use of critical values from previous computations and addition of $e = 9.6$ in. yields the results shown in Table 10-1.1. These values of K and Ke/t are plotted for the $p_t m$ ratio of $(A_{st}/bt)(f_y/0.85f'_c) = [6/(12)(24)][36/(0.85)(3)] = 0.294$ on the $d/t = 0.9$ curves of Fig. 10-1.4 by crosses. Obviously, a number of calculations of this sort were required to produce the family of interaction curves shown in the figure.

Table 10-1.1 Summary of Computations for Fig. 10-1.1

e (in.)	e/t	P_u (kips)	M_u (ft-kips)	Comp	Bal	Tens	K	Ke/t
0	0	664	0	X			1.097	0
2.4	0.1	511	102	X			0.845	0.084
7.0	0.29	354	207	X			0.585	0.171
9.6	0.4	301	241	X		X	0.498	0.199
10.7	0.45	277	248		X		0.458	0.205
14.0	0.58	206	240			X	0.341	0.198
0	∞	0	121			X	0	0.100

A study of such interaction diagrams can provide useful insight into the load-carrying capacity of columns. Some observations are:

1. As the percent of reinforcing steel goes up, the column requires a larger eccentricity to change from compressive to tensile-type failure mode.
2. The actual balance point used in the charts is found by equating compression-type and tension-type failures ($e = 9.6$ in Fig. 10-1.1) rather than by P_b and M_b equations (indicating $e = 10.7$ in. in Fig. 10-1.1).
3. A given column can carry more bending moment if some axial load is present.
4. It appears that for tensile failures, the value of ϕ should vary so that with $e = \infty$, $\phi = 0.9$. See provisions of ACI 9-2.1.2 and the design portion of this chapter for such a procedure.

10-1.6 An Interaction Equation. When actual axial load and column moments are divided by their corresponding allowables, an interaction equation is obtained:

$$\frac{P_u}{P_{\text{allow}}} + \frac{M_u}{M_{\text{allow}}} \leqq 1.0, \tag{10-1.7}$$

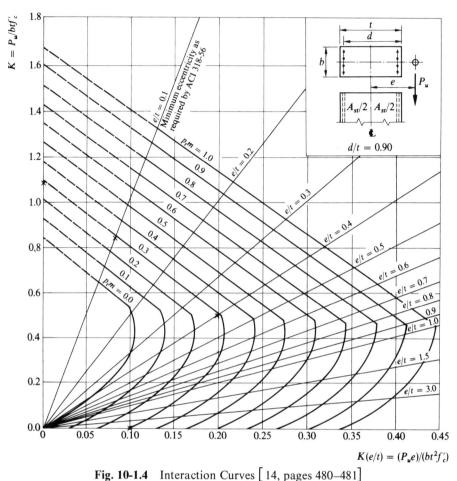

Fig. 10-1.4   Interaction Curves $\left[\, 14,\ \text{pages}\ 480\text{--}481 \,\right]$

(a) Bending and axial load—$d/t = 0.90$; rectangular sections with symmetrical reinforcement

as suggested by Fig. 10-1.5. Without substantial modification, this equation is valid only for compressive-type failure, and the procedure is seldom used directly in conjunction with reinforced concrete. However, an equation of this form is developed in the next section for steel structures.

10-2 DEVELOPMENT OF INTERACTION CURVES FOR STEEL MEMBERS ANALYZED PLASTICALLY

Interaction curves which develop moment curvature relationships as a function of axial thrust and residual stress form the basis for steel beam-column formula-

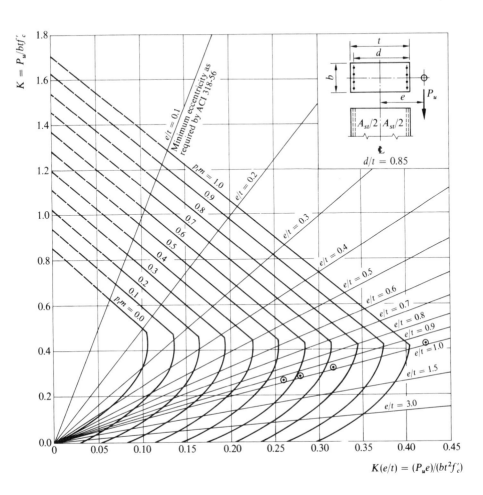

(b) Bending and axial load—$d/t = 0.85$; rectangular sections with symmetrical reinforcement

tions. Results shown in Figs. 10-2.7 and 10-2.10 are dependent upon two main steps.

Step 1 is embodied in secs. 10-2.1 through 10-2.3, and culminates in construction of Fig. 10-2.6, originally presented by Ketter, Kaminsky, and Beedle [1]. First, a method to obtain ϕ, P, and M for a specific member and stress distribution is developed. From these relationships, an auxiliary curve is then plotted. By appropriate use of this plot, desired moment-thrust-curvature interaction is illustrated.

Step 2, sec. 10-2.4, based on work by Galambos and Ketter [2], uses these results to obtain concentrated angle change corresponding to known (or

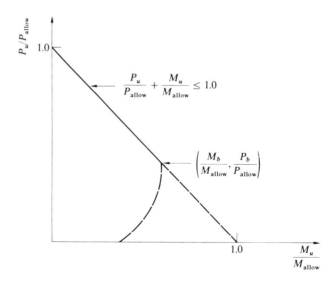

Fig. 10-1.5 Interaction Equation

assumed) bending moments. Then integration by use of a numerical technique finds point slopes and deflections which satisfy member configuration and loading. From point deflections at intermediate points, slope at the beam end is computed. These values plotted for a particular ratio of L/r and P/P_y with specified end condition eventually supplies a critical M/M_y ratio. With a sufficient number of the latter points, the final curves may be plotted.

10-2.1 Relationship of ϕ, P, and M. Basic to the following discussions are assumptions that planes remain plane after bending (the Bernoulli-Navier hypothesis) and that materials are perfectly elasto-plastic.

With these assumptions, member rotation can be determined by relating extreme fiber strains to overall depth (see Fig. 10-2.1) so that for small angles

$$\phi = \tan \frac{(\epsilon_1 - \epsilon_2)}{d} \doteq (\epsilon_1 - \epsilon_2)/d, \qquad (10\text{-}2.1)$$

where $\phi \equiv$ member rotation, $\epsilon_1 \equiv$ *positive* strain in top fiber, $\epsilon_2 \equiv$ *negative* strain in bottom fiber, and $d \equiv$ overall depth of beam

When yielding has occurred, Eq. (10-2.1) must be altered to

$$\phi = (\epsilon_y - \epsilon_2)/(d - y_{cc}), \qquad (10\text{-}2.2)$$

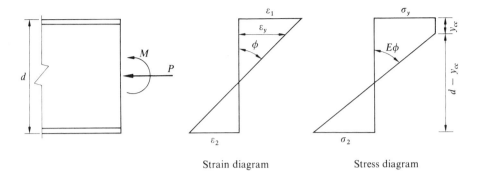

Fig. 10-2.1   Typical Strain and Stress Diagrams

where $y_{cc} \equiv$ depth of compressive yielding. Then, utilizing the relationship $\epsilon = \sigma/E$, a form dependent upon stress rather than strain is obtained:

$$\phi = (\sigma_y - \sigma_2)/[E(d - y_{cc})]. \qquad (10\text{-}2.3)$$

Thus, for any given section with the value of y_{cc} and σ_2 set, ϕ is readily determined. With these values all known, the resultants of the stress blocks can also be easily computed as demonstrated in Chap. 2. With forces summed,

$$P = C - T, \qquad (10\text{-}2.4)$$

where $P \equiv$ concentrated axial thrust, $C \equiv$ resultant of compressive stress block, and $T \equiv$ resultant of tensile stress block. See Fig. 10-2.2.

The distance between P and C can be determined by breaking C into two components C_1 and C_2 equal, respectively, to T and P; Summing moments; we have $T(jd) - P(e'') = 0$ or $e'' = T(jd)/P$. With e'' known, the true eccentricity e and the resulting moment $M = Pe$ can be determined.

In summary, for any given section, when a reasonable stress distribution is assumed, the corresponding values of ϕ, P, and M can be found rather easily.

10-2.2 Auxiliary Curves to Particular Moment Curvature Relationship. Ketter, Kaminsky, and Beedle [1] developed the moment curvature relationship for a W8 x 31. Because of its low shape factor (lower than nearly all other rolled WF sections), this member provides a conservative solution which can be applied universally.

When a W8 x 31 has yielded to one-quarter of its depth ($y_{cc} = d/4$), as shown in Fig. 10-2.3, incrementing the value of σ_2 produces corresponding

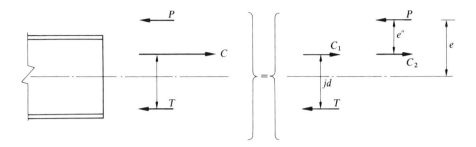

Fig. 10-2.2   Resolution for $P = C - T$ and $M = Pe$

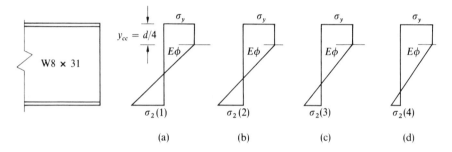

Fig. 10-2.3   Variations of P and M with Constant y_{cc} (a)   (b)   (c)   (d)

values of ϕ, P, and M which can be computed. These values are plotted in curve 3 of Fig. 10-2.4. By using different values of y_{cc}, sufficient values of ϕ, P, and M can be obtained to plot curves that correspond to each particular y_{cc} such as those shown in Fig. 10-2.4. To generalize these curves so that they may apply to other rolled sections, values of ϕ, P, and M are all nondimensionalized. Values of ϕ are divided by ϕ_y (curvature corresponding to initial outer fiber yielding under plastic bending moment, without axial load), P by P_y (axial load corresponding to yield stress over entire section, without bending moment), and M by M_y (bending moment which initiates yield point in extreme fiber, without axial force).

Figure 10-2.4 shows that for any particular ratio of P/P_y, unique values of M/M_y exist which correspond to all possible ratios of ϕ/ϕ_y. In this figure, the $M - \phi$ curve for $P = 0.2P_y$ is illustrated. Broken lines originate from the intersection of the horizontal line $P/P_y = 0.2$ and the five curves which relate P/P_y

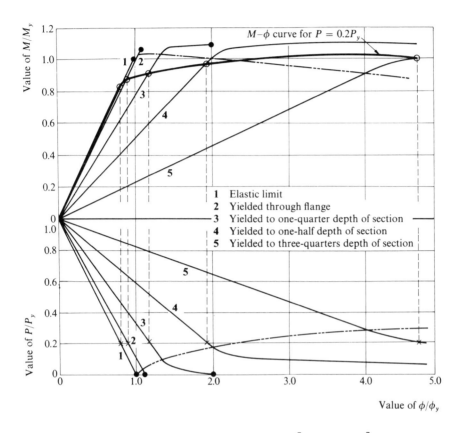

Fig. 10-2.4 Typical Auxiliary Curves [1, page 1038]

and ϕ/ϕ_y for various yield depths (X's on the figure). Intersections of these vertical broken lines and M/M_y–ϕ/ϕ_y curves with corresponding yield depths (heavy circles on the figure) form the locus through which the M–ϕ curve (heavy solid line) is drawn. This single curve is reproduced in Fig. 10-2.5 with intersecting points (shown as circles) corresponding to the circles of Fig. 10-2.4.

10-2.3 Moment–Thrust–Curvature Relationships. The previous procedure was repeated starting at different P/P_y ratios and the curves of Fig. 10-2.6 [2] were developed. When stress diagrams were altered in accordance with residual stresses (for a typical residual stress pattern, see Fig. 3-2.2), the curves represented by broken lines in Fig. 10-2.6 were obtained.

10-2.4 A Plastic Interaction Curve. To construct interaction curves for plastic capacity, it is necessary to assume (1) type of end moments, (2) an L/r ratio,

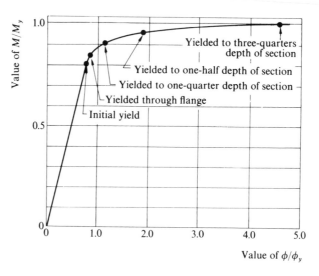

Fig. 10-2.5   Moment Curvature Relationship $(P = 0.2\,P_y)$ [1, page 1039]

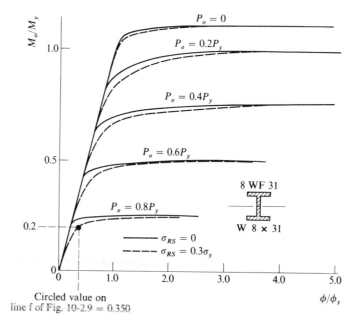

Circled value on
line f of Fig. 10-2.9 = 0.350

Fig. 10-2.6   Moment–Thrust–Curvature Relationships [2, page 3]

and (3) a ratio of P/P_y, for each point to be plotted. Using these parameters, the critical value of M/M_p is obtained by numerical integration.

To illustrate construction of interaction curves such as those of Fig. 10-2.7 for a beam column hinged at one end and subject to a bending moment at the other end, the computations required to obtain the single circled point (0.233, 0.8) are outlined in the following discussion.

The single point circled on Fig. 10-2.7 is obtained as the critical value of M_o/M_y, as shown in Fig. 10-2.8, for a specific configuration with an L/r of 40 and a P_o/P_y ratio of 0.8. The critical value of 0.233 can be found by calculating sufficient points to plot the curve or by a mathematical search technique.

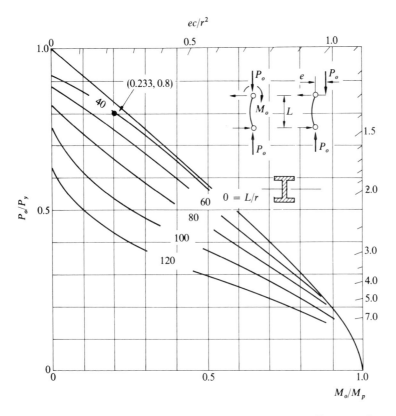

Fig. 10-2.7   Interaction Curve, Plastic, One End Hinged [2, page 8]

Typical computations, in this case leading to the circled point on Fig. 10-2.8 are shown in Fig. 10-2.9. End conditions are stated by moment relationships as $M_o/M_y = 0.2$ at the left-hand end and $M_o/M_y = 0$ on the right-hand end (a hinged end). The ratio P_o/P_y is set at 0.8, L/r_x at 40, and the member is

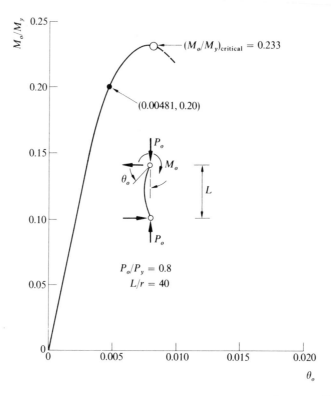

Fig. 10-2.8 Moment versus End Rotation (Typical) [2, page 6]

specified as a W8 **x** 31 (nomenclature of *Steel Construction*, 6th ed. was 8 WF 31).

While not complicated, calculations in Fig. 10-2.9 are tedious. Line a shows the bending moment at each end of eight equal segments of the beam due to an end moment of 0.2 M_y = 181 in.-kips (linear variation). Line b shows a set of *assumed* deflections. The product of this assumed deflection and the axial load causes an additional bending moment at each point, recorded on line c. The total bending moment, line d, is simply the sum of lines a and c, which is reduced to M_x/M_y in line e when divided by M_y = 181/0.2 = 905.

With M_x/M_y known at the end of each segment, corresponding values of ϕ/ϕ_y can be read from Fig. 10-2.6 (made to larger scale). These ϕ/ϕ_y represent concentrated angle change coefficients. For simplicity they are used with $\lambda\phi_y$ (λ is segment length, $L/8$) factored out as noted in the column labeled MF.

Angle changes from line f are accumulated to obtain the slope at each point (line g). The summing sequence is noted on the figure by arrows. Result-

Given: $L/r_x = 40$; $P_o/P_y = 0.8$; W8 × 31 section; $L = 138.8$ in.; $\lambda = 17.35$ in.

Assumed: $M_o/M_y = 0.2$

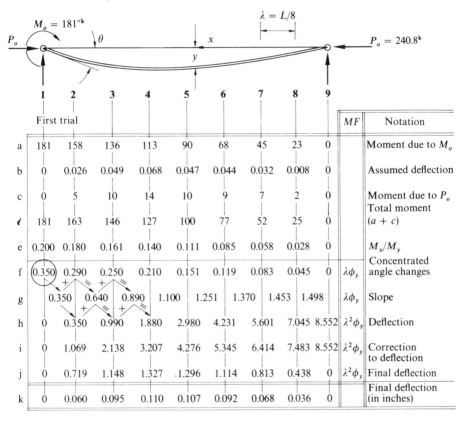

	1	2	3	4	5	6	7	8	9	MF	Notation
	First trial										
a	181	158	136	113	90	68	45	23	0		Moment due to M_o
b	0	0.026	0.049	0.068	0.047	0.044	0.032	0.008	0		Assumed deflection
c	0	5	10	14	10	9	7	2	0		Moment due to P_o
d	181	163	146	127	100	77	52	25	0		Total moment $(a + c)$
e	0.200	0.180	0.161	0.140	0.111	0.085	0.058	0.028	0		M_x/M_y
f	0.350	0.290	0.250	0.210	0.151	0.119	0.083	0.045	0	$\lambda\phi_y$	Concentrated angle changes
g		0.350	0.640	0.890	1.100	1.251	1.370	1.453	1.498	$\lambda\phi_y$	Slope
h	0	0.350	0.990	1.880	2.980	4.231	5.601	7.045	8.552	$\lambda^2\phi_y$	Deflection
i	0	1.069	2.138	3.207	4.276	5.345	6.414	7.483	8.552	$\lambda^2\phi_y$	Correction to deflection
j	0	0.719	1.148	1.327	.1.296	1.114	0.813	0.438	0	$\lambda^2\phi_y$	Final deflection
k	0	0.060	0.095	0.110	0.107	0.092	0.068	0.036	0		Final deflection (in inches)

Fig. 10-2.9 Typical Numerical Integration to Calculate Deflections [2, page 6]

ing deflections are obtained in line h when the slope for each segment is multiplied by the segment length and accumulated again. Mathematically this is numerical intergration; in structural analysis the technique is often called the *conjugate beam method*.

The last value on line h, which is 8.552, should be zero because of a simple hinge at the right support. Correction is made by rigid body rotation by means of the alterations $(1/8)(8.552) = 1.069$, $(2/8)(8.552) = 2.138$, etc., shown on line i. Final deflection coefficients (line j) are the difference between lines i and h. When each coefficient of line j is multiplied by $\lambda^2\phi_y$, resulting deflections are obtained (line k).

Deflection *computed* on line k and the values *assumed* on line b do *not* agree, therefore, the process must be repeated to obtain a second tabular solution by use of the computed values of line k as the *new* assumed values for line b. This entire procedure is repeated again and again until assumed and computed deflections are in reasonable agreement. Galambos and Ketter report that agreement was achieved on the fourth trial for this configuration, with deflections of the first three points: $y_o = 0$ (at $x = 0$), $y_1 = 0.070$ (at $x = \lambda$), and $y_2 = 0.113$ (at $x = 2\lambda$). Slope at the end of the beam-column was then determined by a finite difference equation†

$$\theta = [4(0.070) - 0.113]/2\lambda = 0.00481. \qquad (10\text{-}2.5)$$

The coordinates, $(\theta, M_o/M_y)$ or $(0.00481, 0.20)$ are plotted as a heavy circle on Fig. 10-2.8.

New points on the curve of Fig. 10-2.8 are determined by incrementing M_o/M_y. Thus, with patience and fortitude, the final critical value represented by $M_o/M_y = 0.233$ was determined—representing the single circled point on Fig. 10-2.8.

A moment's reflection will suggest the tremendous effort needed to obtain these interaction curves; their importance will become evident in following sections. Although such curves have been obtained for various end conditions, only one more is reproduced here, in Fig. 10-2.10. Substantial testing has verified relations shown on these interaction curves and they form the basis for both plastic and elastic design formulations.

10-3 INTERACTION EQUATIONS—GENERAL YIELDING

The interaction curve of a beam in single curvature, with $L/r = 0$, can be approximated by a straight line (see Fig. 10-2.10) and expressed as:

$$\frac{M}{M_p} = 1.18(1 - \frac{P}{P_y}), \qquad (10\text{-}3.1)$$

where $M \equiv$ actual bending moment, $M_p \equiv$ maximum allowable plastic bending moment, $P \equiv$ actual axial load, and $P_y \equiv$ load which would cause member to yield without flexure. At any point where the member has sufficient lateral

---

† A simple first forward-difference expression for beam slope would be $(y_1 - y_o)/\lambda$, which for small angles approximates the first derivative of the deflection curve. Substituting numerical values, $\theta = (0.070 - 0)/17.35 = 0.00403$. However, to obtain more accurate results, the second forward-difference expression $\theta = (-y_2 + 4y_1 - 3y_o)/2\lambda$ was used.

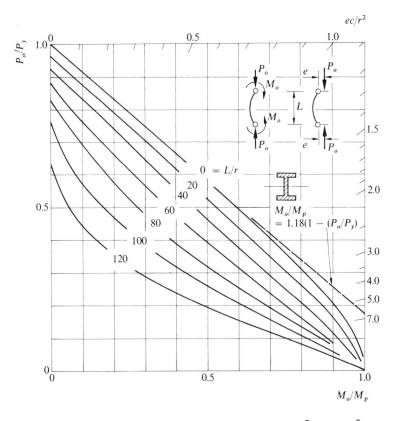

Fig. 10-2.10 Interaction Curve—Single Curvature $\begin{bmatrix}2,\ \text{page } 8\end{bmatrix}$

support ($L/r = 0$), Eq. (10-3.1) can be reordered to plastic design interaction formula 2.4-3 of AISC:

$$\frac{P}{P_y} + \frac{M}{1.18 M_p} \leq 1, \qquad M < M_p, \tag{10-3.2}$$

where both P and M have been multiplied by a load factor of 1.7. Physically M cannot exceed the allowable plastic moment, but this also accounts for the actual curvature which exists in the $L/r = 0$ curve of Fig. 10-2.10, because as M approaches M_p the second term $M/1.18 M_p$ approaches a maximum value of 0.85.

In allowable stress design, the nonlinearity cannot be handled as directly. To be conservative, Eq. (10-3.2) is rewritten as

$$\frac{P}{P_y} + \frac{M}{M_p} \leq 1.0, \tag{10-3.3}$$

which represents a straight line from $P_o/P_y = 0$ to $P_o/P_y = 1$. For use, Eq. (10-3.3) is rearranged further; axial forces are divided by cross-sectional area, flexural moments by section moduli, and denominators now include the factor of safety:

$$\frac{f_a}{F_a} + \frac{f_b}{F_b} \leq 1.0, \tag{10-3.4}$$

where $f_a \equiv$ actual axial stress, $F_a \equiv$ allowable axial stress in absence of bending, $f_b \equiv$ actual flexural stress, and $F_b \equiv$ allowable flexural stress in absence of axial force.

When f_a/F_a is less than 0.15, and by addition of a term to consider biaxial bending, AISC formula 1.6-2 is obtained as:

$$\frac{f_a}{F_a} + \frac{f_{bx}}{F_{bx}} \frac{f_{by}}{F_{by}} \leq 1.0, \qquad \left[\frac{f_a}{F_a} \leq 0.15 \right], \tag{10-3.5}$$

where subscripts refer to the axis about which bending occurs. As f_a/F_a exceeds 0.15, the formula still applies if $L/r = 0$, which occurs at points of lateral support. In such cases, failure is by general yielding rather than instability and F_a is replaced by 0.6 F_y (AISC formula 1.5-1 with $L = 0$). Concurrently, f_b is computed by using the larger (absolute value) of the two moments at the points of lateral support to obtain AISC formula 1.6-1b:

$$\frac{f_a}{0.60 F_y} + \frac{f_{bx}}{F_{bx}} + \frac{f_{by}}{F_{by}} \leq 1.0, \qquad \left[\frac{L}{r} \doteq 0 \right]. \tag{10-3.6}$$

Equations (10-3.2) (plastic design), and (10-3.5) and (10-3.6) (allowable stress design) are used when the possibility of lateral torsional buckling is small. In a crude sense, their use compares to the design of reinforced concrete beam-columns in which slenderness effects can be neglected.

10-4   INTERACTION EQUATIONS—INSTABILITY

As the distance between points of lateral support increase, flexural terms in both AISC and ACI formulations are multiplied by a *moment magnification factor*, called δ in the ACI Code. This factor is composed of two parts:

1. An amplification factor which adjusts the bending moment so that it includes secondary moments generated by the axial force acting through an eccentric "arm" (see Figure 3-1.1 as an example).

2. A reduction factor which adjusts the bending moment so that it (a) includes stability against sidesway and (b) considers nature of loading between points of support.

10-4.1 Amplification Factor. The curves of Fig. 10-2.10 can be represented adequately if Eq. (10-3.2) is rewritten as:

$$\frac{P}{P_{CR}} + \frac{1}{[1 - (P/P_e)]} \frac{M}{M_m} \le 1, \tag{10-4.1}$$

where $P \equiv$ axial load including load factors; $M \equiv$ maximum applied moment including load factor; $P_{CR} \equiv 1.7 \, AF_a$ (critical column load), where $A \equiv$ gross area of member, and $F_a \equiv$ allowable stress for concentrically loaded column [Eqs. (3-2.9) and (3-2.10); AISC formulas 1.5-1 and 1.5-2]. P_e is critical Euler buckling load suitably modified.

If the member is braced in the weak direction,

$$M_m \equiv M_p;$$

if unbraced in the weak direction,

$$M_m \equiv \left(1.07 - \frac{(l/r_y)\sqrt{F_y}}{3160}\right) M_p \le M_p.$$

The amplification factor is

$$\frac{1}{[1 - (P/P_e)]}, \tag{10-4.2}$$

where

$$P_e \equiv \frac{23A \, F'_e}{12},$$

the Euler critical buckling load is equal to $(\pi^2 EA)/(Kl_b/r_b)^2$ and

$$F'_e \equiv \frac{12\pi^2 E}{23(Kl_b/r_b)^2}.$$

For allowable stress design, including a factor of safety and comparing stresses directly, Eq. (10-4.1) is rewritten (see AISC formula 1-6.1a):

$$\frac{f_a}{F_a} + \frac{1}{[1 - f_a/F'_e]} \frac{(f_b)}{(F_b)} \le 1, \tag{10-4.3}$$

where the amplification factor is

$$\frac{1}{1 - \dfrac{f_a}{F'_e}}. \tag{10-4.4}$$

In the ACI formula, the amplification factor is written as

$$\frac{1}{1 - \dfrac{P_u}{\phi P_c}}, \tag{10-4.5}$$

where $P_c \equiv (\pi^2 EI)/(k\ell_u)^2 =$ Euler critical buckling load and $P_u \equiv$ axial load including load factor.

10-4.2 Reduction Factor, C_m. The amplification factor was obtained from Fig. 10-2.10 for a member of single curvature, its maximum deflection occurring near mid-span. For other configurations having reduced eccentricity, the amplification factor would magnify the moments too much. Therefore, C_m is always less than or equal to unity.

For frames not braced against sidesway, C_m is specified as 1.0 by the ACI Code and 0.85 by the AISC Code.

For frames not braced against sidesway which have transverse loading on the span, ACI specifies 1.0 whereas AISC allows either (a) a rational analysis, or (b) $C_m = 0.85$ if ends of members are unrestrained, or (c) $C_m = 1.0$ if member ends are unrestrained.

By rational analysis, AISC defines

$$C_m \equiv 1 + \psi \frac{f_a}{F'_e}, \qquad (10\text{-}4.6)$$

where

$$\psi \equiv \frac{\pi^2 \delta_o EI}{M_o L^2} - 1, \qquad (10\text{-}4.7)$$

and $\delta_o \equiv$ maximum deflection due to transverse load and, $M_o \equiv$ maximum moment between supports due to transverse loading.

The short ensuing discussion, as far as the repeat of Eq. (10-4.7), gives some insight into theoretical development of these equations. Timoshenko (*Elastic Stability*, 1.7 and 1.11) shows that, within 2% error for $P/P_{CR} < 0.6$,

$$\delta = \frac{\delta_o}{1 - P/P_{CR}}, \qquad (10\text{-}4.8)$$

where $\delta_o \equiv$ deflection due to lateral loads only; $\delta \equiv$ deflection due to lateral loads plus moment produced by axial load $= P\delta$; $P \equiv$ axial load; $P_{CR} \equiv$ Euler critical buckling load. On this basis, then [3]

$$M_{max} = M_o + P\delta = M_o + \frac{P\delta_o}{1 - P/P_{CR}}, \qquad (10\text{-}4.9)$$

where M_o is the moment due to lateral loads only, and P is the moment due to axial load at eccentricity δ. By rearrangement of Eq. (10-4.9) and the use of a preset expression for ψ,

$$M_{max} = M_o \left(1 + \frac{P\delta_o/M_o}{1 - (P/P_{CR})}\right) = \frac{M_o}{1 - (P/P_{CR})} \left(1 - \frac{P}{P_{CR}} + \frac{P\delta_o}{M_o}\right)$$
$$= \frac{M_o}{1 - (P/P_{CR})} \left(1 + \frac{\psi P}{P_{CR}}\right) \qquad (10\text{-}4.10)$$

the following expression for ψ is obtained:

$$1 - \frac{P}{P_{CR}} + \frac{P\delta_o}{M_o} = 1 + \psi \frac{P}{P_{CR}}$$

so

$$\psi = -1 + \frac{P_{CR}\delta_o}{M_o}. \tag{10-4.11}$$

In Eq. (10-4.10) the denominator corresponds to the amplification factor of Eq. (10-4.2). Then, the reduction factor is defined from the remainder of Eq. (10-4.10) when P is replaced by f_a and P_{CR} by F'_e of Eq. (10-4.4). Thus,

$$C_m = 1 + \psi \frac{P}{P_{CR}} = 1 + \psi \frac{f_a}{F'_e} \tag{10-4.6}$$

and with $F'_e = (\pi^2 EI)/L^2$ (no safety factor here) from Eq. (10-4.11)

$$\psi = -1 + \frac{\pi^2 EI\delta_o}{L^2 M_o} = \frac{\pi^2 \delta_o EI}{M_o L^2} - 1. \tag{10-4.7}$$

As an example of a typical calculation for ψ, consider a simply supported beam with a lateral load Q at mid-span and axial load P. Maximum moment due to lateral load only is $M_o = QL/4$. Maximum deflection caused by only Q is $QL^3/48EI$. Therefore

$$\psi = \frac{\pi^2 (QL^3/48EI)EI}{(QL/4)L^2} - 1 = 0.822 - 1 = -0.178.$$

In table C.1.6.1.2 of the AISC Commentary (*Steel Construction*, p. 5-133), where several loading cases are shown, this value is rounded to -0.2.

For frames not braced against sidesway which have no transverse loading, ACI Code specifies $C_m = 0.6 + 0.4 M_1/M_2 \geq 0.4$ and AISC Code specifies $C_m = 0.6 - 0.4 M_1/M_2 \geq 0.4$. Signs differ because ACI uses beam sign convention whereas AISC uses the frame sign convention (See sec. E-5). In both cases, M_1 has the smaller absolute value. To eliminate error note that C_m is greater than 0.6 when the member is in single curvature, less than 0.6 when in double curvature.

10-4.3 Magnification Factor. To summarize, the magnification factor δ is the product of an amplification factor and a reduction factor. Amplification factors are:

$$\frac{1}{1 - \dfrac{P}{P_e}} \quad \text{for plastic design in steel;} \tag{10-4.2}$$

$$\frac{1}{1 - \dfrac{f_a}{F'_e}} \text{ for allowable stress design, steel;} \qquad [10\text{-}4.4]$$

and

$$\frac{1}{1 - \dfrac{P_u}{\phi P_e}} \text{ for reinforced concrete by USD,} \qquad [10\text{-}4.5]$$

if $K\ell_u/r \geq 22$ for unbraced frames or if $K\ell_u/r \geq 34 - 12\, M_1/M_2$ for braced frames. (M_2 is the absolute value of the larger end moment).

The reduction factor C_m is defined as follows:

	By ACI	*By AISC*
Unbraced frames:	1.0	0.85
Braced frames:		
lateral load between supports:	1.0	$1 + \phi \dfrac{f_a}{F'_e}$
or, conservatively, member ends restrained:	—	0.85
or, conservatively, member ends unrestrained:	—	1.0
no lateral loads between supports:	$0.6 + 0.4\, M_1/M_2$ ≥ 0.4 (beam sign convention)	$0.6 - 0.4\, M_1/M_2$ ≥ 0.4 (frame sign convention)

For single curvature $C_m > 0.6$.
For double curvature $C_m < 0.6$.

10-4.4 Use of the Magnification Factor. ACI and AISC employ the magnification factor whenever instability failure is possible.

Equation (10-3.2), plastic design of steel, is modified to AISC formula 2.4-2 by the following:

1. changing P_y to P_{CR} (stability criterion);
2. conservatively dropping the 1.18 coefficient so that M can reach M_u;
3. replacing M_p by M_m to recognize the possibility of buckling about either axis;
4. adding the magnification factor to the bending term, as

$$\frac{P}{P_{CR}} + \frac{C_m M}{\left(1 - \dfrac{P}{P_e}\right) M_m} \leq 1, \qquad (10\text{-}4.12)$$

where terms are defined with Eq. (10-4.1) and C_m is specified in sec. 10-4.3. For allowable stress analysis, a similar transformation yields AISC formula 1.6-1a:

$$\frac{f_a}{F_a} + \frac{C_{mx}f_{bx}}{\left(1 - \dfrac{f_a}{F'_{ex}}\right)F_{bx}} + \frac{C_{my}f_{by}}{\left(1 - \dfrac{f_a}{F'_{ey}}\right)F_{by}} \leqq 1.0, \qquad (10\text{-}4.13)$$

where F_a is governed by maximum slenderness ratio, F'_{ex} is governed by slenderness ratio in the plane of bending (about X axis), F'_{ey} is governed by slenderness ratio in the plane of bending (about Y axis), f_b(x and y) is the largest flexural stress (absolute value) within the unsupported span, and F_b(x and y) use $C_b = 1$.

A reinforced concrete column is designed to carry P_u and M_u (both including load factors) so long as the effects of slenderness can be neglected ($K\ell_u/r \geqq 22$ or $34 - 12\, M_1/M_2$ for unbraced or braced frames, respectively). When slenderness must be considered, the column is designed for P_u and δM_u, where δ is the magnification factor summarized in sec. 10-4.3.

Considerable effort is being directed toward fostering more precise analytical techniques for frames so that results can be used without these modifications or considerations. Although such methods are mathematically practical when using electronic computers, two factors mitigate against widespread acceptance at an early date. One reason concerns design costs that might be disproportionately high and the other concerns physical parameters (particularly for reinforced concrete structures) that vary too widely to insure more reliable overall results (see secs. 2-8.2 and 2-8.3). At present, most office design is based upon the magnification factor concept.

Before either nomograph (Fig. 3-3.2) can be used to determine effective length for individual members in frames, it is first necessary to decide whether the frame is laterally supported or not. The ACI Commentary [4, page 41] suggests that a frame has sufficient lateral support from a shear wall, shear truss, etc., if these bracing elements have a total stiffness greater than six times the stiffness of the corresponding columns. By use of this concept, a building can be designed as if it were laterally supported, with final dimensions of the shear wall (or other stiffening device) established as the final step. For instance, if a certain floor has eighteen 14 in. x 16 in. columns and fourteen 16 in. x 20 in. columns (all long dimensions parallel), the shear wall(s) would require a stiffness parallel to the long column dimension of six times

$$18[14(16)^3/12] + 14[16(20)^3/12] = 235{,}400 \text{ in.}^4$$

or parallel to short column dimensions, six times

$$18[16(14)^3/12] + 14[20(16)^3/12] = 161{,}400 \text{ in.}^4$$

Assuming two parallel 8 in. thick shear walls (with heights identical to column lengths), the required wall length is found by equating shear wall I to six times column I:

$$2(8t^3/12) = (235,400)(6)$$

so $t = 102$ in. In the other direction, required length for two parallel 8 in. thick shear walls is 90 in. (rounded to nearest inch).

By ACI 10.11.5.1 it is required that a magnification factor δ be determined independently for each column and also for all columns acting as a unit for a floor. The larger of these two values is then used for each particular column. Floor δ is determined by replacing P_u and P_o by ΣP_u and ΣP_o, respectively in Eq. (10-4.5). By this method, both total floor and individual column stability are assured. Although not required by AISC, a similar technique could be readily applied to Eqs. (10-4.2) and (10-4.4).

10-5   STEEL BEAM-COLUMN DESIGN

An adequately designed beam-column which has $f_a/F_a \leqq 0.15$ must satisfy

$$\frac{f_a}{F_a} + \frac{f_{bx}}{F_{bx}} + \frac{f_{by}}{F_{by}} \leq 1.0, \qquad [10\text{-}3.5]$$

where F_a is computed from the maximum KL/r; f_b (x and/or y) is calculated from largest moments present in the respective place of bending; and F_b (x and/or y) is allowable stress as if bending moment existed in the respective plane only, and without axial force.

When f_a/F_a exceeds 0.15, appropriate formulas are:

$$\frac{f_a}{F_a} + \frac{C_{mx}f_{bx}}{\left(1 - \dfrac{f_a}{F'_{ex}}\right)F_{bx}} + \frac{C_{my}f_{by}}{\left(1 - \dfrac{f_a}{F'_{ey}}\right)F_{by}} \leqq 1.0 \qquad [10\text{-}4.13]$$

and

$$\frac{f_a}{0.6F_y} + \frac{f_{bx}}{F_{bx}} + \frac{f_{by}}{F_{by}} \leqq 1.0. \qquad [10\text{-}3.6]$$

The first of these equations satisfies stability criteria and so must use the largest moments present; the second equation satisfies yield criterion and must use the larger of the two moments found at points of lateral support. In the case of biaxial bending and axial load, this could require use of each equation twice, even if points of lateral support coincide in each plane:

1. Maximum x and y moments may not occur at the same location, requiring use of the first equation twice.

2. Maximum x moment might be at one support with maximum y moment at the other, requiring use of the second equation twice.

Although both allowable stresses F_a and F_e depend upon unbraced length, the former is based on column action while the latter is always figured with regard to only its respective plane of bending.

10-5.1 Design Aids. In practice, when columns are designed without computer assistance (AISC has developed such a program), two general methods are employed. When the beam-column is predominantly "column," bending moments are manipulated into equivalent axial loads, and a column is chosen for such increased load. When combined stresses occur in a member which is predominantly "beam," axial loads are converted into equivalent bending moments, and design proceeds as for a beam. Rubinsky [5] presents efficient methods of attack for both procedures which work well for different combinations of practical loadings, including biaxially loaded beam-columns. Also, Ting [6] has developed a calculator to increase speed in the calculation of interaction equations. It is recommended that both references be consulted before embarking on any project requiring extensive design of beam-columns. In this text, the equivalent column approach of *Steel Construction*, page 3-7 ff, and a simplistic equivalent beam formulation are demonstrated for a beam-column bending about only one axis.

10-5.2 Equivalent Column Method. In *Steel Construction*, the basic equations have been modified so that tabulated values can be used to facilitate design. The derivation of these equations is considered to be relatively unimportant because it is expected that final selection will be analyzed by AISC Specification equations *un*modified.

These equations convert bending moments into an equivalent axial load P'. When P' is added to actual axial load P, the result P^T is used to enter concentric load column tables for choice of a trial section (as in Chap. 3). The modified equations for uniaxial bending are:

$$P^T = P + [B_x M_x (F_a/F_{bx})] \qquad (f_a/F_a \leq 0.15), \qquad (10\text{-}5.1)$$

$$P^T = P + [B_x M_x C_{mx}(F_a/F_{bx})\{a_x/[a_x - P(K\ell)^2]\}] \qquad (f_a/F_a > 0.15), \qquad (10\text{-}5.2)$$

and

$$P^T = P(F_a/0.6F_y) + [B_x M_x(F_a/F_{bx})], \qquad (10\text{-}5.3)$$

where $P \equiv$ actual axial load (kips), $M_x \equiv$ actual bending moment about X axis (in.-kips), $F_a \equiv$ allowable axial stress (ksi), $F_{bx} \equiv$ allowable flexural stress about X axis (ksi), $F_y \equiv$ yield stress of steel (ksi), $K\ell \equiv$ effective length in plane of bending (in.), $C_{mx} \equiv$ reduction factor evaluated from column geometry, and a_x and B_x are coefficients defined by member size. (*Note.* If bending is about the Y axis, change M_x, F_{bx}, C_{mx}, a_x, and B_x to M_y, F_{by}, C_{my}, a_y, and B_y, respectively.)

A column loaded as shown in Fig. 10-5.1 is used to indicate the design procedure.

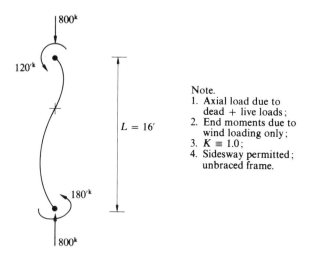

Note.
1. Axial load due to dead + live loads;
2. End moments due to wind loading only;
3. $K \equiv 1.0$;
4. Sidesway permitted; unbraced frame.

Fig. 10-5.1   Combined Bending and Axial Force

1. Design for axial load only as per Chap. 3. $P = 800$ kips, $K = 1.0$, and $L = 16$ ft. Therefore, using Table 1, page 3-14 of *Steel Construction*, select W14 x 150, good for 816 kips at $KL = 16.0$.

2. For W14 x 150, compute f_a and F_a: $f_a = P/A = (\frac{3}{4})(800)/44.1 = 13.6$ ksi. The fraction $\frac{3}{4}$ is used to reduce the load so that allowable stresses can be used at their nominal values, which is equivalent to using unreduced loads at 4/3 nominal allowable stress permitted. With $KL/r = (12)(16)/3.99 = 48$, (*Steel Construction*, page 5-84) $F_a = 18.53$ ksi. Therefore, with combined loads including wind, $f_a/F_a = 0.734 > 0.15$ so Eqs. (10-5.2) and (10-5.3) must be used in lieu of Eq. (10-5.1).

3. Compute P_T using Eq. (10-5.2). The bending moment is also reduced, $(\frac{3}{4})(180) = 135$ ft-kips, to account for the one-third increase in stress allowed for load combinations involving wind. This is normal procedure in practice because it facilitates elimination of noncritical load combination without calculation.

From page 5-84 for a W14 x 150, $B_x = 0.184$ and $a_x = 267$ x 10^6. For an unbraced frame, $C_m = 0.85$. Section modulus for the member is 240; so, $f_{bx} = 12(135)/240 = 6.75$ ksi. From step 2, $F_a = 18.53$, so $F_a/0.6 F_y = 18.53/22 = 0.843$ and $P(KL^2) = 600$ x $(12$ x $16)^2 = 22.2$ x 10^6. By the column table, $L_c = 16.4$; because $16.4 > 16$, $F_{bx} = 24$ ksi and $F_a/F_{bx} = 18.53/24.0 = 0.772$.

Substitution into Eq. (10-5.2) gives

$$P_1^T = 600 + (0.184)(12 \times 135)(0.85)(0.772)[267/(267 - 22.2)] = 813 \text{ kips}$$

and

$$P_2^T = 600(0.843) + 0.184(12 \times 135)(0.772) = 736 \text{ kips.}$$

4. Use the largest P^T, 813 kips, and recheck adequacy by column tables. In this case, a W14 x 150 is still adequate.

5. Because the modified equations, in general, are conservative, it is reasonable normally to try a size below that predicted by the modified equations. Of course, such choice cannot be smaller than required by pure axial load.

6. Prove by analysis using the basic AISC Specification equations as shown in Table 10-5.1. Do *not* use modified equations.

Efficient, error-free work usually is enhanced by tabular calculations.

10-5.3 Equivalent Beam Method. In the equivalent beam method, Eq. (10-4.13) (uniaxial bending only) is written using magnification factor δ as

$$\frac{f_a}{F_a} + \delta \frac{f_b}{F_b} \leq 1.0 \tag{10-5.4}$$

Solution of the inequality for F_b yields

$$F_b \geq \frac{\delta f_b}{1 - (f_a/F_a)}. \tag{10-5.5}$$

Then, multiply both sides by section modulus,

$$M^M = \frac{\delta M_o}{1 - (f_a/F_a)} = \frac{C_m M_o}{[1 - (f_a/F_e')][1 - (f_a/F_a)]}, \tag{10-5.6}$$

where $M^M \equiv$ magnified bending moment and $M_o \equiv$ actual bending moment.

The procedure is to assume a section, determine f_a, F_a, and F_e', and solve for M^M. The member is then checked by the "Allowable Moments in Beams"– curve of *Steel Construction*. If the assumed section appears to agree reasonably well with M^M, an analysis is performed. Otherwise, a larger or smaller section is chosen for a subsequent trial.

Two primary difficulties are encountered when using this procedure: (1) If f_a/F_a is quite small, M^M changes rapidly in the iteration process making final selection difficult. (2) A limited number of members are shown in the curves so that a desirable section might be overlooked. In the latter case, the method can still be used by estimating F_b and determining a required section modulus.

10-5.4 Analysis. A complete analysis of the choice obtained in sec. 10-5.1 is presented in Table 10-5.1.

Table 10-5.1   Analysis for Combined Bending and Axial Load

Loads (16' column $K = 1.0$ $C_m = 0.85$)

Dead and Live 800 kips (axial)
Wind $120^{/k}$ (top)
              $180^{/k}$ (bottom)

Design Loads

Without Wind $P = 800$ kips
With Wind $\left. \begin{array}{l} P = \frac{3}{4} \times 800 = 600 \text{ kips} \\ M = \frac{3}{4} \times 180 = 135^{/k} \end{array} \right\}$ No further stress adjustment

Section Properties   W 14 × 150   A36

$A = 44.1$ in.^{2}
$Z = 240$ in.^{3} $F_b = 24$ ksi $L_u = 16.8 > 16.0$
$r_x = 6.37$ $F_y = 36$ ksi
$KL/r_x = 12(16)/6.37 = 30.1$ $F'_e = 164.9$ AISC p. 5-94 based on 30.1.
$r_y = 3.99$
$KL/r_y = 12(16)/3.99 = 48.1$ $F_a = 18.52$ ksi. AISC p. 5-84 based on 48.1.

Stresses   *D and L*   D and L and Wind

$f_a = \dfrac{800}{44.1} = 18.2 < 18.5 \therefore$ OK $f_a = \dfrac{600}{44.1} = 13.61$ ksi

$f_b = \dfrac{12(135)}{240} = 6.75$ ksi

$\dfrac{f_a}{F_a} + \dfrac{C_m f_b}{(1 - f_a/F'_e)F_b}$ $= \dfrac{13.6}{18.52} + \dfrac{0.85(6.75)}{\left(1 - \dfrac{13.6}{164.9}\right)24} = 0.734 + 0.261$

$= 0.995 < 1$ OK

$\dfrac{f_a}{0.6F_y} + \dfrac{f_b}{F_b}$ $= \dfrac{13.6}{22.0} + \dfrac{6.75}{24.0} = 0.618 + 0.281 = 0.899 < 1$ OK

Note. Even though the last two steps can be done in one operation with a mini-calculator, it is better to show the magnitude of the two elements so that effects of axial force and bending can be mentally visualized.

Note that r_x is used in the determination of F'_e because of bending direction. On the other hand, r_y is used to determine F_a. That is, the radius of gyration and the effective length must be computed with reference to the *bending axis* to obtain F'_e. But, the correct F_a is for the combination of effective length and radius of gyration *producing the smallest* allowable stress.

Stress proofing for each loading condition is shown independently: $D + L$ without alteration, $D + L +$ wind values having been reduced by three-quarters to account for the allowable stress increase of one-third. In actual practice, columns must often be designed for several loading combinations; because of this, factoring of M and P rather than allowable stresses is very common.

10-6 REINFORCED CONCRETE DESIGN

Most reinforced concrete columns are designed using either computer programs, such as one available from the Portland Cement Association (PCA) or extensive tables such as the CRSI Handbook [3] and various others [7] available from PCA. In the absence of such aids, or when loads are outside tabulated ranges, reasonable progress can be made by the interaction curves such as those of Fig. 10-1.4. Final proof is by basic relationships typical of those used in sec. 10-1.

Table 10-6.1 shows tabular computations leading to the design of a rectangular column having t restricted to 20 in., $f'_c = 4000$ psi, and $f_y = 60,000$ psi. Ultimate axial load is 290.5 kips and ultimate bending moment is 479.5 ft-kips. Both must be divided by $\phi = 0.7$ and the latter must be multiplied by magnification factor δ, in this case assumed to be unity.

As a first trial, column width b is assumed to be 12 in. Eccentricity exceeds minimums and is used directly. Multiplicative factor 12 converts $\delta M_u/\phi$ to inch-kips. Coefficients for the charts are then computed. With $1\frac{1}{2}$ in. cover, assuming #4 ties and, say, #11 bars, d/t is about 0.85. Coordinates of coefficients are plotted on Fig. 10-1.4, the first point being beyond an acceptable range.

In trial 2, $b = 16$ in., $p_t m$ is read from the chart as 0.74. Required steel area is computed by rearranging $p_t m = (A_s/bt)(0.85f'_c/f_y)$.

Although b of 16 in. and 18 in. are adequate, steel placement is difficult. The fourth trial with $b = 20$ in. is satisfied by 4-#11 bars in the two faces that resist bending.

Ties are #4 bars spaced at the minimum of

1. 48 tie diameters $= 48 \times \frac{1}{2} = 24$ in.;
2. 16 longitudinal bar $\phi = 16 \times 11/8 = 22$ in.;
3. least column dimension $= 20$ in.

For this rather simple case, because e/t remains constant, the coefficients plot in a straight line. Normally, several trials are necessary, and when the d/t ratio does not coincide with a particular set of curves, it is advantageous to interpolate between charts. An additional complication is encountered when the value of δ changes with each change in column size for slender members.

Table 10-6.1   RC Column Design

$P_u = 290.5^k$ $M_u = 479.5^{\prime k}$		$P_u/\phi = 415^k$ $\delta M_u/\phi = 685^{\prime k}$		$f_c' = 4$ ksi $f_y = 60$ ksi	
t (in.)	20	20	20	20
b (in.)	12	16	18	20
$e = 12\delta M_u/P_u \geq 0.1t \geq 1$ (in.)	19.81	19.81	19.81	19.81
$(P_u/\phi)/(btf_c')$	0.432	0.324	0.288	0.259
$[(12\delta M_u)/\phi]/(bt^2 f_c')$	0.428	0.321	0.285	0.257
$p_t m$	off	0.74	0.63	0.55
$A_{ST} = (p_t m)bt(0.85 f_c')/f_y$		13.42	12.85	12.47
$p = A_{ST}/bt$		0.042	0.036	0.031

10-7 DERIVATION OF ULTIMATE STRENGTH DESIGN COLUMN EQUATIONS

The fundamental column equation, Eq. (3-5.3), as discussed in Chap. 3, is an experimentally verified equation which expresses $\sigma = P/A$ or $P = \sigma_c A_c + \sigma_s A_s$. Equations for combined bending and axial load have their bases in this same expression. Because columns subject to bending may fail by either over-extension of the tensile reinforcement or by overstressing of the concrete in compression, two different types of equations are required. The bifurcation load at which failure can be either by tension or compression failure is denoted balanced load and is represented symbolically by P_b and M_b.

10-7.1 Computation of c_b. This balanced loading condition is defined by two criteria: concrete is strained to 0.003 in./in. and tensile steel is just at yield. A graphical representation of such a configuration is seen in Fig. 10-7.1 where strain in steel reinforcing is equated to yield stress by Hooke's law. By similar triangles

$$\frac{c_b}{d} = \frac{0.003}{0.003 + f_y/E} \tag{10-7.1}$$

which when rearranged with $E = 29(10^6)$ psi shows the distance from the neutral axis to the extreme fiber in compression for balanced load:

$$c_b = 87,000\, d/(87,000 + f_y). \tag{10-7.2}$$

Concrete does not actually fail until strain exceeds 0.003 in./in. However, in keeping with the philosophy (sec. 2-5.6) that sudden explosive-type failures incurred by overstressing of concrete without prior yielding of tensile steel are to be minimized, this conservative value seems reasonable.

10-7.2 Basic Equations for P_u and $P_u e$. ACI-63 formulas 19-1 and 19-2 are obtained directly by statics. Figure 10-7.2 shows that summation of forces

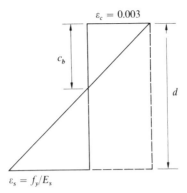

Fig. 10-7.1 Strain Diagram—Balanced Load

parallel to P_u yields

$$-P_u + C_s + C_c - T = 0 \qquad \text{or} \qquad P_u = C_s + C_c - T. \qquad (10\text{-}7.3)$$

Summation of moments about the tensile steel gives

$$-P_u e' + C_c(d - a/2) + C_s(d - d') = 0. \qquad (10\text{-}7.4)$$

Replacement of C_s by its equivalent value $A'_s f_y$, C_c by $0.85 f'_c ab$, and T by $A_s f_s$ and addition of the ϕ term yields ACI-63 formula 19-1:

$$P_u = \phi [0.85 f'_c ab + A'_s f_y - A_s f_s], \qquad (10\text{-}7.5)$$

where $b \equiv$ width of column, $A_s \equiv$ area of tensile steel, $f_y \equiv$ yield stress of tensile steel, $A'_s \equiv$ area of compressive steel, and $f_s \equiv$ actual stress in compressive steel.

ACI-63 formula 19-2 is developed by the same substitutions into Eq. (10-7.4):

$$P_u e' = \phi [0.85 f'_c ab (d - a/2) + A'_s f_y(d - d')], \qquad (10\text{-}7.6)$$

where e' is an eccentricity measured from the tensile steel, related to true eccentricity e, as previously discussed, with respect to Fig. 10-1.2.

10-7.3 Balanced Load and Balanced Moment. The equation for balanced load is obtained directly from Eq. (10-7.5) by replacing f_s by f_y and a by $0.85 c_b$. The latter substitution is based on the assumption that Whitney's equivalent stress block extends downward a distance a_b approximately equal to 0.85 of the distance to the neutral axis.

Summation of moments about the plastic centroid, as sketched in Fig. 10-7.3, yields ACI-63 formula 19-3:

$$M_b = P_b e_b \qquad (10\text{-}7.7)$$
$$= \phi [0.85 f'_c b a_b (d - d'' - a_b/2) + A'_s f_y(d - d' - d'') + A_s f_y d''], \qquad [10\text{-}1.5]$$

where $d'' \equiv$ distance from plastic centroid to centroid of tensile steel (typically $e' = e + d''$) and $e_b \equiv$ true eccentricity for balanced load.

10-7.4 Tension Failure. Figure 10-7.2 also can be used to derive an equation for tensile failure. However, the basic equations (10-7.5) and (10-7.6) are modified by deducting the area of compressive reinforcement from the area of compressive concrete. With this change, the resultant of the compression reinforcement stress block is

$$C_s = A'_s(f_y - 0.85\,f'_c) \tag{10-7.8}$$

so that

$$P_u = \phi\left[0.85f'_c ab + A'_s(f_y - 0.85f'_c) - A_s f_y\right] \tag{10-7.9}$$

and

$$P_u e' = \phi\left[0.85f'_c ab(d - a/2) + A'_s(f_y - 0.85f'_c)(d - d')\right] \tag{10-7.10}$$

with e' measured from tensile steel.

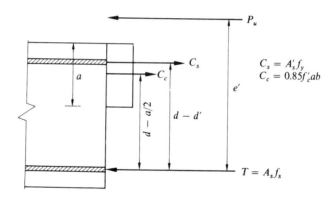

Fig. 10-7.2   General Column—Axial and Bending

Multiplication of the right-hand side of Eq. (10-7.9) by e' and equating it to the right-hand side of Eq. (10-7.10) results in

$$0.85f'_c abe' + A'_s(f_y - 0.85f'_c)e' - A_s f_y e' = \tag{10-7.11}$$
$$0.85f'_c ab(d - a/2) + A'_s(f_y - 0.85f'_c)(d - d'),$$

which can be rewritten in the form of a quadratic in a as

$$[0.85f'_c b/2]\,a^2 + [0.85f'_c b(e' - d)]\,a$$
$$+ [A'_s(f_y - 0.85f'_c)(e' - d + d') - A_s f_y e'] = 0. \tag{10-7.12}$$

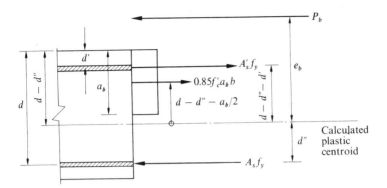

Fig. 10-7.3 Balanced Moment $M_b = P_b\, e_b$

Therefore, using the quadratic formula, we find that

$$a = \frac{-0.85f_c'b(e'-d)}{2(0.85f_c'b/2)}$$

$$\pm\ \frac{\sqrt{[0.85f_c'b(e'-d)]^2 - 4(0.85f_c'b/2)\left[A_s'(f_y - 0.85f_c')(e'-d+d') - A_sf_ye'\right]}}{2(0.85f_c'b/2)}$$

(10-7.13)

Cancelling $0.85f_c'b$ throughout the right-hand side, then dividing both sides by d, we get

$$\frac{a}{d} = \frac{d-e'}{d} \pm \sqrt{\left(\frac{e'-d}{d}\right)^2 - 2\frac{\left[A_s'(f_y - 0.85f_c')(e'-d+d') - A_sf_ye'\right]}{0.85f_c'bd^2}}$$

(10-7.14)

Rearrange and substitute as follows:

1. $(d-e')/d$ into $1 - e'/d$;
2. $[(e'-d)/d]^2$ as $(-1 + e'/d)^2 = (1 - e'/d)^2$;
3. p for A_s/bd;
4. p' for A_s'/bd;
5. m for $f_y/0.85f_c'$.

Equation (10-7.14) then reduces to

$$\frac{a}{d} = 1 - e'/d \pm \sqrt{(1 - e'/d)^2 - 2\left[(mp' - p')\left(\frac{e'-d+d'}{d}\right) - \frac{mpe'}{d}\right]}.$$

(10-7.15)

Reorder terms and let $m' = m - 1$, so $mp' - p' = p'm'$:

$$a/d = 1 - e'/d \pm \sqrt{(1 - e'/d)^2 + 2[e'/d(mp - p'm') + p'm'(1 - d'/d)]}.$$

$$(10-7.16)$$

To make use of this expression for a/d, factor $\phi(0.85 f'_c bd)$ from all terms of Eq. (10-7.9) to obtain

$$P_u = \phi(0.85 f'_c bd) \left[(a/d) + \frac{A'_s f_y - A'_s 0.85 f'_c - A_s f_y}{0.85 f'_c bd} \right], \quad (10-7.17)$$

where the last three terms are equivalent to $p'm - p' - pm = p'(m - 1) - pm = p'm' - pm$, and the expression reduces to

$$P_u = \phi(0.85 f'_c bd)[(a/d) + p'm' - pm]. \qquad (10-7.18)$$

Finally, ACI-63 formula 19-4 is obtained by substituting the value of a/d from Eq. (10-7.16) into Eq. (10-7.18):

$$P_u = \phi \Big\{ 0.85 f'_c bd \Big[p'm' - pm + (1 - e'/d)$$

$$+ \sqrt{(1 - e'/d)^2 + 2[(e'/d)(pm - p'm') + p'm'(1 - d'/d)]} \Big] \Big\}. \quad (10-7.19)$$

This equation for a rectangular reinforced concrete column designed to preclude a tension failure can be simplified if reinforcement is symmetrical. Then, because $p' = p$, the expression $p'm' - pm$ becomes $p'm - p' - pm = -p'$ and ACI-63 formula 19-5 is obtained as

$$P_u = \phi \Big\{ 0.85 f'_c bd \Big[-p + 1 - e'/d + \sqrt{(1 - e'/d)^2 + 2p[e'/d + m'(1 - d'/d)]} \Big] \Big\}.$$

$$(10-7.20)$$
$$[10-1.3]$$

In the rather unusual case where a column has no compressive reinforcement, $p' = 0$, ACI-63 formula 19-6 is evident readily from reduced Eq. (10-7.19) as

$$P_u = \phi \Big\{ 0.85 f'_c bd \Big[-pm + 1 - e'/d + \sqrt{(1 - e'/d)^2 + 2e'pm/d} \Big] \Big\}. \quad (10-7.21)$$

10-7.5. Compressive Failure—Direct Solution.

The equation for a compressive failure is developed semi-theoretically. Although based on the static equations, in latter steps a reasonable numerical constant replaces a term which apparently is dependent upon several parameters.

Tests indicate (see sec. 2-5.6) that maximum ultimate moment as carried by concrete in compression is $f'_c bd^2/3$. Ultimate moment resisted by the section is computed, therefore, by taking moments about the tensile steel with

$\frac{1}{3}f'_c bd^2$ used in place of $C(jd)$ for the concrete in compression, and we see that

$$P_u e' = A'_s f_y(d - d') + f'_c bd^2/3; \qquad (10\text{-}7.22)$$

or, dividing through by e' and multiplying the second term by t/t, we have

$$P_u = \frac{A'_s f_y}{e'/(d - d')} + \frac{f'_c bt}{3e't/d^2}. \qquad (10\text{-}7.23)$$

If the plastic centroid is assumed at column cross-section centerline—reasonable for a compression-type failure—e' may be expressed as (illustrated in Fig. 10-7.4)

$$e' = (2e + d - d')/2 \qquad (10\text{-}7.24)$$

and

$$e' = e + d - t/2. \qquad (10\text{-}7.25)$$

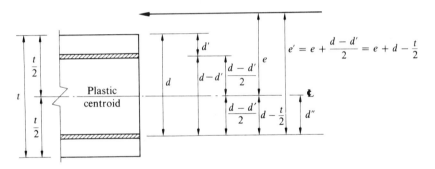

Fig. 10-7.4   Geometry of Eccentricity

When e' in the first term of Eq. (10-7.23) is replaced by Eq. (10-7.24) and e' in the second term by Eq. (10-7.25), P_u becomes

$$P_u = \frac{A'_s f_y}{(2e + d - d')/2(d - d')} + \frac{f'_c bt}{(3et + 3dt - 3t^2/2)/d^2}. \qquad (10\text{-}7.26)$$

The denominator of the first term reduces to $e/(d - d') + \frac{1}{2}$, and the denominator of the second term may be converted to $[(3te)/d^2] + [(6td - 3t^2)/2d^2]$. Equation (10-7.26) then takes the form

$$P_u = \frac{A'_s f_y}{e/(d - d') + \frac{1}{2}} + \frac{f'_c bt}{[(3te)/d^2] + [(6td - 3t^2)/2d^2]}. \qquad (10\text{-}7.27)$$

When the eccentricity e is zero, Eq. (10-7.27) reduces to

$$P_u = 2A'_s f_y + \frac{f'_c bt}{(6td - 3t^2)/2d^2}, \qquad (10\text{-}7.28)$$

which is equivalent to the fundamental equation for P_o [ACI-63 formula 19-7, or Eq. (10-1.1)] only if $(6td - 3t^2)/2d^2 = 1.18$ (that is, $1/0.85$). Although this term varies with t and d and approaches 1.5, the constant 1.18 is used because P_u then correlates well with actual test results. By substitution of this constant, Eq. (10-7.27) reduces to ACI-63 formula 19-10:

$$P_u = \phi \left[\frac{A'_s f_y}{(e/d - d') + \frac{1}{2}} + \frac{btf'_c}{(3te/d^2) + 1.18} \right]. \qquad \begin{array}{c} (10\text{-}7.29) \\ [10\text{-}1.2] \end{array}$$

10-7.6 Compression Failure—Load Moment Equations. The plot of Fig. 10-1.1 shows a linear relationship between P_u and M_u when failure mode is compressive. The upper ordinate of this line is defined by P_o, Eq. (10-1.1), and the lower P_u ordinate by P_b, Eq. (10-1.4). The lower M_b abscissa is equal to $P_b e_b$, Eq. (10-1.5). By use of these coordinates two additional equations are obtained by geometry. see Fig. 10-7.5, in which

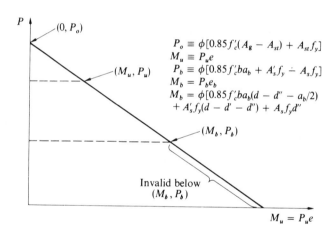

Fig. 10-7.5 Linear Relationships in Compression

$$P_o = \phi [0.85f'_c(A_g - A_{st}) + A_{st}f_y], \qquad P_b = \phi [0.85f'_c ba_b + A'_s f_y - A_s f_y],$$
$$M_u = P_u e, \qquad \qquad M_b = P_b e_b,$$
$$M_b = \phi [0.85f'_c ba_b(d - d'' - a_b/2) + A'_s f_y(d - d' - d'') + A_s f_y d''].$$

By similar triangles

$$\frac{P_o - P_u}{P_o - P_b} = \frac{M_u}{M_b}, \qquad (10\text{-}7.30)$$

which yields ACI-63 formula 19-9 when solved for P_u:

$$P_u = P_o - (P_o - P_b)(M_u/M_b). \qquad (10\text{-}7.31)$$

Replacement of M_u by $P_u e$ and M_b by $P_b e_b$ allows Eq. (10-7.31) to be written

$$P_u = P_o - (P_o - P_b)(P_u e / P_b e_b) \qquad (10\text{-}7.32)$$

or

$$P_u \left[1 + \frac{P_o - P_b}{P_b} \left(\frac{e}{e_b} \right) \right] = P_o, \qquad (10\text{-}7.33)$$

so that ACI-63 formula 19-8 is obtained as

$$P_u = P_o / \left[1 + (P_o / P_b - 1)(e / e_b) \right]. \qquad (10\text{-}7.34)$$

Equations for members of other cross sections are obtained in an analogous manner.

10-8 THE SECANT FORMULA

For a time, the secant formula gained wide acceptance because it could take into account "accidental" eccentricities in what appeared to be a rational manner. With acceptance of the tangent modulus theory embodying alterations imposed by residual stresses, the popularity of the secant formula has diminished. Because of its historical importance and continued use in some codes, however, a development seems appropriate.

Choice of a coordinate system with positive x measured downward from column mid-depth through the line of action of external load (see Fig. 10-8.1) shows lateral displacement is

$$y = \Delta \cos \pi x / \lambda, \qquad (10\text{-}8.1)$$

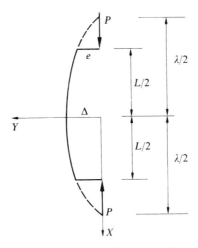

Fig. 10-8.1 Eccentrically Loaded Column

where Δ = mid-span lateral displacement (maximum); λ = half of extended column length, that is, distance from mid-depth to the point where the column would, if extended, intersect the line of action of external load; and y = lateral displacement at x. At x = $L/2$, lateral displacement y is equal to the eccentricity e and Δ can be expressed as

$$\Delta = e/(\cos \pi L/2\lambda), \qquad (10\text{-}8.2)$$

where L = true length of column and e = eccentricity of external load at end of column.

Substitution of this specific value of Δ, as obtained from Eq. (10-8.1) in Eq. (10-8.2), into the original displacement equation yields

$$y = \frac{e[\cos(\pi x/\lambda)]}{\cos(\pi L/2\lambda)}. \qquad (10\text{-}8.3)$$

Based on Euler buckling criteria discussed in Chap. 3, if the column did extend to the full length λ, the critical buckling load would be

$$P = \pi^2 EI/\lambda^2, \qquad (10\text{-}8.4)$$

where E = modulus of elasticity, I = moment of inertia, $P \equiv P_{CR}$, critical load. For such a situation,

$$\lambda = \pi \sqrt{EI/P}. \qquad (10\text{-}8.5)$$

With this value of λ substituted into Eq. (10-8.3),

$$y = \frac{e \cos(x \sqrt{P/EI})}{\cos(L \sqrt{P/EI}/2)}. \qquad (10\text{-}8.6)$$

successive derivatives of y, taken with respect to x, are

$$y' = \frac{-\sqrt{P/EI}\, e \sin(x \sqrt{P/EI})}{\cos(L \sqrt{P/EI}/2)}. \qquad (10\text{-}8.7)$$

and

$$y'' = \frac{-Pe}{EI}\left[\frac{\sin(x \sqrt{P/EI})}{\cos(L \sqrt{P/EI}/2)}\right]. \qquad (10\text{-}8.8)$$

Because maximum moment occurs at mid-height of the column, where $x = 0$,

$$M_{max} = EIy''$$

or, by use of Eq. (10-8.8) for y'',

$$M_{max} = -Pe \sec(L \sqrt{P/EI}/2). \qquad (10\text{-}8.9)$$

An expression for the maximum stress based on both this maximum bending moment and axial load (see Chap. 9) is

$$\sigma_{max} = \frac{P}{A} \pm \frac{M_{max}c}{I} \qquad (10\text{-}8.10)$$

or

$$\sigma_{max} = \frac{P}{A} + \frac{Pe \sec (L\sqrt{P/EI}/2)c}{I}, \qquad (10\text{-}8.11)$$

where $c \equiv$ distance from neutral axis to extreme fiber.

Using the definition of radius of gyration, $r = \sqrt{I/A}$, (or $I = Ar^2$), in Eq. 10-8.11 yields the classical secant formula:

$$\sigma_{max} = \frac{P}{A}\left[1 + \frac{ec}{r^2}\sec\left(\frac{L}{2r}\sqrt{P/EA}\right)\right]. \qquad (10\text{-}8.12)$$

To overcome computational difficulties in using the secant formula, Dr. Richard H. J. Pian has developed efficient design aids. [8] Such aids are almost a necessity when designing under codes specifying the secant formula.

As used in the AASHO 1969 Code [9, page 331] the equation is written for permissible average stresses as:

$$f_s = \frac{f_y/\eta}{1 + [0.25 + e_gc/r^2]\ B \csc \Phi} = \frac{P}{A}, \qquad (10\text{-}8.13)$$

where e_g = eccentricity of applied load at end of column having the maximum moment;

0.25 = factor added to the total eccentricity term to account for accidental eccentricity due to such things as (a) initial crookedness and (b) load not applied as assumed, etc.;

$B = \sqrt{a^2 - 2a \cos \Phi + 1}$; a term to account for unequal loading eccentricities or end moments;

$\Phi = (L/r)\sqrt{(\eta f_s)/E}$. Note that f_s must be known here in order to figure f_s from the equation!

η = factor of safety;

e_s = eccentricity at end opposite to e_g;

$a = \{[(e_sc)/r^2] + 0.25\}/\{[(e_gc)/r^2] + 0.25\}$.

For additional information refer to Young's paper [10] and a bulletin written by Stephenson, Henson, and Cloninger[11].

PROBLEMS

10-1 given: Cross section _____ [(a), (b), or (c) of Fig. 10.P.1] with 8- _____ (#8, #9, #10, or #11) bars, $t = D =$ _____ (18,

20, or 22 in.), $f_y =$ _____ (40, 50, or 60 ksi), $f_c' =$ _____ (3, 4, or 5 ksi).

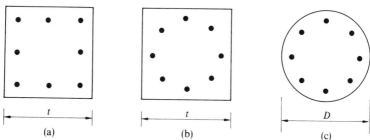

Fig. 10-P.1   Column Cross Sections

develop: P_u, e, and M_u relationships.

plot: Interaction curve.

10-2 given: System of Fig. 10-P.2 with $b_1 =$ _____ (10, 12, 14, or 16 in.), $b_2 =$ _____ (12, 16, or 20 in.), $t_1 =$ _____ (same as b_1), $t_2 =$ _____ (same as b_2), $t_3 =$ _____ (22, 24, 26, or 30 in.), $t_4 =$ _____ (26, 28, 32, or 34 in.).

find: Minimum dimensions of two shear walls so that columns can be considered laterally supported in E–W direction.

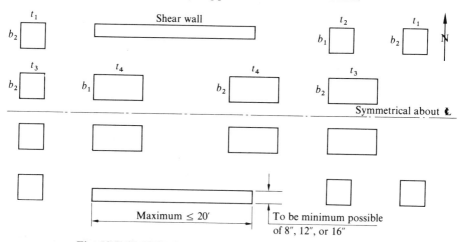

Fig. 10-P.2   Column and Shear Wall Floor Support System

10-3 given: Column of Fig. 10-P.3 used in _____ (braced, unbraced) frame with axial load _____ (40, 50, or 60%) dead, balance live load. Moments due to wind only: $P =$ _____ (530, 570, or 610 kips), $M_T =$ _____ (80, 90, or 100 ft-kips), $M_B =$ _____ (140, 150, or 160 ft-kips), $F_y =$ _____ (36 or 50 ksi).

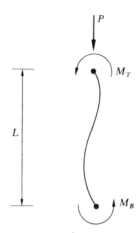

Fig. 10-P.3   Column with End Moments

design: Most economical steel rolled section.

present: Analysis proof.

10-4 given: Same as in problem 10-3 except $f_y = $ _____ (40, 50, or 60 ksi),
$f'_c = $ _____ (3, 4, or 5 ksi).

design: Reinforced concrete column.

present: Analysis proof.

REFERENCES

1. Ketter, Robert L., Edmund L. Kaminsky, and Lynn S. Beedle, "Plastic Deformation of Wide-Flange Beam-Columns," *Transactions of American Society of Civil Engineers*, V. 120 (1955), pp. 1028–1061.

2. Galambos, Theodore V., and Robert L. Ketter, "Columns Under Combined Bending and Thrust," *Proceedings of American Society of Civil Engineers, Journal of the Engineering Mechanical Division*, V. 85, No. EM2, Part 1 (April 1959), pp. 1–30; *Transactions*, V. 126 (1961), pp. 1–25.

3. *CRSI Handbook*, Concrete Reinforcing Steel Institute, 228 North LaSalle Street, Chicago, Illinois, 60601 (1972).

4. *Commentary on Building Code Requirements for Reinforced Concrete* (ACI 318–71), American Concrete Institute, Detroit, Michigan (1971).

5. Rubinsky, Moe A., "Rapid Selection of Beam Columns," *Engineering Journal*, V. 5, No. 3 (July 1965), pp. 100–122.

6. Ting, Ea-Tu, "Calculator for Beam-Column Design," *Engineering Journal*, V. 7, No. 2 (April 1970), pp. 62–63.

7. "Ultimate Load Tables for Circular Columns" (IS017.01D); "Ultimate Load Tables for Spirally Reinforced Square Columns" (IS044.01D); "Ultimate Load Tables for Tied Columns" (IS078.01D); "Ultimate Strength Design of Reinforced Concrete Columns" (EB009.01D); Portland Cement Association, Old Orchard Road, Skokie, Illinois 60076.

8. Pian, Richard H. J. "Column Design Simplified," *Proceedings of the American Society of Civil Engineers*, Paper No. 1231 (1957).

9. *Standard Specifications for Highway Bridges*, 10th ed., The American Association of State Highway Officials, 341 National Press Building, Washington, D.C. 20004 (1969).

10. Young, D. H., "Stresses in Eccentrically Loaded Steel Columns," *Mémoires International Association for Bridge and Structural Engineering*, V. 1, Zurich (1932), pp. 507–517.

11. Stephenson, Henson K. and Cloninger, K., *Stress Analysis and Design of Steel Columns*, Texas Engineering Experiment Station, Bulletin 129, College Station, Texas (February 1955).

ADDITIONAL HELPFUL READINGS

12. Massonnet, Charles. "Stability Considerations in the Design of Steel Columns," *Proceedings of the American Society of Civil Engineers, Journal of the Structural Division*, V. 85, No. ST7, Part 1 (September 1959), pp. 75–111.

13. Ketter, Robert L. "Further Studies of the Strength of Beam-Columns," *Proceedings of the American Society of Civil Engineers, Journal of the Structural Division*, V. 87, No. ST6 (August 1961), pp. 135–152.

14. Whitney, Charles S., and Edward Cohen. "Guide for Ultimate Strength Design of Reinforced Concrete," *Journal of American Concrete Institute*, V. 28, No. 5 (November 1956), *Proceedings* V. 53, Title No. 53–25, pp. 455–490.

15. Ferguson, Phil M. "Simplification of Design by Ultimate Strength Procedures," *Proceedings of the American Society of Civil Engineers, Journal of Structural Division*, V. 82, No. ST4 (July 1956), pp. 1–25.

16. Reese, Raymond C., et al. "Explanatory Notes on Appendix (Ultimate Strength Design) to "Building Code Requirements for Reinforced Concrete (ACI 318–56)", *Journal of American Concrete Institute*, V. 29, No. 3 (September 1957), *Proceedings* V. 54, Title No. 54–12, pp. 197–204.

17. Archibald, Raymond, et al. "Report of ASCE-ACI Joint Committee on Ultimate Strength Design," *Proceedings of the American Society of Civil Engineers, Structural Division*, V. 81, No. 809 (October 1955), pp. 1–68.

CHAPTER 11

STABILITY-RELATED AISC
DESIGN EQUATIONS

11-1 FUNDAMENTAL CRITERIA

Most equations of the AISC Specifications, as published in *Steel Construction*, are developed from stability criteria related to torsion, column, and plate action. Basic to this study is the work of Konrad Basler[1, 2, 3] which is used as a primary reference throughout most of this chapter. The second of these references is based on his dissertation.

Necessary aspects of torsion are developed in Chap. 9 and column action is discussed in Chaps. 3 and 10. Plate equations for critical buckling stress, σ_{CR}, developed as Eqs. (C-4.17) and (C-4.18) in Appendix C, are

$$\sigma_{CR} = k \frac{\pi^2 \sqrt{EE_T}}{12(1 - v^2)} \left(\frac{h}{b}\right)^2 \tag{11-1.1}$$

and

$$\sigma_{CR} = k \frac{\pi^2 E}{12(1 - v^2)} \left(\frac{h}{b}\right)^2. \tag{11-1.2}$$

Equation (11-1.1) (inelastic formulation) degenerates to Eq. (11-1.2) when the mode of failure is linearly elastic. In these equations, $E \equiv$ modulus of elasticity, $E_T \equiv$ tangent modulus in inelastic region, $v \equiv$ Poisson's ratio, $h \equiv$ thickness of plate, $b \equiv$ width of plate, $k \equiv$ the plate buckling coefficient which relates (1) width and length of plate and (2) edge support conditions of plate.

Plate buckling coefficient k is evaluated and discussed in Appendix C for a plate simply supported on all four sides. Additional values for k are presented in Fig. 11-1.1 for rectangular plates simply supported along their loaded edges. Coefficient k varies from a low of 1.00, when longitudinal edges are free, to a high of 6.97, when longitudinal edges are fixed. Two special cases applicable

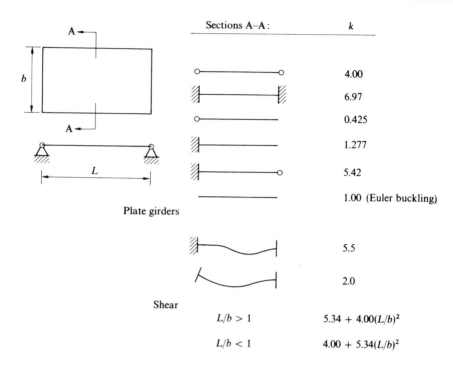

Fig. 11-1.1   Plate Buckling Coefficient k

to webs of plate girders are also shown, $k = 5.5$, when the compression flange is restrained against rotation and $k = 2.0$ when the compression flange can rotate. When a plate is subject to shear, Bleich[4] develops

$$k = 5.34 + 4.00 \, (L/b)^2 \qquad (L/b > 1) \qquad (11\text{-}1.3)$$

and

$$k = 4.00 + 5.34 \, (L/b)^2 \qquad (L/b < 1) \qquad (11\text{-}1.4)$$

in a manner similar to typical developments in Appendix C.

11-2   WIDTH–THICKNESS RATIOS

To insure that no local buckling of plate elements of rolled or built-up steel sections can occur, specifications cite permissible width-to-thickness ratios. Mathematically, such local plate buckling is precluded when buckling stress σ_{CR} is greater than yield point stress F_y. Thus, $F_y/\sigma_{CR} \leqq 1.0$. To obtain an additional factor of safety against this highly undesirable type of failure, the relationship actually used is:

$$\sqrt{F_y/\sigma_{CR}} \lesssim 0.7. \tag{11-2.1}$$

Replacement of critical stress by Eq. (11-1.2) shows that protruding elements will not buckle if:

$$\sqrt{\frac{F_y}{\{(k\pi^2 E)/[12(1 - v^2)]\}(t/b)^2}} \leq 0.7. \tag{11-2.2}$$

To transform this to a more useful equation, E and v are replaced, respectively, by 29,000 and 0.3. With coefficients rounded slightly, then

$$b/t \lesssim 113.3\sqrt{k/F_y}. \tag{11-2.3}$$

This equation gives maximum permissible ratios of b/t dependent upon only type of steel and end support conditions. In specification equations, plate buckling coefficients have been evaluated with numerical coefficients rounded slightly.

11-2.1 Basic Ratios. Section 1.9.1 of the AISC. Code specifies width-thickness ratios of unstiffened projecting elements under compression. These values are determined typically as follows: It is assumed that one leg of a single angle in compression can be represented by a plate hinged on one edge and free on the other, with a k of 0.425 (from Fig. 11-1.1). Therefore $b/t \leq 113.3\sqrt{0.425/F_y}$ or $b/t \leq 73.9/\sqrt{F_y}$, which is rounded to specification equation (Specification sec. 1.9.1.2).

$$b/t \leq 76.0/\sqrt{F_y}. \tag{11-2.4}$$

If F_y is 36 ksi, then $b/t \leq 12.7$ in., which agrees with the limit shown in AISC specification, Appendix A, P.5-72, for A36 steel.

By contrast, the stems of T sections are assumed to be fixed to a rigid flange. For a fixed-free plate, $k = 1.277$ and $b/t \leq 127/\sqrt{F_y}$. Intermediate between hinged-free and fixed-free plate conditions, a k value of 0.70 is assumed applicable to projecting flanges for $95.0/\sqrt{F_y}$.

When compression elements are supported along two edges, k varies between a low of 4.00, if both edges act as hinges, to 6.97, if both edges give fixed support. Therefore, in AISC Specifications, sec. 1.9.2, k for *all other compressed stiffened elements* is assumed at the intermediate value 5.0, restricting the b/t ratio to $253/\sqrt{F_y}$, or 42.2 for A36 steel. For two special cases, variations are used:

$238/\sqrt{F_y}$ for flanges of square and rectangular box sections to provide a more conservative value because of negligible torsional restraint between members.

$317/\sqrt{F_y}$ for unsupported width of cover plates perforated with a succession of access holes to account for increased torsional restraint.

In unusual cases where it seems desirable to exceed these limits, equations are provided in AISC, Appendix C, which allow for reasonable effective areas and stiffnesses.

11-2.2 Compact Sections. AISC Specification sec. 1.5.1.4.1 defines an increased allowable flexural stress for compact sections, that is, those sections which can develop a plastic hinge without local buckling. Such members have a larger attainable shape factor and have their allowables increased by 10% to $0.66F_y$.

Formulas for web-depth ratios are developed by two different criteria: (a) $14,000/\sqrt{F_y(F_y + 16.5)}$, developed in sec. 11-4.3 and (b) those formulated by requiring plastic hinge formation without buckling, treated here.

Figure 11-2.1 shows results obtained from theoretical plate analysis (broken line) compared to a simpler bilinear curve. As analyzed, certain parameters were assumed fixed: $A/A_w = 2$, $d/d_f = 1.05$, $\epsilon_{max}/\epsilon_y = 4.0$, and $\sigma_y = 33$ ksi (A7 steel).

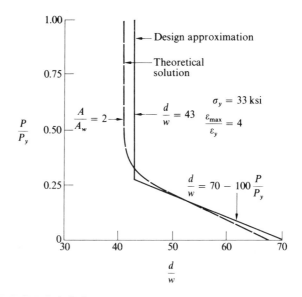

Fig. 11-2.1 Limiting Slenderness Ratios for the Web Plate of Wide-Flange Shapes Under Combined Bending and Axial Compression [5, page 51]

From this,

$$d/w = 70 - 100 \, P/P_y \quad \text{if} \quad P/P_y \leq 0.27 \qquad (11\text{-}2.5)$$

and

$$d/w = 43 \qquad\qquad \text{if} \quad P/P_y > 0.27. \qquad (11\text{-}2.6)$$

These expressions were used in the AISC Specifications p. 5-54, of 1963 because plastic design was limited to A7 and A373 (both 33 ksi) and A36 (36 ksi).

Variations are not linear with the reciprocal of $\sqrt{F_y}$ but more theoretical expressions are complex and contain parameters which are not as yet precisely defined. Therefore, AISC-1969, has modified Eq. (11-2.5) to (*Steel Construction*, p. 5-61)

$$\frac{d}{t} = \frac{412}{\sqrt{F_y}} \left(1 - 1.4 \frac{P}{P_y}\right) \quad \text{if} \quad P/P_y \leq 0.27, \qquad (11\text{-}2.7)$$

which for A36 still reduces to $68.7 - 96.1 \, P/P_y$, agreeing closely with Eq. (11-2.5). The lower limit for A36, $257/\sqrt{F_y} = 42.8$, compares to 43 of Eq. (11-2.6).

Allowable stress formulations, AISC, sec. 1.5.1.4.1d, Supplement No. 1, are obtained by eliminating factored loads from Eq. (11-2.7), that is, $0.27/1.7 \doteq 0.16$ and $1.4(1.7) \doteq 2.33$.

Table 11-2.1 compares minimum requirements for allowable stress design with those for compact sections and plastic design. AISC Code equation or section numbers appear in parentheses below respective equations. The only deviation, for unstiffened projecting elements by plastic design, has the tabulated value of 8.5 for 36 ksi (AISC, sec. 2.7) converted to $51/\sqrt{F_y}$ for ease of comparison. Table 11-2.2 is similar but limited to A36 steel.

From these tables, it is apparent that an allowable of $0.66 \, F_y$ for compact sections is based on plastic design requirements.

11-3   LATERAL BUCKLING

AISC Specification formulas 1.5-6a, 1.5-6b, and 1.5-7 intuitively present an anomaly. The correct value is the *larger* of these, and *not* the smaller, except that no value can be larger than $0.60 \, F_y$ unless distance between points of lateral support is equal to or less than (AISC 1.5.1.4.1e)

$$76.0 \, b_f/\sqrt{F_y} \qquad (11\text{-}3.1)$$

or

$$20{,}000 \, A_f/d \, (F_y). \qquad (11\text{-}3.2)$$

The basis for the development of these equations rests heavily on material previously discussed in relation to torsion and column action.

Table 11-2.1 Width-Thickness Ratios with AISC References

	Allowable Stress Design		Plastic Design
	Minimum	Compact Section	
Projecting Elements			
Unstiffened	$\dfrac{b}{t} = \dfrac{b_f}{2t_f} \leqq \dfrac{95.0}{\sqrt{F_y}}$	$\dfrac{b_f}{2t_f} \leqq \dfrac{52.2}{\sqrt{F_y}}$	$\dfrac{b_f}{2t_t} \leqq \dfrac{51}{\sqrt{F_y}}$ ($F_y = 36$)
	(1.9.1.2)	(1.5.1.4.1b)	(2.7)
Stiffened	$\dfrac{b_f}{2t_f} \leqq \dfrac{317}{\sqrt{F_y}}$	$\dfrac{b_f}{2t_f} \leqq \dfrac{190}{\sqrt{F_y}}$	$\dfrac{b_f}{2t_f} \leqq \dfrac{190}{\sqrt{F_y}}$
	(1.9.2.2)	(1.5.1.4.1c)	(2.7)
Web	$\dfrac{14,000}{\sqrt{F_y(F_y + 16.5)}}$	$\dfrac{d}{t} = \dfrac{412}{\sqrt{F_y}}\left(1 - 2.33\dfrac{f_a}{F_y}\right)$ if $\dfrac{f_a}{F_y} \leqq 0.16$	$\dfrac{d}{t} = \dfrac{412}{\sqrt{F_y}}\left(1 - 1.4\dfrac{P}{P_y}\right)$ if $\dfrac{P}{P_y} \leqq 0.27$
		$\dfrac{d}{t} = \dfrac{257}{\sqrt{F_y}}$ if $\dfrac{f_a}{F_y} > 0.16$	$\dfrac{d}{t} = \dfrac{257}{\sqrt{F_y}}$ if $\dfrac{P}{P_y} > 0.27$
	(1.10.2)	(1.5-4a and 1.5-4b, Supplement No. 1) (Formulas of Sec. 1.5.1.4.1d)	(2.7-1a and 2.7-1b) (Formulas of Sec. 2.7)

Table 11-2.2 Width-Thickness Ratios for A36 Steel

| | Allowable Stress Design | | Plastic Design |
	Minimum	Compact Section	
Projecting Elements			
Unstiffened	$\dfrac{b_f}{2t_f} \leq 15.8$	$\dfrac{b_f}{2t_f} \leq 8.7$	$\dfrac{b_f}{2t_f} \leq 8.5$
Stiffened	$\dfrac{b_f}{2t_f} \leq 52.8$	$\dfrac{b_f}{2t_f} \leq 31.7$	$\dfrac{b_f}{2t_f} \leq 31.7$
Web	$\dfrac{d}{t_w} \leq 322$	$\dfrac{d}{t_{w(max)}} = 68.7 \quad \left(\dfrac{f_a}{F_A} \equiv 0\right)$	$\dfrac{d}{t_{w(max)}} = 68.7 \quad \left(\dfrac{P}{P_y} \equiv 0\right)$
	For web minimum, d is clear distance between flanges	$\dfrac{d}{t_{w(min)}} = 42.8$	$\dfrac{d}{t_{w(min)}} = 42.8$
		For web minimum compact section and for web minimum plastic design, d is depth of beam	

11-3.1 Basic Differential Equation. Fig. 11-3.1 depicts a typical rolled steel section subject to a uniform bending moment M. (This action may be observed convincingly by bending commercially available I sections made of balsa wood.) If the angle β of lateral rotation is small, (lateral displacement assumed negligible) then (sec. A-3)

$$EI_z \frac{d^2 v}{dx^2} = M \sin \beta = \beta M \tag{11-3.3}$$

and

$$EI_y \frac{d^2 w}{dx^2} = M \cos \beta = M. \tag{11-3.4}$$

From Eq. (9-9.9),

$$GC_t \frac{d\beta}{dx} - EI_w \frac{d^3 \beta}{dx^3} = M^T = -\frac{dv}{dx} M. \tag{11-3.5}$$

The torsional moment term can be visualized most readily in the plan view of Fig. 11-3.1. Bending moment M is represented as a vector parallel to Y'; its component in the negative direction of the x axis is of a magnitude equal to $M \times \sin$ (lateral angle); so that the torsional moment is almost equal to $-M(dv/dx)$.

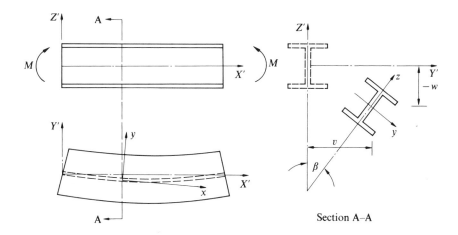

Fig. 11-3.1   Lateral Buckling Geometry

Differentiation of Eq. (11-3.5) with respect to x yields

$$GC_t \frac{d^2\beta}{dx^2} - EI_w \frac{d^4\beta}{dx^4} = -\frac{d^2v}{dx^2} M = -\frac{\beta M^2}{EI_z}, \qquad (11\text{-}3.6)$$

where the last term is secured by replacing d^2v/dx^2 by its equivalent from Eq. (11-3.3). Reordering, we have

$$\frac{d^4\beta}{dx^4} - \left(\frac{GC_t}{EI_w}\right)\left(\frac{d^2\beta}{dx^2}\right) - \frac{\beta M^2}{E^2 I_z I_w} = 0. \qquad (11\text{-}3.7)$$

Two constants are introduced to simplify the solution:

$$\frac{GC_t}{EI_w} \equiv \frac{1}{A^2} \quad \text{and} \quad \frac{M^2}{E^2 I_z I_w} \equiv \frac{1}{B^4}. \qquad (11\text{-}3.8)$$

Then Eq. (11-3.7) may be written in operator form as

$$\left(\mathscr{D}^4 - \frac{1}{A^2}\,\mathscr{D}^2 - \frac{1}{B^4}\right) \beta = 0. \qquad (11\text{-}3.9)$$

A solution to the operator quartic is

$$\mathscr{D}^2 = \frac{1}{2A^2} \pm \sqrt{\frac{1}{4A^4} + \frac{1}{B^4}} \qquad (11\text{-}3.10)$$

Or, by use of the definitions,

$$m \equiv \sqrt{-\frac{1}{2A^2} + \sqrt{(1/4A^4) + (1/B^4)}} \qquad (11\text{-}3.11)$$

and

$$n \equiv \sqrt{\frac{1}{2A^2} + \sqrt{(1/4A^4) + (1/B^4)}}, \qquad (11\text{-}3.12)$$

it is seen that $\mathcal{D} = mi; -mi; n; -n$; so

$$\beta = C_1 \sin mx + C_2 \cos mx + C_3 e^{nx} + C_4 e^{-nx}. \qquad (11\text{-}3.13)$$

11-3.2 Critical Buckling Moment. Four arbitrary constants of integration appearing in Eq. (11-3.13) are evaluated by use of appropriate boundary conditions. At the left end, considering simple supports, both rotation and flange bending moment (torsional moment) are zero (see Fig. 9-10.1). Using the first condition, at $x = 0$, $\beta = 0$, we have

$$\beta = 0 = C_1(0) + C_2(1) + C_3(1) + C_4(1). \qquad (11\text{-}3.14)$$

Invoking the second condition, at $x = 0$, $d^2\beta/dz^2 = 0$, we have

$$d^2\beta/dz^2 = 0 = -m^2 C_1(0) - m^2 C_2(1) + n^2 C_3(1) + n^2 C_4(1). \qquad (11\text{-}3.15)$$

If Eq. (11-3.14) is multiplied by m^2 and the result added to Eq. (11-3.15), then

$$0 = C_3(m^2 + n^2) + C_4(m^2 + n^2), \qquad (11\text{-}3.16)$$

which shows that

$$C_3 = -C_4 \qquad (11\text{-}3.17)$$

and therefore from Eq. (11-3.14)

$$C_2 = 0. \qquad (11\text{-}3.18)$$

Thus Eq. (11-3.13) can be written in simpler form as

$$\beta = C_1 \sin mx + 2C_3 \left(\frac{e^{nx} - e^{-nx}}{2} \right) \qquad (11\text{-}3.19)$$

or

$$\beta = C_1 \sin mx + 2C_3 \sinh nx. \qquad (11\text{-}3.20)$$

At the opposite end of the beam, rotation and flange bending moment also are zero, or at $x = L$, $\beta = 0$, and $d^2\beta/dz^2 = 0$, so that

$$\beta = 0 = C_1 \sin mL + 2C_3 \sinh nL \qquad (11\text{-}3.21)$$

and

$$d^2\beta/dz^2 = 0 = -m^2 C_1 \sin mL + 2n^2 C_3 \sinh nL. \qquad (11\text{-}3.22)$$

In these two equations, neither m nor n can be zero [Eqs. (11-3.8) through

(11-3.12) and C_1 and C_3 cannot be zero concurrently. Because sinh nL can be zero only if $nL = 0$, the two possible zero terms are sin mL and the constant C_3. But, if sin $mL = 0$, then

$$m = p\pi/L \qquad p = 1, 2, 3, \ldots \qquad (11\text{-}3.23)$$

By use of the smallest nontrivial value of p, $m = \pi/L$ or [Eq. (11-3.11)]

$$\left(\frac{\pi}{L}\right)^2 = m^2 = -\frac{1}{2A^2} + \sqrt{\frac{1}{4A^4} + \frac{1}{4B^4}}, \qquad (11\text{-}3.24)$$

so that

$$\left[\left(\frac{\pi}{L}\right)^2 + \frac{1}{2A^2}\right]^2 = \left(\frac{\pi}{L}\right)^4 + \left(\frac{\pi}{L}\right)^2 \frac{1}{A^2} + \frac{1}{4A^4} = \frac{1}{4A^4} + \frac{1}{B^4} \qquad (11\text{-}3.25)$$

Cancelling the common terms and replacing A and B by their equivalents from Eq. (11-3.8), we get

$$\frac{\pi^4}{L^4} + \left(\frac{\pi^2}{L^2}\right)\left(\frac{GC_t}{EI_w}\right) = \frac{M^2}{E^2 I_z I_w} \qquad (11\text{-}3.26)$$

or

$$M^2 = \frac{\pi^4 E^2 I_z I_w}{L^4} + \frac{\pi^2 GC_t EI_z}{L^2}. \qquad (11\text{-}3.27)$$

Finally, critical bending moment for lateral buckling may be expressed as

$$M_{CR} = \frac{\pi}{L}\sqrt{EI_z GC_t}\sqrt{1 + (\pi^2 EI_w)/(L^2 GC_t)}. \qquad (11\text{-}3.28)$$

11-3.3 Critical Lateral Buckling Stress. Critical stress is determined by dividing both sides of Eq. (11-3.28) by section modulus S. If effective compressive area is taken as the area of one flange plus one-sixth the area of the web, which seems reasonable,

$$A_{\text{eff}} = A_f + \frac{1}{6}A_w = w_f t_f \left(1 + \frac{A_w}{6A_f}\right). \qquad (11\text{-}3.29)$$

Then section modulus is computed as transfer moment of intertia (I_0 neglected):

$$I = 2A_f \left(1 + \frac{A_w}{6A_f}\right)\frac{d_1^2}{4}$$

divided by $c = d_2/2$ distance from centroid of section to extreme fiber, so that

$$S = \frac{I}{c} = \frac{d_1^2 t_f w_f [1 + (A_w/6A_f)]}{d_2},$$

where $w_f \equiv$ width of flange, $t_f \equiv$ thickness of flange, $d_1 \equiv$ distance from centroid of section to centroid of effective area. Assuming that $d_1 = d_2 = d$ results in an approximate expression for section modulus which tends to compensate for neglecting I_0 by using a somewhat smaller value for c:

$$S = dt_f w_f [1 + (A_w/6A_f)] . \tag{11-3.30}$$

The definitions given for shear modulus, torsional constant, and warping constant are:

$$G = E/2(1 + v), \qquad\qquad [\text{B-5.11}]$$

$$C_t = \frac{1}{3}\Sigma bt^3 = \frac{1}{3}(2w_f t_f^3) + \frac{1}{3}dt_w^3, \qquad [9\text{-}4.36] \left.\begin{array}{c} \\ \\ \\ \\ \\ \end{array}\right\} \tag{11-3.31}$$

$$I_w = I_f d_3^2/2 = t_f w_f^3 d^2/24, \qquad [9\text{-}9.7]$$

where $2d_3$ is depth of web excluding flange thickness which is approximated by d as used in Eq. (11-3.30).

The moment of inertia about the weak axis is

$$I_z = 2(t_f w_f^3/12). \tag{11-3.32}$$

If I_z, I_w, G, and C_T are replaced by these equivalent forms in Eq. (11-3.27), and if both composite terms are divided by S^2 as expressed in Eq. (11-3.30), the resulting equation for critical stress is:

$$\sigma_{CR}^2 = \frac{\pi^4 E^2}{L^4} \frac{t_f w_f^3 d^2}{24} \left(\frac{2(t_f w_f^3)}{12d^2 t_f^2 w_f^2 (1 + A_w/6A_f)^2} \right)$$

$$+ \frac{\pi^2 EE (2w_f t_f^3 + dt_w^3)}{L^2 2(1 + v)3} \left(\frac{2t_f w_f^3}{12d^2 t_f^2 w_f^2 (1 + A_w/6A_f)^2} \right) \tag{11-3.33}$$

or

$$\sigma_{CR}^2 = \left(\frac{\pi^2 E}{L^2} \frac{t_f w_f^3}{12 w_f t_f (1 + A_w/6A_f)} \right)^2$$

$$+ \frac{(\pi^2 E^2)/[18(1 + v)]}{(L^2 d^2)/(t_f^2 w_f^2)} \left(\frac{(2w_f t_f^3 + dt_w^3)/(2w_f t_f^3)}{[1 + (A_w/6A_f)]^2} \right), \tag{11-3.34}$$

where $(t_f w_f^3)/(t_f^2 w_f^2)$ is rewritten as $(1/w_f t_f^3)/(1/t_f^2 w_f^2)$, both being equal to w_f/t_f. The last term may be set equal to 1 if $A_w \ll A_f$ or $t_f = 1.73t_w$. This is more obvious if rearranged as follows:

$$\frac{1 + (t_w^2/2t_f^2)(A_w/A_f)}{(1 + A_w/6A_f)^2} \doteq 1, \tag{11-3.35}$$

where $A_w = dt_w$ and $A_f = w_f t_f$.
The second part of the first term of Eq. (11-3.24) may be replaced by r, the radius of gyration, because

$$r^2 = \frac{I_f}{A} = \frac{t_f w_f^3}{12 w_f t_f (1 + A_w/6A_f)},$$ (11-3.36)

while if $v = 0.3$,

$$\frac{\pi^2}{18(1 + v)} = 0.422 \quad \text{or} \quad 0.65^2.$$ (11-3.37)

By use of these relationships, Eq. (11-3.34) reduces to

$$\sigma_{CR} = \sqrt{\left[\frac{\pi^2 E}{(L/r)^2}\right]^2 + \left[\frac{0.65E}{(Ld/t_f w_f)}\right]^2}$$ (11-3.38)

11-3.4 Relationship of Critical Stresses. The first term of Eq. (11-3.38) suggests Euler column buckling, actually induced by nonuniform torsion, and the second term suggests St. Venant torsion. Thus, Eq. (11-3.38) can be written symbolically as

$$\sigma_{CR} = \sqrt{[\sigma_{CR}(\text{col})]^2 + [\sigma_{CR}(\text{St. Venant})]^2}.$$ (11-3.39)

This form is represented graphically by a right triangle in Fig. 11-3.2.

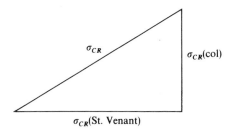

Fig. 11-3.2 σ_{CR} Related

These two instability types yield different relative stress magnitudes for identical span lengths. Typically, however, critical stress for column-type buckling is more pronounced when that for St. Venant torsion is relatively insignificant, and vice versa. Lateral buckling occurs when stress reaches σ_{CR}, which is always greater than either σ_{CR} (col) or σ_{CR} (St. Venant). Hence, a conservative approach is to set σ_{CR} equal to the greater of either σ_{CR}(col) or σ_{CR}(St. Venant), anticipating that the lower critical stress is probably much smaller than the higher stress value used. Of course, allowable stress is still limited by a percentage of yield strength.

This explains the logic behind provisions for allowable flexural compressive stress of AISC Specification 1.5.1.4.5 when $\sigma_{CR}(\text{col})$ and $\sigma_{CR}(\text{St. Venant})$ are re-expressed in specification form.

11-3.5 AISC Formula 1.5-7. The second term of Eq. (11-3.38) reduces to AISC formula 1.5-7 if divided by a 1.57 factor of safety, E is replaced by 29,000, the product $t_f w_f$ by flange area A_f, and C_b is included. Thus

$$F_b = \frac{0.65(29,000)C_b}{1.57\ L(d/A_f)} = \frac{12,000\,C_b}{Ld/A_f}. \tag{11-3.40}$$

In Chapter 2, C_b is discussed and is expressed by Eq. (2-6.11).

11-3.6 AISC Formula 1.5-6b. When failure is by elastic buckling, the Eulerean term is simply multiplied by C_b and divided by a safety factor of about 1.68, so

$$\sigma_{CR(\text{col})} = \frac{\pi^2 E}{(L/r)^2}\frac{C_b}{\text{F.S.}} \Rightarrow F_b = \frac{170,000\,C_b}{(L/r_T)^2}. \tag{11-3.41}$$

Cutoff point for this equation is when σ_{CR} exceeds $F_y/3$, expressed as

$$\frac{L}{r_T} \geq \sqrt{510,000\,C_b/F_y}. \tag{11-3.42}$$

11-3.7 AISC Formula 1.5-6a. When L/r_T is less than Eq. (11-3.42), the Eulerean term is replaced by a more representative function which allows for

1. distribution of bending moment (C_b),
2. end support with respect to vertical movement,
3. end restraint with respect to rotation about the horizontal axis.

The critical stress for a column with any KL/r ratio is expressed as

$$\sigma_{CR} = \sigma_y - \frac{\sigma_y^2}{4\pi^2 E}\left(\frac{KL}{r}\right)^2. \tag{3-2.6}$$

If K is defined as 1, C_b is included, and appropriate factors of safety are used. Then this equation makes a reasonable replacement for the Eulerean term when allowable stresses are between $0.6\,F_y$ and $1/3\,F_y$, limits expressed by AISC as

$$\sqrt{[102(10^3)\,C_b]/F_y} \leq L/r_T \leq \sqrt{[510(10^3)\,C_b]/F_y}. \tag{11-3.43}$$

To obtain the final result, the column equation for critical stress is modified as

$$\frac{\sigma_{CR}}{\text{F.S.(1)}} = \sigma_b = \sigma_y\left(\frac{1}{\text{F.S.(1)}} - \frac{\sigma_y(L/r_T)^2}{\text{F.S.(2)}4\pi^2 E}\right), \tag{11-3.44}$$

which reduces to the code equation 1.5-6a if F.S.(1) = 1.5 and F.S.(2) $\doteq$ 1.33:

$$F_b = \left(\frac{2}{3} - \frac{F_y(L/r_T)^2}{1,530,000\,C_b} \right) F_y. \tag{11-3.45}$$

11-4   WEB CRIPPLING AND BUCKLING

AISC Specification, sec. 1-10.2, restricts depth-thickness ratios of plate girder webs to resist buckling; sec. 1.10.10.2 has other criteria against buckling measured as an allowable bearing (compressive) stress. Code specification 1.10.10.1 and 1.15.5 prescribe means to resist web crippling.

11-4.1 Web Crippling. Critical position for bearing failure in the web of a rolled section is at the toe of the fillet, found as the k value in tables of Dimensions for Detailing of *Steel Construction* for structural shapes. Load is assumed to distribute through the member at a 45° angle (Fig. 11.4.1) making bearing area equal to "length over which concentrated load is acting (N) plus k or $2k$," all multiplied by web thickness. Using the basic relationship $\sigma = P/A$ for interior loads (Specification formula 1.10-8), we have

$$\frac{P}{t_w\,(N\,+\,2k)} \leqq 0.75\,F_y \tag{11-4.1}$$

and for end reactions (Specification formula 1.10-9),

$$\frac{R}{t_w\,(N\,+\,k)} \leqq 0.75\,F_y, \tag{11-4.2}$$

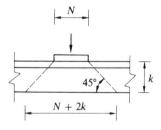

Fig. 11-4.1   Web Crippling

where $P \equiv$ concentrated load (kips), $R \equiv$ reaction (kips), $t_w \equiv$ web thickness (in.), $N \equiv$ length of bearing ($\geq k$ for end reactions), and $k \equiv$ distance from toe of fillet to outermost fiber of flange (in.). Reduction in length for end reaction from $N + 2k$ to $N + k$ results because no distribution is possible over a nonexistent end portion of a beam. If these stresses are exceeded, then bearing stiffeners must be used.

11-4.2 Compressive Stress—Plate Girder Webs. Critical stress for plate girder webs is defined as (refer to beginning of this chapter)

$$F_{CR} = k_1 \underbrace{\frac{\pi^2 E}{12(1 - v^2)} \left(\frac{t}{h}\right)^2}_{\text{vertical}} + k_2 \underbrace{\frac{\pi^2 E}{12(1 - r^2)} \left(\frac{t}{a}\right)^2}_{\text{transverse}}, \qquad (11\text{-}4.3)$$

where $k_1 \equiv 5.5$ if flange is restrained against rotation and 2.0 if it is not so restrained (Fig. 11-1.1), $h \equiv$ clear distance between flanges, $a \equiv$ clear distance between transverse stiffeners and $t \equiv$ thickness of web. Factoring and re-ordering, we see that Eq. (11-4.3) becomes

$$F_{CR} = \left[k_1 + k_2 \left(\frac{h}{a}\right)^2 \right] \frac{\pi^2 E}{12(1 - v^2)} \left(\frac{t}{h}\right)^2. \qquad (11\text{-}4.4)$$

Setting $k_2 = 4.0$ for hinged ends, $E = 29,000$, $v = 0.3$, and adding a 2.62 factor of safety, we have

$$\sigma_w \leqq \left(k_1 + \frac{4}{(a/h)^2} \right) \frac{10,000}{(h/t)^2} \text{ ksi.} \qquad (11\text{-}4.5)$$

This corresponds to Specification formulas 1.10-10 and 1.10-11 when the proper value of k_1 is inserted (either 5.5 or 2).

11-4.3 Web Buckling. Figure 11-4.2 depicts a segment encompassing two web stiffeners of a plate girder deforming under load. Angle of rotation (compare to sec. A-3) is defined as

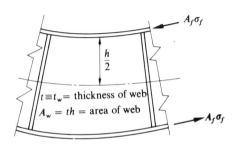

Fig. 11-4.2 Web Buckling

$$\rho \doteq (\epsilon_{\mathrm{flg}}) dx/(h/2), \tag{11-4.6}$$

where $\epsilon_{\mathrm{flg}} \equiv$ strain in flange and $h \equiv$ depth of section.

Vertical component of flange force is expressed by

$$f_f = A_f \sigma_f \sin \rho = 2 A_f \sigma_f \epsilon_f dx/h, \tag{11-4.7}$$

because $\sin \rho \doteq \rho$, and where $A_f \equiv$ flange area, $\sigma_f \equiv$ flange stress, $\epsilon_f \equiv \epsilon_{\mathrm{flg}} \equiv$ strain in flange.

The web has a resisting force f_w, equal to Euler buckling stress (Fig. 11-1.1 for $k = 1$) acting over an area of web thickness t_w (or simply t) and length dx:

$$f_w = \frac{\pi^2 E}{12(1 - v^2)} \left(\frac{t}{h}\right)^2 t\, dx. \tag{11-4.8}$$

Although equilibrium requires that $f_w = f_f$, to preclude the possibility of buckling the component of vertical flange force must always be equal to or less than the resisting web force. Thus Eqs. (11-4.7) and (11-4.8) are related:

$$2 A_f \sigma_f \epsilon_f \frac{dx}{h} \leq \frac{\pi^2 E}{12(1 - v^2)} \left(\frac{t}{h}\right)^2 t\, dx. \tag{11-4.9}$$

By rearrangement of Eq. (11-4.9)—note that $t(h)$ equals web area A_w—an expression for depth to thickness is obtained:

$$\left(\frac{h}{t}\right)^2 \leq \frac{\pi^2 E}{24(1 - v^2)} \frac{A_w}{A_f} \frac{1}{\sigma_f \epsilon_f}. \tag{11-4.10}$$

Equation (11-4.10) defines the limiting ratio of depth to thickness of a web so that buckling will not occur. However, in this form it is too complex for everyday use. Simplification is obtained by (1) replacing v by 0.3 and E by 29,000, (2) setting the ratio A_w/A_f to a reasonable minimum of 0.5, (3) replacing flange stress σ_f by σ_y (desired at failure), and finally, (4) flange strain ϵ_f is defined as $(\sigma_y + \sigma_{RS})/E$, where σ_y is the yield stress and σ_{RS} the residual stress (considered to be 16.5 ksi; see Chap. 3). Equation (11-4.10) reduced and rounded is the equation of AISC, sec. 1.10.2:

$$\frac{h}{t} \leq \frac{14{,}000}{\sqrt{\sigma_y(\sigma_y + 16.5)}}. \tag{11-4.11}$$

11-5   WEB SHEAR AND INTERMEDIATE STIFFENERS

A plate girder web does not become useless when it reaches its buckling load because of tension field action, with the plate girder web acting as tension diagonals of a Pratt truss. For this reason, intermediate stiffeners must provide sufficient strength to perform as truss verticals with end panels designed to provide anchorage.

11-5.1 Geometry of Tension Field. The upper part of Fig. 11-5.1 indicates symbols used to define a tension field between two intermediate stiffeners located distance a apart. Web thickness is denoted by t, tensile stress by σ_t,

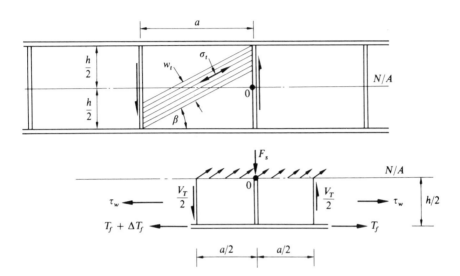

Fig. 11-5.1 Web Shear and Intermediate Stiffners

and width of tensile field by w_t. Shear resisted by the tension field, symbolized as V_T, is

$$V_T = w_t \sigma_t t \sin \beta. \tag{11-5.1}$$

The magnitude of w_t is related geometrically to web depth h and length between stiffeners a as

$$w_t = h \cos \beta - a \sin \beta. \tag{11-5.2}$$

Substitution of this into the previous equation, after multiplication, yields

$$V_T = \sigma_t t (h \sin \beta \cos \beta - a \sin^2 \beta),$$

which can be simplified by trigonometric identities to:

$$V_T = \sigma_t t \left[\frac{h}{2} \sin 2\beta - \left(\frac{a}{2} - \frac{a \cos 2\beta}{2} \right) \right]. \tag{11-5.3}$$

Equation (11-5.3) is differentiated and equated to zero to obtain relationships between a, h, and β which produce maximum tension. Thus

$$dV_T/d\beta = \sigma_t t (h \cos 2\beta - a \sin 2\beta) = 0 \tag{11-5.4}$$

so that ($\sigma \neq 0$ and $t \neq 0$)

$$h/a = \tan 2\beta. \tag{11-5.5}$$

The value of β is found by trigonometric expansion of $\tan 2\beta$ and solution of the resulting quadratic equation. In expanded form, then $h/a = (2 \tan \beta)/(1 - \tan^2 \beta)$, which reduces to quadratic form

$$\tan^2 \beta + \frac{2a}{h} \tan \beta - 1 = 0,$$

and yields

$$\tan \beta = \sqrt{1 + (a/h)^2} - (a/h). \tag{11-5.6}$$

Equation (11-5.5) is represented by a triangle in Fig. 11-5.2, from which it is seen that $\sin 2\beta = 1/\sqrt{1 + (a/h)^2}$.

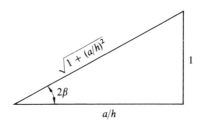

Fig. 11-5.2   Representation of Eq. (11-5.5)

11-5.2 Ultimate Shear. Ultimate shear is the sum of both beam V_B and tension field shear V_T resistances, pictured in the lower half of Fig. 11-5.1:

$$V_u = V_B + V_T. \tag{11-5.7}$$

Summation of moments about point 0 in this figure gives

$$V_T = h(\Delta T_f/a), \tag{11-5.8}$$

where ΔT_f is the change in flange force. By statics ΔT_f is equal to a component of the total tension field force:

$$\Delta T_f = (\sigma_t ta \sin \beta) \cos \beta. \tag{11-5.9}$$

Combining these last two equations and representing $\sin 2\beta$ in terms of the aspect ratio a/h from the previous section,

$$V_T = \sigma_t th \sin 2\beta/2 = \sigma_t th/[2\sqrt{1 + (a/h)^2}]. \tag{11-5.10}$$

To complete the derivation, two terms must be defined:

$V_P \equiv$ shear force causing web plastification (corresponds to plastic hinge in flexure), and $= \tau_y ht$.

$\tau_y \equiv \sigma_y/\sqrt{3}$, the yield point in shear (see sec. E-3).

The total beam shear is represented as

$$V_B = \tau_{CR} ht, \tag{11-5.11}$$

where $\tau_{CP} \equiv$ critical shearing stress, or

$$V_B = V_P(\tau_{CR}/\tau_y). \tag{11-5.12}$$

Equation (11-5.12) uses the relation

$$V_P = \tau_y ht, \tag{11-5.13}$$

where $ht = V_P/\tau_y$.

By combining Eqs. (11-5.12) and (11-5.10), $V_u = V_B + V_T$ is obtained.

$$V_u = V_P \left[\frac{\tau_{CR}}{\tau_y} + \frac{\sqrt{3}}{\sigma_y}\left(\frac{\sigma_t}{2}\right)\frac{1}{\sqrt{1 + (a/h)^2}} \right], \tag{11-5.14}$$

where, in the second term, ht is replaced by V_P/τ_y and subsequently τ_y by $\sqrt{3}/\sigma_y$.

Based on Von Mises' yield condition, principal stresses σ_1 and σ_2 are related to yield stress σ_y (sec. E-3) by

$$\sigma_1 = \sigma_y + (\sqrt{3} - 1)\sigma_2. \tag{E-3.4}$$

For the limiting case, with $\tau_{CR} \equiv$ critical shear stress; $\sigma_t \equiv$ uniform tensile stress in web cross section; $\tau_y = \sigma_y/\sqrt{3}$; $\sigma_1 = \tau_{CR} + \sigma_t$; and $\sigma_2 = -\tau_{CR}$; Eq. [E-3.4] becomes

$$\frac{\sigma_t}{\sigma_y} = 1 - \frac{\tau_{CR}}{\tau_y}. \tag{11-5.15}$$

Substitution of this expression for σ_t/σ_y into Eq. (11-5.14) yields

$$V_u = V_P \left[\frac{\tau_{CR}}{\tau_y} + \frac{1 - (\tau_{CR}/\tau_y)}{1.15 \sqrt{1 + (a/h)^2}} \right]. \tag{11-5.16}$$

This equation is re-expressed as AISC formulas 1.10-1 and 1.10-2 by:

1. replacing τ_{CR}/τ_y by C_V,
2. dividing both sides by ht,
3. which alters V_P to τ_y. This, in turn, is replaced by $\sigma_y/\sqrt{3}$.
4. Division of the right-hand side by a 5/3 factor of safety.

The result is:

$$F_V = \frac{F_y}{2.89} \left[C_V + \frac{\langle 1 - C_V \rangle}{1.15 \sqrt{1 + (a/h)^2}} \right] \leq 0.4 \, F_y, \tag{11-5.17}$$

where $\langle 1 - C_V \rangle$ means to use the result only if it is positive; otherwise the quantity is zero and $a/h \leq 3$ and $\leq [260/(h/t)]^2$.

11-5.3 Evaluation of C_V. The value of C_V is found by expanding terms by previous definitions:

$$C_V = \tau_{CR}/\tau_y = \tau_{CR}(\sqrt{3}/\sigma_y) \tag{11-5.18}$$

and then squaring to obtain

$$C_V^2 = 3\tau_{CR}^2/\sigma_y^2.$$

In order to correspond favorably to test results [2, page 169], two definitions are used for τ_{CR}:

$$\tau_{CR} = \tau_{CRI} \quad \text{when} \quad \tau_{CRI} \leq 0.8\,\tau_y \tag{11-5.19}$$

and

$$\tau_{CR} = \sqrt{0.8\,\tau_y\,\tau_{CRI}} \quad \text{when} \quad \tau_{CRI} > 0.8\,\tau_y, \tag{11-5.20}$$

where τ_{CRI} is defined by the plate buckling equations:

$$\tau_{CRI} \equiv k\,\frac{\pi^2 E}{12(1 - v^2)}\left(\frac{t}{h}\right)^2 \tag{11-1.2}$$

and

$$k = 5.34 + 4.00/(a/h)^2 \quad \text{when} \quad a/h > 1.0 \tag{11-1.3}$$

or

$$k = 4.00 + 5.34/(a/h)^2 \quad \text{when} \quad a/h < 1.0. \tag{11-1.4}$$

To evaluate C_V for these two conditions, v is defined as 0.3 and E as 29,000 ksi. Using Eq (11-5.18) and (11-5.19), the first AISC formula for C_V is obtained as

$$C_V = \frac{\sqrt{3}}{\sigma_y}\left(\frac{k\pi^2 29{,}000}{12(1 - 0.3^2)(h/t)^2}\right) \doteq \frac{45{,}000\,k}{F_y(h/t)^2}, \tag{11-5.21}$$

where $t \equiv$ web thickness (in.) and $h \equiv$ clear distance between flanges (in.)

The second AISC formula for C_V is obtained by use of Eq. (11-5.20) as

$$C_V^2 = \frac{3(0.8)\tau_y}{\sigma_y^2}\left(\frac{k\pi^2 29{,}000}{12(1 - 0.3^2)(h/t)^2}\right). \tag{11-5.22}$$

This reduces, when τ_y is replaced by $\sigma_y/\sqrt{3}$, to

$$C_V \doteq \frac{190}{h/t}\sqrt{\frac{k}{F_y}}. \tag{11-5.23}$$

Note that σ_y is redefined by F_y to agree with AISC Specifications.

11-5.4 Area of Stiffeners. Formula 1.10-3 of AISC. Specifications is obtained by summing forces vertically from Fig. 11-5.1 so that

$$F_s = \sigma_t\, ta \sin^2 \beta \qquad (11\text{-}5.24)$$

represents compressive force in a stiffener. Using a trigonometric substitution for $\sin^2 \beta$ and the triangle of Fig. 11-5.2

$$\sin^2 \beta = \frac{1}{2}(1 - \cos 2\beta) = \frac{1}{2}\left(1 - \frac{a/h}{\sqrt{1 + (a/h)^2}}\right). \qquad (11\text{-}5.25)$$

Substitution of this expression for $\sin^2 \beta$ into Eq. (11-5.24) results in the following equation. (Note that the expression for $\sin^2\beta$ has been multiplied by a/h, a from Eq. (11-5.24), and that the total expression has been multiplied by h to maintain the equality.)

$$F_s = \frac{\sigma_t}{2}\left(\frac{a}{h} - \frac{(a/h)^2}{\sqrt{1 + (a/h)^2}}\right) ht. \qquad (11\text{-}5.26)$$

Equation (11-5.26) is rearranged to correspond to AISC formula 1.10-3 by dividing by σ_y, replacing σ_t/σ_y by $1 - (\tau_{CR}/\tau_y)$ from Eq. (11-5.15) and subsequently by $1\text{-}C_V$ using Eq. (11-5.18), and introducing the following two additional terms:

$$Y \equiv \frac{\text{yield point of web steel}}{\text{yield point of stiffener steel}}$$

$$D \equiv 1.0 \text{ for stiffeners furnished in pairs,}$$
$$\equiv 1.8 \text{ for single angle stiffeners,}$$
$$\equiv 2.4 \text{ for single plate stiffeners.}$$

Then

$$A_{st} = \frac{1 - C_V}{2}\left[\frac{a}{h} - \frac{(a/h)^2}{\sqrt{1 + (a/h)^2}}\right] YDth. \qquad (11\text{-}5.27)$$

11-6 OTHER FORMULAS

Flange stress reductions of AISC Specification 1.10.6 do not normally result in major variations. In effect, it allows for a section as shown in Fig. 11-6.1 to account for buckling of the web under flexure.

AISC formulas 1.5.5a and 1.5.5b (Supplement No. 1) supply straight-line variations in stress, the former from 0.60 to 0.66 and the latter from 0.60 to 0.75. This is easily verified by substituting extreme limits of $b_f/2t_f$ from $95.0/\sqrt{F_y}$ to $52.2/\sqrt{F_y}$.

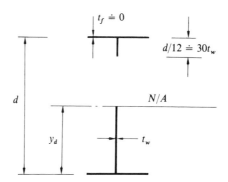

Fig. 11-6.1 Modified Section for Flange Stress Reduction

No reduction in stresses due to combined tension and shear is required unless bending tensile stresses exceed $0.6\,F_y$. When flexural tensile stresses exceed three-quarters of their allowable, a reduction is required by Code formula 1.10-7:

$$(0.825 - 0.375\,f_V/F_V)F_y.$$

This linear equation has its limiting value when $f_V/F_V = 1$, resulting in $0.450\,F_y$ which is equivalent to $(\tfrac{3}{4})\,0.6\,F_y$.

Formulas related to column action or combined bending and column action are discussed in Chaps. 3 and 10. Bearing stiffeners are designed as columns.

For more detail of material discussed in this chapter, references 1, 2, and 3, by Basler, are essential.

REFERENCES

1. Basler, Konrad and Bruno Thurlimann, "Strength of Plate Girders in Bending," *Journal of Structural Division*, American Society of Civil Engineers, V. 87, ST 6 (August 1961), pp. 153–181.
2. Basler, Konrad, "Strength of Plate Girders in Shear," *Journal of Structural Division*, American Society of Civil Engineers, V. 87, ST 7 (October 1961), pp. 151–180.
3. Basler, Konrad. "Strength of Plate Girders Under Combined Bending and Shear," *Journal of Structural Division*, American Society of Civil Engineers, V. 87, ST 7 (October 1961), pp. 181–197.
4. Bleich, Friedrich and Lyle B. Ramsey, *Buckling Strength of Metal Structures*, McGraw-Hill, New York (1952).
5. *Commentary on Plastic Design in Steel*, ASCE Manual of Engineering Practice No. 41, Welding Research Council and ASCE (1961).

CHAPTER **12**

EPILOGUE

In a very real sense, this chapter might well be titled *Prologue*, because it is intended to relate what you will be doing tomorrow with what you have learned using this textbook. You have probably reached the stage of considering the desirability of graduate study in the light of your own abilities and financial status. To help you evaluate these possibilities, the following sections of this chapter are presented: (1) The Structural Engineer in Practice, (2) A Structural Engineer's Training, and (3) The State of the Art.

12-1 THE STRUCTURAL ENGINEER IN PRACTICE

A specialized educational background does not guarantee that you will use such knowledge throughout your professional career (in fact, education appears to have astonishingly little correlation to career). For this reason, many educational institutions have gone more and more toward a common "core," stressing overall engineering competence with corresponding de-emphasis on specialization—particularly at the undergraduate level. For this reason, it is important to realize that the caption for this section presupposes that such a person might consider himself to be first, an engineer; second, a civil, architectural, aeronautical, agricultural, or construction engineer; and third, a structural engineer.

But, whether a specialist or generalist first, a structural engineer normally will find himself engaged in one of the following work situations: (1) Consulting, (2) Government, (3) Petrochemical, (4) Aerospace, (5) Education, (6) Research, and/or (7) Construction.

Before consideration of each of these, however, it is important to classify structural projects into three broad categories that permeate the entire construction industry: architectural structures, engineering structures, and "packaged" structures.

Architectural structures are those which have public habitation or utilization as their primary purpose. Such structures normally are desired or required by a person (or persons) with the financial ability to carry out the entire operation. Such a person is typically called a *client*.

A client, hopefully after some investigation, chooses an architect to be in charge of the design and on-site inspection of a structure. Such an architect is usually referred to as the *principal*. Occasionally, laws are written in such a way that a structural engineer can be a principal.

The principal organizes a team composed of architects and engineers who conceive and develop drawings which define the structure. Such plans consider usage, structural adequacy, heating and ventilation, and electrical circuitry. As these plans being to "jell", the client is appraised of proposed design solutions. At this stage, a client can usually be satisfied so technical work leading to final plans can be completed without undue alteration.

Once plans are complete, the job is *let to bid*. That is, contractors are advised that a structure is to be built and given plans from which they can estimate the *as-built* cost. In general, the lowest bid is accepted.

This contractor then subcontracts parts of the job to other contractors and fabricators. For instance, a steel fabricator is chosen (probably during bid stage) who prepares *shop drawings* for the steel portion of the structure from *engineering drawings*. Shop drawings present sufficient detail to be used by technicians and laborers, who fabricate individual pieces, and include erection details for field crews. When shop drawings are completed, they are returned to the engineer for his approval. When all is in readiness, the contractor commences the job, hopefully under the supervision of a representative of the engineer. Upon completion of the job, a final inspection is made and a check is made to be sure that all bills have been paid. Then the structure is turned over to the client.

Engineering structures are built by a similar process, but the architect, if any, normally is hired by the engineer who is the principal. Engineering structures include bridges, industrial buildings, dams, etc.

Packaged structures are those which are designed and built by the same firm. Although package dealers are accepted in the petrochemical area, they have often been considered suspect when bidding architectural structures or industrial buildings, because the client does not have an unbiased principal to look out for his interests. However, proponents suggest that "turnkey" construction is the only way to obtain true economy and enhance innovation.

Traditionally, in the United States, engineers seldom change sides; that is, they work either as agents of the principal or of a contractor. Therefore, when you take your first job, you will probably be deciding your future line of work.

If you choose to enter into *consulting*, you may be in a two-man office or a very large one. Your office may restrict itself to one certain type of job or specialize the work that you do. On the other hand, you and/or your firm may be committed to a wide variety of tasks. In any case, the work is usually challenging—constantly subject to the time-versus-money squeeze. Typically your firm will work on a percentage of total structure cost. This means that the *better* and *more economical* a job, the less money your company will make! Survival in such circumstances is best enhanced by competence and efficiency. Because of this environment, you can expect to gain rapidly in both experience and responsibility. Conversely, even the best new graduate engineer will probably cost† his employer for the first six months, earn his way the second half-year, repay during the third half-year what he cost in the first six months, and finally begin to be a positive asset during the last half of the second year. In general, you can expect your pay to reflect this fact and the fact that your employer expects you to become his competitor one day. To offset both, he may have a built-in way for you to become an associate to share in the overall profits and motivate you to stay with the firm. If you look forward to being your own boss, this is an excellent way to go.

Work in *government* is found with cities, counties, states, and the federal agencies. In general, financial return is competitive, particularly if fringe benefits are included. Biggest objections to working for government are the need to cut red tape and the nonobjectivity of decisions (and decision-makers). Some good engineers have left governmental agencies soon after joining them because they felt that they were not used to their fullest ability nor were they sufficiently challenged. Such occurrences have led to charges of laziness and boondoggling among engineers in government. Interestingly enough, while such charges may be true in certain offices, personal observation has shown a usually high calibre of engineer in government jobs. If you want security and close to regular hours, then you may be looking for government work. (If you are industrious and would like to see our tax money wisely spent, then I *hope* you are looking for government work.)

Work with the *petrochemical* industry seems to be rewarding, provides high salaries and interesting challenges. The only drawback seems to be lack of individual interest in this specialized area. Most engineers contacted either found petrochemical work exciting and rewarding or they had long since left it.

The *aerospace* industry has traditionally offered top dollar and fringe benefits (including extra paid schooling) for new graduates. Particular drawbacks seem to be loss of feelings of individuality, relatively early ceilings on income, and some insecurity. In general, the man who goes into aerospace is advised to obtain graduate work *before* entering this area of employment.

---

† An experienced engineer usually can do a new engineer's job completely in less time than is actually spent developing and molding the new engineer's talents.

The *educator* of today is paid much better than in the past, but based on training and aptitude, less than he could make in industry. My advice to any would-be engineering teacher is to get substantial industrial experience and a Ph.D. before starting a teaching career. Though it is costly in terms of time and money, it is of inestimable value both in and out of the classroom. It is costly because too few institutions will recognize industrial experience in your rank or paycheck. But, if you really want to *teach*, then your first interest must be the student. Although mathematics and sciences can be taught well without practical experience, it is doubtful that this is true of engineering.

If your bent is *research*, by all means do not stop until you have a Ph.D. With a doctoral degree you can find rewarding work, in private and governmental organizations, that will pay well and give you a real sense of accomplishment.

Maybe you really want to build—Then the place for you is *construction.* Your pay will be above average, and so will the hours you put in. You will find a future of travel, adventure, and challenge. Also, you can soon be your own boss. If you stay in construction upward of five years, you will probably stay in it for the rest of your life—in fact, if you don't like it, you had best get out early in your career. Those men who have chosen construction appear to be happy, to live well, and to have a justifiably tremendous pride in their accomplishments and abilities.

12-2 A STRUCTURAL ENGINEER'S TRAINING

If you really want to be a competent, efficient *structural engineer*, you still have a long way to go. Hopefully, having studied this textbook, you will recognize that you have only begun to "see." Additional courses that will interest you fall into four main categories: mathematics, mechanics, related courses, and structures.

You will find an ability to *use mathematics* a very real asset. Study of linear algebra and advanced calculus broaden your scope and give you the ability to read and comprehend at a much more mature level. Become accomplished in solution of differential and partial differential equations in both closed form and by numerical techniques. For the latter, you will need competence in computer programming.

In addition, you can anticipate that probability theory and statistics will appear more frequently in your future work. Another interesting and important study is of mathematical techniques needed for automated, optimum-cost design.

As ability to solve more complex expressions evolves, anticipate attendant use of more precise mathematical models using such newer techniques. In fact, it is almost certain that you will never "complete" your study of mathematics.

Courses in *mechanics* deal almost entirely with idealized mathematical models for which solutions exist. But, until mechanics are mastered, additional progress is stifled. Names of subjects included in this field are theory of linear and nonlinear elasticity and plasticity, theory of dynamics and vibrations, and theory of stability. These courses provide means for translating mathematical solutions into solutions of real-world models.

Related courses are of two broad types: technical and nontechnical. Of the technical, you may consider such related areas as aerodynamics, fluid mechanics, thermodynamics, etc. Or, conversely, you might consider the relational fields of transportation, environmental, and soil sciences. The non-technical material relates to the humanities and the social sciences. Hope-fully, as the electronic computer frees us from drudgery and allows us time to contemplate—our knowledge can be applied more equitably to the solution of socio-economical problems in society.

In *structures* you will find advanced courses in (1) analysis, including matrix methods and finite elements, (2) design, including architectural struc-tures, industrial structures, bridges, and dams, (3) material-oriented courses, concrete structures, etc., and (4) more advanced theory such as applications courses in dynamics of structures, stability of structures, and plates and shells.

These deal with the real world, and rely heavily on mathematics and mechanics in your background. Mastery of these integrating courses provides tools you need to look forward to each new challenge with confidence and pleasure.

12-3 STATE OF THE ART

Material quoted in this section is taken from a report, written nearly ten years prior to this book's copyright, by the Committee on Research, Structural Division, ASCE, published under the title "Research Needs in Structural Engineering for the Decade 1966–1975." [1]

In summary, the breakdown of structural research needs forecast for that decade were as follows:

Methods of Analysis and Design	10-year Cost (Millions of $)
General Methods of Analysis	12
General Methods of Design	12
Use of Small Scale Models	5
Analysis and Design for Dynamic Loads	6
Design of Towers	1
Special Building Problems	1
Experimental Verification of Design Analyses	3
	40

Methods of Analysis and Design	10-year Cost (Millions of $)	
Electronic Computation	25	
"... a complete man-machine system for the design of structures."		
Masonry and Reinforced Concrete		
Composite Construction	1	
Reinforced Concrete Slabs	2	
Folded Plate Construction	1	
Limit Design	3	
Precast Structural Concrete	5	
Prestressed Concrete	2	
Reinforced Concrete Columns	3	
Masonry Design and Practice	2	
Shear and Diagonal Tension	5	
Cracking of Concrete	2	
Behavior of Complete Structure	2	
Bond and Anchorage of Reinforcement	1	
Torsion of R/C Members	1	
Tensile Strength of Concrete	1	
Reinforcing Materials and Methods	1	
High Strength Concrete Structures	1	
Deflection Predictions	0.5	
Maintenance and Modifications	0.5	
		33
Metal Structures		
Compression Members	5	
Reticulated Structures	1	
Light Gage Metal Structures	1	
Light Weight Alloys	2	
Plastic Design	1	
"Core and Skin" Structures	1	
Flexural Members	5	
Structural Connections	5	
Orthotropic Plate Structures	1	
Tubular Structures	1	
Fatigue	7	
Fracture	7	
Suspended Structures	2	
		40
Structural Plastic		
Time and Temperature Effects	2	
Behavior under Long Duration Loads	2	
Fatigue Behavior	2	

Methods of Analysis and Design	10-year Cost (Millions of $)	
Properties of Structural Elements	2	
Connections	4	
Anchorage of Prestressing Strands	0.5	
Use of Plastics in Composite Construction	2	
Structural Use of Brittle Materials	1	
Development of Design Procedures	2.5	
		18
Wood Structures		
Composite Construction	1.5	
Laminated Elements	2	
Flexural Members	1	
Plywood Panels	0.5	
Wood Shell Roofs	1	
Stability of Wood Members	1	
Fire Retardant Treatments	1	
Connections	1	
		9
Buried Structures		
Free-Field Effects	2	
Soil-Structure Interaction	10	
Dynamic Analysis of Soil-Structure System	2.5	
Shock Analyses	5	
Structures in Ice and Snow	0.5	
		20

Structures in Outer Space

... Structures in outer space are no longer figments of wild imaginations; they will be realities within a few years. Hence, it is essential that structural engineers begin to consider such structures—their requirements and the nature of the environment in which they will exist. At the moment, an extensive program in this area probably is not justified, but certainly work must be begun. It is essential that close contact be maintained with the space scientists to obtain the necessary information concerning the environments in which such structures may exist (be it zero-gravity space or on the surface of the moon or a planet), and to assimilate this information as rapidly as it becomes available to develop structural forms, structural materials, and means of fabrication consistent with these environments and the functional requirements of the needed structures. [1, page 308]

9

Methods of Analysis and Design	10-year Cost (Millions of $)
Underwater Structures	9
New Structural Materials and Forms	20
Implementation of New Knowledge	5
	Total 228

Typical interesting excerpts follow:

Technology has now advanced to the point that such simplifications are no longer necessary or justifiable, and it is proper that methods of analysis and design be updated to reflect more completely the behavior of the entire structure as the dynamic system that it really is.[1, page 288]

... development of adhesives which might be expected ultimately to replace welding, riveting, and bolting as connection methods for metal structures.[1, page 288]

... account must be taken of the statistical and probabilistic aspects of material properties, structural loadings, fabrication techniques, and workmanship quality and uniformity in the design of structures. Structural optimization techniques ... must be developed.[1, page 293]

... of the infinitive variety of shell forms that are available, only a comparative few of the simpler types have been studied to any significant extent.[1, page 293]

The mechanism of load transfer from reinforcement to the surrounding concrete is not yet fully understood.[1, page 299].

The estimates and quotations seem to speak for themselves. We still have not arrived, but as engineers, we are where we are because we have fought the timeless battle of safety versus economy. This battle is never-ending! As an engineer, this will become your battle too—a battle replete with personal excitement, challenge, and reward—a battle that consistently makes our world a better place in which to live and grow.

REFERENCE

1. Committee on Research, "Research Needs in Structural Engineering for the Decade 1966–1975," *Journal of the Structural Division*, American Society of Civil Engineers, V. 92, ST-5 (October 1966).

UNIVERSALS

This elaboration of sec. 1-2.6 contains highly important, although possibly redundant, material. It is referred to throughout the text to correlate what might otherwise be viewed as disjointed information.

A-1 MOHR'S CIRCLE

Mohr's circle is a graphical representation of a tensor of second order. As such, it may be applied to problems involving stresses, strains, and moments of inertia related to two-dimensional space. In all these cases coordinates representing horizontal and vertical ordinates (H, V) are expressed in the form

$$H = T + C \cos 2\theta - K \sin 2\theta \qquad (A\text{-}1.1)$$

and

$$V = C \sin 2\theta + K \cos 2\theta. \qquad (A\text{-}1.2)$$

Squaring each of these, we have

$$(H - T)^2 = C^2 \cos^2 2\theta - 2CK \sin 2\theta \cos 2\theta + K^2 \sin^2 2\theta \quad (A\text{-}1.3a)$$

and

$$V^2 = C^2 \sin^2 2\theta + 2CK \sin 2\theta \cos 2\theta + K^2 \cos^2 2\theta \qquad (A\text{-}1.3b)$$

which, when summed, yield a single equation:

$$(H - T)^2 + V^2 = C^2 + K^2. \qquad (A\text{-}1.4)$$

This equation is for a circle displaced positively T units from the origin along the horizontal axis, and with radius of $\sqrt{C^2 + K^2}$.

Geometry for a general point (H, V) is shown in Fig. A-1.1. From this sketch ($\sin \alpha = K/\sqrt{C^2 + K^2}$),

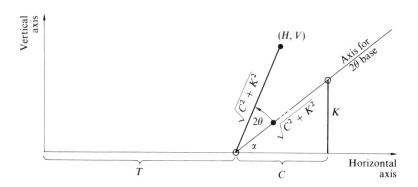

Fig. A-1.1   Geometry of Point (H, V)

$$
\begin{aligned}
H &= T + \sqrt{C^2 + K^2} \cos(2\theta + \alpha) \\
&= T + \sqrt{C^2 + K^2}\,(\cos 2\theta \cos \alpha - \sin 2\theta \sin \alpha) \\
&= T + \sqrt{C^2 + K^2}\left[\cos 2\theta\,(C/\sqrt{C^2 + K^2}) - \sin 2\theta\,(K/\sqrt{C^2 + K^2})\right]
\end{aligned}
$$

which reduces to

$$H = T + C \cos 2\theta - K \sin 2\theta. \qquad [\text{A-1.1}]$$

This indicates that angle 2θ is measured from an axis defined by a radius rotated through an angle α. End points of this radius have coordinates (T, O) and $(T + C, K)$. This axis is more commonly defined as a diameter having end points (H_x, K) and $(H_y, -K)$, as indicated in Fig. A-1.2. With H_x, H_y, and K known, T is found as $(H_x + H_y)/2$ and C as $(H_x - H_y)/2$. Given information about angle 2θ, any (H, V) coordinates can be evaluated. Conversely, angles 2θ can be found for locations of:

Maximum H, the point where righthand side of circle crosses horizontal axis, and coordinates are $(T + \sqrt{C^2 + K^2}, 0)$.

Minimum H, the point where left-hand side of circle crosses horizontal axis, and coordinates are $(T - \sqrt{C^2 + K^2}, 0)$.

Maximum V, the peak of the circle of coordinates $(T, \sqrt{C^2 + K^2})$.

Although Mohr's circle yields results by measuring from scaled drawings, it usually is used in roughly sketched form as an aid to efficient, error-free calculation. This procedure eliminates the need to either memorize or look up complex formulas, and gives a better picture of the actual physical phenomena.

A-1.1 Application of Mohr's Circle to Stress.   Figure A-1.3 shows a typical

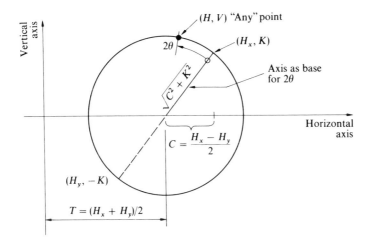

Fig. A-1.2   Generalized Mohr's Circle

elemental planar section of any generalized structural member under stress. Another free body of this section is shown with resultants S_n and S_s acting on any plane θ. Finally, sign conventions are shown for both normal and shearing stresses.

Summation of forces (1) parallel to S_n and (2) parallel to S_s yields the following equations if (a) $S_{yx} \equiv S_{xy}$ (sec. A-4), (b) $dx/ds \equiv \sin\theta$, (c) $dy/ds \equiv \cos\theta$, (d) $\sin^2\theta \equiv (1 - \cos 2\theta)/2$, (e) $\cos^2\theta \equiv (1 + \cos 2\theta)/2$, and (f) $\sin 2\theta \equiv 2\sin\theta\cos\theta$:

$$S_n = \frac{S_x + S_y}{2} + \frac{S_x - S_y}{2}\cos 2\theta - S_{xy}\sin 2\theta; \qquad (A\text{-}1.5)$$

$$S_s = \frac{S_x - S_y}{2}\sin 2\theta + S_{xy}\cos 2\theta; \qquad (A\text{-}1.6)$$

where $S_x \equiv$ normal (axial) horizontal stress, $S_y \equiv$ normal (axial) vertical stress, $S_{xy} \equiv$ shearing (tangential) stress (vertical faces), $S_{yx} \equiv$ shearing (tangential) stress (horizontal faces), $S_n \equiv$ normal force on any plane at angle θ, $S_s \equiv$ shearing force on any plane at angle θ.

Equations (A-1.5) and (A-1.6) for S_n and S_s, or their Mohr's circle representations, show stresses only on a particular plane. Since an infinite number of planes can pass through any point, with an unlimited number of θ angles, an infinite number of S_n and S_s values exist at any particular point. The equations, then, tell what is happening on only one particular plane at a time.

These equations are of the same form as Eqs. (A-1.1) and (A-1.2), with T replaced by $(S_x + S_y)/2$, C by $(S_x - S_y)/2$, and K by S_{xy}. By comparison to Fig. A-1.2, then, coordinates to the diameter of Mohr's circle are (S_x, S_{xy}) and $(S_y, -S_{xy})$. Angles 2θ measured from an axis through this diameter correspond to angles θ on the physical structure measured from the X axis.

In these equations θ is the only variable; therefore, maximum and minimum values are found by setting derivatives with respect to θ equal to zero. For the former, $dS_n/d\theta$ results in $\tan 2\theta = -2S_{xy}/(S_x - S_y)$. This expression is related to a right triangle with horizontal leg $(S_x - S_y)/2$, vertical leg S_{xy}, and hypotenuse $\sqrt{[(S_x - S_y)/2]^2 + S_{xy}^2}$ then,

$$\sin 2\theta = -S_{xy}/\sqrt{[(S_x - S_y)/2]^2 + (S_{xy})^2},$$

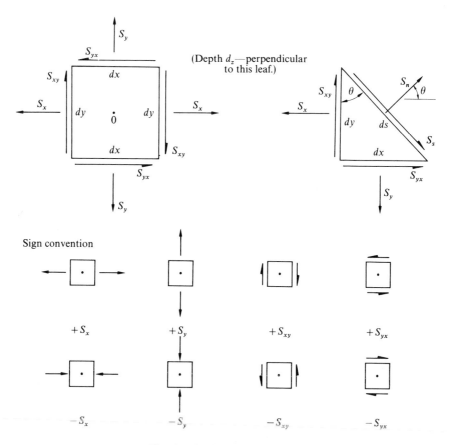

Fig. A-1.3 Stress at a Point

and

$$\cos 2\theta = (S_x - S_y)/2\Big/\sqrt{[(S_x - S_y)/2]^2 + (S_{xy})^2}.$$

Substitution of these expressions into Eqs. (A-1.5) and (A-1.6), after reduction, results in

$$S_{nm} = \frac{S_x + S_y}{2} \pm \sqrt{[(S_x - S_y)/2]^2 + (S_{xy})^2}. \tag{A-1.7}$$

In a similar way, with $\tan 2\theta = + S_{xy}/\sqrt{(S_x - S_y)}/2$, obtained by setting the derivative of Eq. (A-1.6) to zero,

$$S_{sm} = \pm\sqrt{[(S_x - S_y)/2]^2 + (S_{xy})^2}, \tag{A-1.8}$$

where $S_{nm} = $ maximum normal (axial) stress and $S_{sm} = $ maximum shearing stress.

This same result is obtained from Mohr's circle in the following manner (refer to Fig. A-1.4)

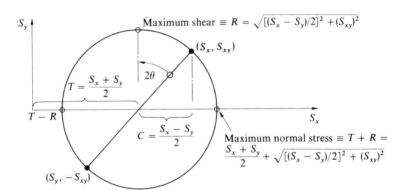

Fig. A-1.4   Maximum Shear and Normal Stress

1. Plot (S_x, S_{xy}) and $(S_y, - S_{xy})$.
2. Connect these points to obtain diameter of Mohr's circle.
3. Sketch circle.
4. Find distance to center of circle $T = (S_x + S_y)/2$.
5. Find length of horizontal projection of radius on S_x axis $\equiv C = (S_x - S_y)/2$.
6. Find radius by Pythagorean theorem:

$$R = \sqrt{[(S_x - S_y)/2]^2 + (S_{xy})^2.}$$

7. Maximum shear, equal to R, occurs at top of circle with $\tan 2\theta = 2S_{xy}/(S_x - S_y)$; Eq. (A-1.8) agrees with R.

8. Maximum stress occurs at

$$T + R = \frac{S_x + S_y}{2} + \sqrt{[(S_x - S_y)/2]^2 + (S_{xy})^2,}$$

corresponding to Eq. (A-1.7).

As a numerical example, calculations are shown for a certain point in a particular prestressed concrete beam with vertical shear of 1000 psi, horizontal compression due to prestress tendons of 5000 psi, and horizontal tension due to bending moment of 2000 psi. Principal tensile stress, compressive stress, shearing stress, and the angle each of these makes with the neutral axis of the beam is determined by: (see Fig. A-1.5).

1. $S_x = -5000 + 2000 = -3000$ psi (compression).

2. $S_y = 0$ (no normal stress acts perpendicular to the neutral axis).

3. $S_{xy} = 1000$ psi.

4. Plot the extremities of the chord as (S_x, S_{xy}) $(-3000, 1000)$; and $(S_y, -S_{xy})$ $(0, -1000)$. (See Fig. A-1.5.)

5. Center of the circle is at -1500 determined as $(-3000 + 0)/2$.

6. Radius of the circle is $\sqrt{1500^2 + 1000^2} = 1800$ psi.

7. Maximum values of axial stress lie along the horizontal axis at extremities of the circle. The maximum shearing stress is parallel to the vertical axis and equal to radius of the circle.

8. To determine the angle at which these stresses occur, compute 2θ as $\tan^{-1}(1000/1500) = 33.6°$. A plane $\theta = 16.8°$ (half of 33.6) counterclockwise from the neutral axis has the 3300 psi compressive stress while 300 psi tensile stress acts on a perpendicular plane. These stresses are depicted to scale in Fig. A-1.6.

9. Maximum shearing stress occurs at an angle (rotated clockwise from the neutral axis) of $(90 - 33.6)/2 = 28.2°$. Corresponding normal stress is 1500 psi compression.

A-1.2 Application of Mohr's Circle to Moments of Inertia. Necessary basic relations from statics are summarized as:

$$I_x = \int y^2 dA, \text{ moment of inertia about } X \text{ axis,} \tag{A-1.9}$$

$$I_y = \int x^2 dA, \text{ moment of inertia about } Y \text{ axis,} \tag{A-1.10}$$

$$I_{xy} = \int xy dA, \text{ product of inertia about centroidal axis,} \tag{A-1.11}$$

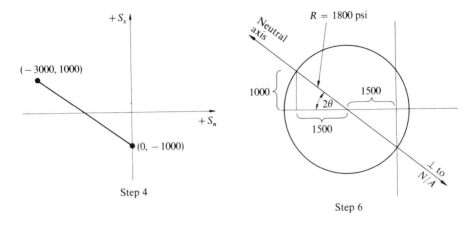

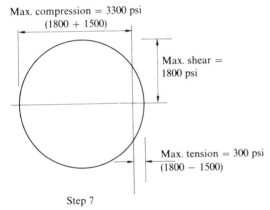

Fig. A-1.5 Solution by Mohr's Circle

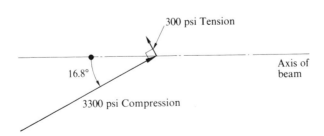

Fig. A-1.6 Maximum Stresses

$$I_p = \int p^2 dA, \text{ polar moment of inertia,} \tag{A-1.12}$$

where $p^2 = x^2 + y^2$ so

$$I_p = I_x + I_y, \tag{A-1.13}$$

(*Note.* This is an invariant, that is, the sum of any two moments of inertia taken about mutually perpendicular axes is equal to a constant polar moment of inertia.)

$$I_x = I_0 + Ad^2, \text{ transfer moment of inertia} \tag{A-1.14}$$

and $r = \sqrt{I/A}$, where $r \equiv$ radius of gyration, $I \equiv$ moment of inertia, and $A \equiv$ cross-sectional area.

Although moment of inertia can be determined about any axis, axes XY and $X'Y'$ (Fig. A-1.7) are taken as a companion set of centroidal axes. Then

$$I'_y = \int (x')^2 dA = \int (y \sin \theta + x \cos \theta)^2 \, dA, \tag{A-1.15}$$

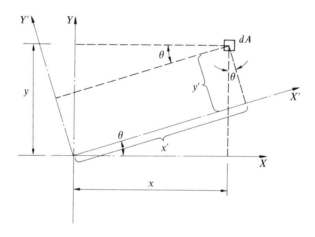

Fig. A-1.7   Rotation of Axes—Effect on Moment of Inertia

so integrating and using the basic relations from statics summarized above,

$$I'_y = \frac{I_x + I_y}{2} - \frac{I_x - I_y}{2} \cos 2\theta + I_{xy} \sin 2\theta. \tag{A-1.16}$$

Similarly,

$$I'_x = \frac{I_x + I_y}{2} + \frac{I_x - I_y}{2} \cos 2\theta - I_{xy} \sin 2\theta, \tag{A-1.17}$$

and

$$I'_{xy} = \frac{I_x - I_y}{2} \sin 2\theta + I_{xy} \cos 2\theta. \qquad \text{(A-1.18)}$$

These equations are also in the form of Eqs. (A-1.1) and (A-1.2), so Mohr's circle is directly applicable.

As an example, maximum and minimum moments of inertia for a L 9 x 4 x 1, using data from *Steel Construction*, are found, where $I_x = 97.0$ in.[4], $I_y = 12.0$ in.[4], $R_{min} = 0.834$, $\tan \theta = 0.203$, $A = 12.0$ in.[2].

1. From $\tan \theta = 0.203$, $\theta = 11.5°$ and $2\theta = 23°$.

2. For this angle of rotation, maximum and minimum moments of inertia I'_x and I'_y are obtained with $I'_{xy} = 0$. Reordering Eq. (A-1.18) with I'_{xy} set at zero,

$$I_{xy} = \frac{I_y - I_x}{2} \tan 2\theta \qquad \text{(A-1.19)}$$

or

$$I_{xy} = \frac{12 - 97}{2} \tan 23 = -18 \text{ in.}^4.$$

3. Diameter end points (I_x, I_{xy}) and $(I_y, -I_{xy})$ are plotted in Fig. A-1.8(a). Base of the shaded triangle is computed as $(I_x - I_y)/2$ and its hypotenuse (radius of Mohr's circle) by the Pythagorean theorem.

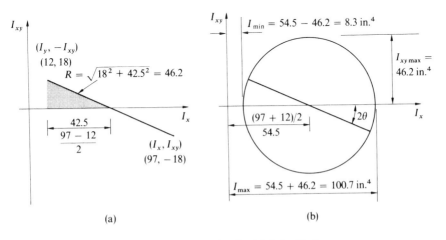

(a)   (b)

Fig. A-1.8 Solution by Mohr's Circle

4. Maximum and minimum moments of inertia are computed in part (b) of the figure, extremes on Mohr's circle. Maximum value for product of inertia is simply the radius.

5. Using the invariant as a check, we have

$$I_p = I_x + I_y = I_{max} + I_{min},$$

or

$$97 + 12 = 8.3 + 100.7 = 109.$$

6. As a secondary check, $2\theta = \tan^{-1}(18/42.5) = 23°$, so that $\theta = 11.5°$ and $\tan\theta = 0.203$, which checks with the value shown in *Steel Construction*.

A-1.3 Summary. Table A-1.1 correlates uses of Mohr's circle as illustrated herein. In addition, three points are restated for emphasis:

1. The sign of I_{xy} (or S_{xy}) always goes with I_x (or S_x); conversely, the opposite sign of I_{xy} (or S_{xy}) is associated with I_y (or S_y).

2. Radius to (I_x, I_{xy}) [or (S_x, S_{xy})], in general, is directly related to centroidal or neutral axis of member.

3. (a) 2θ rotation of Mohr's circle corresponds to (b) θ measured on physical structure; both rotations in same direction from respective axes.

Table A-1.1 Comparison of Coordinates

	Mohr's Circle	Stress	Moment of Inertia
Equations	(A-1.1)	(A-1.5)	(A-1.17)
	(A-1.2)	(A-1.6)	(A-1.18)
Horiz. X Coord.	H_x	S_x	I_x
Horiz. Y Coord.	H_y	S_y	I_y
Vert. Coord.	K	S_{xy}	I_{xy}
Coord. of any point	H	S_n	I'_x
	V	S_s	I'_{xy}

A-2 TABULATION OF A, y, Ay, $Ay^2 + I_0$

Frequent need to compute centroids, moments of inertia, and section moduli for irregular sections signifies efficient techniques are very useful. (As evident throughout this text, the approach is much more general than it might first appear.) When an attempt is made to determine a suitable cross section by trial and error, the A, y, Ay, $Ay^2 + I_0$ tabular form is almost mandatory. The technique requires the following steps:

1. Choose a working axis. For slide-rule computations, as this axis nears the actual centroid, the results become more accurate. Except as influenced by efficiency and number size, *any* axis will do.

2. Keep this working axis unchanged throughout computations involving any cross section.

3. Break the cross section into simple elements (generally rectangular or triangular) and compute their respective areas A and arms y. These arms are measured between the *centroid of the element* and the *working axis of the cross section* (positive when upward from the centroidal axis, negative when downward).

4. Compute moment of inertia I of each element about its own centroidal axis.

5. Compute Ay and Ay^2 for each element.

6. Sum all A and all Ay, then compute $\bar{y} = \Sigma Ay / \Sigma A$ which represents the distance between *true centroidal axis of the cross section* and the *assumed working axis*.

7. Sum all Ay^2 and I of every element to obtain the total I *about the working axis*.

8. Because $I = I_0 + Ay^2$ (transfer theorem), $I_0 = \Sigma (Ay^2 + I$ of elements) $- (\Sigma Ay)\bar{y}$. Therefore, *subtract* the product of ΣAy times $\bar{y}$ from I obtained with respect to the working axis to get the (true, minimum) moment of inertia of the entire cross section with respect to its true centroidal axis.

Case 1. A typical tabulation for the L 9 x 4 x 1 shown in Fig. A-2.1 appears in Table A-2.1. (The working axes X and Y have been chosen from among several reasonable ones included in the infinite possibilities.)

The figures shown in brackets in the table normally are not shown, although it is a good idea to write the values on a separate line so that I is easy to total. Note that $\bar{x} = 6.0/12 = 0.5$ whereas $\bar{y} = 0/12 = 0$. The quantity -3.00 subtracted from 100.00 represents $(\bar{x})(\Sigma Ax)$, or $(0.5)(6.0)$, and is always negative because the moment of inertia is a minimum about a centroidal axis.

Case 2. An alternate orientation for the angle is shown in Fig. A-2.2. For this case, the tabular solution is in Table A-2.2. This comparison indicates special characteristics of problems involving products of inertia. First, product of inertia is always zero about any axis of symmetry. Second, product of inertia about the principal axis of an entire section involves the transfer function Axy [in this case, $12(-1.0)(-1.5) = 18.0$] which is *always* taken as negative.

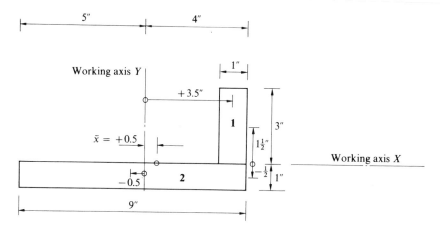

Fig. A-2.1   L9 × 4 × 1 Angle (*Case 1*)

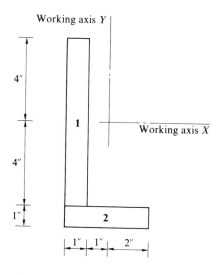

Fig. A-2.2   L9 × 4 × 1 Angle (*Case 2*)

Third, a change in orientation of the axes changes the sign for the product of inertia.

Case 3. A WF beam with cover plates, Fig. A-2.3, is used as a final example. Typically, it is required to find the section moduli (top and bottom) for several possible combinations of the beam and plate. Section modulus Z is defined as I/y. The calculations are presented in tabular form in Table A-2.3. Note that (1) all calculations are performed with respect to working axis; (2) y_t and y_b

Table A-2.1 Computations for Fig. A-2.1

Point	A	x	Ax	Ax² + I_y	y	Ay	Ay² + I_x	Axy
1	3	3.5	10.5	36.75	1.5	4.5	6.75	15.75
			[3(1³/12)] =	0.25		[1(3³/12)] =	2.25	2.25
2	9	−0.5	−4.5	2.25	−0.5	−4.5	2.25	
			[1(9³/12)] =	60.75		[9(1³/12)] =	0.75	
	12	x̄ = 0.5	6.0	100.00	ȳ = 0	0	12.00	18.00
			−3.00			−0	0 [12(0.5)0]	
			97.00			12.00	18.00	

$x \longrightarrow$

Table A-2.2 Computations for Fig. A-2.2.

Point	A	x	Ax	Ax² + I_y	y	Ay	Ay² + I_x	Axy
1	8.0	−1.5	−12.0	18.0	0	0	0	0
			[8(1³/12)] =	0.7		[1(8³/12)] =	42.7	
2	4.0	0	0	0	−4.5	−18.0	81.0	0
			[1(4³/12)] =	5.3		[4(1³/12)] =	0.3	
	12.0	x̄ = −1.0	−12.0	24.0	ȳ = −1.5	−18.0	124.0	−18.0 [12.0(−1.0)(−1.5)]
			−12.0				−27.0	−18.0
			12.0				97.0	

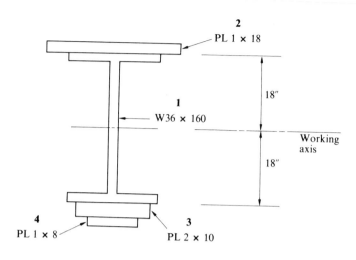

Fig. A-2.3   Cover-Plated Beam (*Case 3*)

Table A-2.3   Computations for Fig. A-2.3

Part	A	y	Ay	$Ay^2 + I$	y_t	Z_t	y_b	Z_b
1 W 36 × 160	47.1	0	0	9760	18.00	542	18.00	542
2 PL 1 × 18	18.0	18.5	333	6160				
3 PL 2 × 10	20.0	− 19.0	− 380	7220				
4 PL 1 × 8	8.0	− 20.5	− 164	3360				
1 + 2	65.1	5.12	333	15920				
				− 1710				
				14210	13.88	1023	23.12	615
1 + 3	67.1	− 5.66	− 380	16980				
				− 2150				
				14830	23.66	627	14.34	1034
All	93.1	− 2.26	− 211	26500				
				− 480				
				26020	21.26	1224	18.74	1388

Note: The I component of each plate about its own centroid is computed as: for **2**, $18(1)^3/12 = 1.5$; for **3** $10(2)^3/12 = 6.67$; and for **4** $8(1)^3/12 = 0.67$. These values were *consciously* neglected because they were so small in comparison to the other quantities.

must extend to the extremity of the plate. For instance, for **1** + **2**, with a plate at the top, $y_t = 18 + 1 - 5.12 = 13.88$ while $y_b = 18 + 5.12 = 23.12$. For all parts acting, $y_t = 18 + 1 + 2.26 = 21.26$ while $y_b = 18 + 2 + 1 - 2.26 = 18.74$.

In the design process, when cover plate size must be determined, quickest results are obtained by (1) maintaining a fixed working axis, (2) adding new sections or partial sections as new parts (quantities above double lines in Table A-2.3), referred to working axis, and (3) computing combinations independently. These procedures are even more important when attempting to "cut in" a prestressed concrete beam where it is essential to visualize the pattern of changes in A, I, and Z due to small iterative changes in cross section.

A-3 $Ely'' = M$

Figure A-3.1(a) shows a beam before bending and (b) shows a segment of that same beam after it has deflected due to applied loading M_{ext}. Although the applied moment deflects this beam into a variable arc, for a very small segment with ds (distance along the neutral axis of the curved member) nearly equal to dx (distance parallel to the original undeflected beam axis), curvature essentially is that of a circular arc or, for this very small length, a straight line.

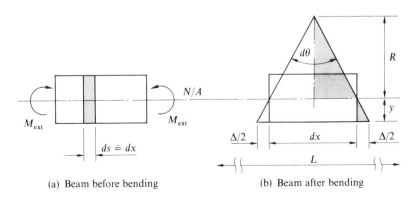

(a) Beam before bending   (b) Beam after bending

Fig. A-3.1   Segment of a Beam Before and After Bending

A-3.1 Elastic Equation. Geometry of the larger shaded area in Fig. A-3.1(b) shows for small θ:

$$\tan \frac{d\theta}{2} \doteq \frac{d\theta}{2} = \frac{dx/2}{R}, \tag{A-3.1}$$

where $d\theta \equiv$ angle which subtends arc length $ds \doteq dx$; $dx \equiv$ arc length $\doteq$ true arc length ds; and $R \equiv$ radius of curvature.

Similarly, from the smaller shaded triangle,

$$\tan \frac{d\theta}{2} \doteq \frac{d\theta}{2} = \frac{\Delta/2}{y},$$ (A-3.2)

where $\Delta \equiv$ extension of dx due to curvature; and $y \equiv$ distance measured from neutral axis to specified fiber; so that $\Delta/2$ and y vary in proportion to their distance from the neutral axis (N/A).

Equating these expressions for $d\theta/2$, we have

$$\frac{dx/2}{R} = \frac{\Delta/2}{y},$$ (A-3.3)

which when solved for Δ/dx yields the equation $\Delta/dx = y/R$. The term Δ/dx represents elongation per unit length which is, by definition, strain and denoted by ϵ. Therefore

$$\epsilon = \frac{\Delta}{dx} = \frac{y}{R}.$$ (A-3.4)

But by Hooke's law, $S = E\epsilon$ or $\epsilon = S/E$. Thus,

$$\epsilon = \frac{y}{R} = \frac{S}{E},$$ (A-3.5)

where $S \equiv$ stress (psi), $E \equiv$ modulus of elasticity (psi), and $\epsilon \equiv$ strain (in./in.).

To simplify, the *elastic equation* is obtained:

$$\frac{1}{R} = \frac{S}{Ey}.$$ (A-3.6)

A-3.2 Curvature Equation. With the beam in static equilibrium, external moments are equal to internal moments at any given cross section, that is, ($M_{ext} = M_{int}$). Internal moments are defined by couples having components of force and distance. Incremental force ΔF acting on a small incremental area is expressed as SdA (lb/in.^{2} times in.^{2} = lb) for a beam of unit width. Distance to this incremental area measured from the neutral axis is called the *arm* and is represented in Fig. A-3.2 by y (which varies between the limits of c_1 and c_2, respective distances to the bottom and top extreme fibers as measured from the neutral axis). Thus incremental internal moment can be expressed as $(y)\Delta F$ or $y(SdA)$, which yields total resisting moment as

$$M_{int} = \int_{c_1}^{c_2} y(SdA).$$ (A-3.7)

By substitution (1) $M_{ext} = M_{int}$ and (2) $S = Ey/R$ [the elastic Eq. (A-3.6)], Eq. (A-3.7) takes the form

$$M_{ext} = \int_{c_1}^{c_2} y\frac{Ey}{R}\,dA = \frac{E}{R}\int_{c_1}^{c_2} y^2 dA, \qquad \text{(A-3.8)}$$

where E and R are removed from the integral sign because they are constants. The remaining integral of $y^2 dA$ is simply moment of inertia I, and results in the *curvature equation*:

$$\frac{1}{R} = \frac{M}{EI}. \qquad \text{(A-3.9)}$$

A-3.3 Flexure Formula. Although only indirectly related to the objective of this section, it is interesting to combine the elastic and static equations,

$$\frac{1}{R} = \frac{S}{Ey} = \frac{M}{EI}.$$

From this, the *flexure formula* is readily obtained as

$$S = \frac{My}{I}. \dagger \qquad \text{(A-3.10)}$$

A-3.4 Curvature for Any Planar Curve. In both elastic and static equations an expression for radius, or more exactly reciprocal of radius, occurs. It is important to recognize that this expression is directly related to the definition of a curve as given in analytics.

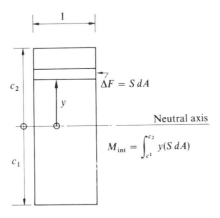

Fig. A-3.2   Beam Cross Section

---

† In the Theory of Elasticity and in the design of metal structures, tension is usually regarded as positive. In the design of concrete structures, compression is usually considered positive. Thus the sign of this expression is plus or minus depending upon the sign convention used.

Figure A-3.3 illustrates any general curve AB with tangents drawn to points P and Q. The tangent to point P makes an angle of B with the X axis and the tangent drawn to point Q makes an angle B plus ΔB with the X axis. Therefore, radian angular difference between these two tangents is represented by angle ΔB. By definition, curvature is the limiting value of the average curvature as Δs approaches zero. Thus

$$K = \lim_{\Delta s \to 0} \frac{\Delta B}{\Delta s} = \frac{dB}{ds}. \tag{A-3.11}$$

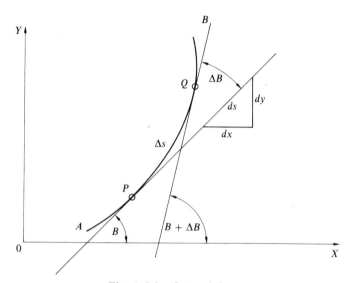

Fig. A-3.3   General Curve

The tangent of angle B may be expressed as the differential dy/dx (often written y'), so that

$$B = \tan^{-1} y',$$

and taking the derivative of B with respect to x, yields

$$\frac{dB}{dx} = \frac{d(\tan^{-1} y')}{dx} = \frac{y''}{1 + (y')^2}. \tag{A-3.12}$$

But by the Pythagorean theorem, $dx^2 + dy^2 = ds^2$, so

$$\frac{ds}{dx} = \left[1 + (y')^2\right]^{\frac{1}{2}} \tag{A-3.13}$$

Division of Eq. (A-3.12) by (A-3.13) results in

$$\frac{dB}{ds} = \frac{y''}{[1 + (y')^2]^{3/2}}. \tag{A-3.14}$$

Because $1/R = K = dB/ds$, $y' = dy/dx$, and $(y')^2 = (dy/dx)^2$, the final expression is often written as

$$K = \frac{1}{R} = \frac{d^2y/dx^2}{[1 + (dy/dx)^2]^{3/2}}. \tag{A-3.15}$$

A-3.5 Equation of Elastic Curve. Equation (A-3.15) expresses curvature for any planar curve. However, the term $(dy/dx)^2$ is the square of beam slope and is much less than 1 for most real beams. Therefore this equation is usually simplified to:

$$\frac{1}{R} = y'' = \frac{d^2y}{dx^2}. \tag{A-3.16}$$

Equating the curvature equation $1/R = M/EI$ to the reduced form of basic curvature equation $1/R = y''$, we get the final equation for the elastic curve:

$$EIy'' = M \tag{A-3.17}$$

The relationships between deflection, slope, bending moment, shear, and load, obtained by differentiation or integration, are summarized in Table A-3.1. These equations also show (by definitions of calculus, but possibly with need for some negative signs in order to agree with the particular sign convention actually used):

Change in Δ between two points = Area of θ diagram between same two points;

Change in θ between two points = Area of M/EI diagram between same two points;

Change in M between two points = Area of V diagram between same two points;

Change in V between two points = Area of load diagram between same two points.

Magnitude of Load at any point = slope of V diagram at same point;

Magnitude of V at any point = Slope of M diagram at same point;

Magnitude of M/EI at any point = Slope of θ diagram at same point;

Magnitude of θ at any point = Slope of Δ diagram at same point.

Table A-3.1   Fundamental Relationships

Deflection	$= \Delta =$	y	$= y$
Slope	$= \theta =$	y'	$= dy/dx$
Moment	$= M = Ely''$		$= EI(d^2y/dx^2)$
Shear	$= V = Ely'''$		$= EI(d^3y/dx^3)$
Load	$= w = Ely''''$		$= EI(d^4y/dx^4)$

The student should be aware that certain incongruities occur because of reducing $[1 + (y')^2]^{3/2}$ to 1. This for instance, is the reason why Euler's equations (Chap. 3) do not lead to values for lateral displacements. Another example is that of a simply supported weightless beam having a constant EI, loaded only by two equal and opposite end moments. Such a beam has a constant bending moment and, therefore, is deflected into a circular arc [Eq. (A-3.9); constant radius R]. However, any method of deflection calculation which is based either directly or indirectly on $Ely'' = M$ yields a computed deflection equation of a parabola. Typically, straight beams deflect in a very flat arc, slope approaches zero, and the differences between parabolic and circular curvature is of no physical consequence.

A-4   SHEARING STRESS

Figure A-4.1 represents an incremental length of beam where C_1 and C_2 are compressive resultants of stress blocks on two opposite faces.

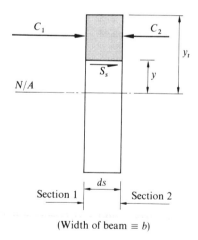

(Width of beam $\equiv b$)

Fig. A-4.1   Shear on Elemental Section

If S_s is shearing stress at a distance y from the neutral axis, then

$$S_s(ds)b = C_2 - C_1, \tag{A-4.1}$$

where $b \equiv$ beam width and $ds \equiv$ length of increment, and it is assumed that moment arms to the centroid of stress blocks are essentially equal.

The stress block resultant C_2 is computed as

$$C_2 = \int_y^{yt} \sigma_2 dA = \int_y^{yt} \frac{My}{I} dA, \tag{A-4.2}$$

where $dA \equiv$ differential area of element which is b wide; and $\sigma_2 \equiv$ flexural stress. Equating these two expressions (including one similar to the above for C_1), we have

$$S_s(ds)b = \int_y^{yt} (M_2 - M_1)y \frac{dA}{I}, \tag{A-4.3}$$

where $M_2 - M_1 = dM$, the change in bending moment between the two cross sections. Rearranging, we see that

$$S_s = \frac{dM}{ds} \left(\frac{1}{Ib}\right) \int_y^{yt} ydA, \tag{A-4.4}$$

where $dM/ds = V$ (shear) and

$$\int_y^{yt} ydA = A\bar{y}$$

is defined as Q. By simplification an equation for both horizontal and vertical shearing stress is obtained:

$$S_s = \frac{VQ}{Ib}. \tag{A-4.5}$$

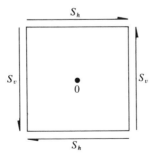

Fig. A-4.2   Shear on Square Element

Figure A-4.2 presents an elemental square with vertical shear S_v and horizontal shear S_h. To maintain static equilibrium, summation about point 0 must be zero. For this reason $S_v = S_h$; the derivation for horizontal stress is equally valid for vertical shear. When width is not included, Eq. (A-4.5) reduces to

$$S_{sf} = \frac{VQ}{I}, \tag{A-4.6}$$

where shear flow S_{sf} is given in units of pounds per inch.

Direction of shear flow usually can be determined by a simple two-step procedure (Fig. A-4.3):

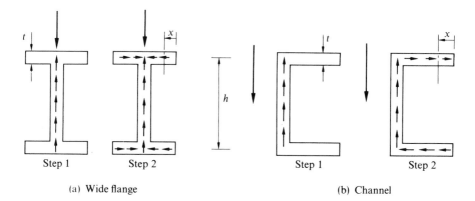

| Step 1 | Step 2 | | Step 1 | Step 2 |

(a) Wide flange   (b) Channel

Fig. A-4.3 Sketching Direction of Shear Flow (a) Wide Flange (b) Channel

1. Show shearing stress in element(s) of cross section which *directly* oppose load.
2. Show balance of stresses to maintain consistent flow.

To obtain the magnitude of the shear flow, define Q so that it is zero at a point where, by inspection, shear stress is zero. Shear flow for points x distance from flange ends in Fig. A-4.3 are:

$$\frac{V}{I_x}Q = \frac{V}{I_x}(\text{area beyond point})(\text{arm to such area}) = \frac{V}{I_x}(tx)\frac{h}{2}. \tag{A-4.7}$$

Total shear *force* in the top flange of the channel is

$$\begin{aligned} F_{flg} &= \left(\frac{VQ}{I_x}\right)_{avg}(b) = [(VQ/I_x)_{max}/2]\,(b) \\ &= \frac{V}{I_x}(tb)(h/2)\frac{1}{2}(b) = \frac{Vtb^2h}{4I_x}. \end{aligned} \tag{A-4.8}$$

PROBLEMS

A-1 given: _____ (L9 x 4 x $\frac{1}{2}$, L9 x 4 x $\frac{5}{8}$, L8 x 6 x 1, L8 x 6 x $\frac{1}{2}$, L8 x 4 x 1, or L8 x 4 x $\frac{9}{16}$

    verify: Properties for designing, shown in *Steel Construction*, page 1–58.

    find: Principal moments of inertia.

A-2 given: Composite section of Fig. A-P.1 with b = _____ (40, 44, or 50 in.), t = _____ (4, 5, or 6 in.), p = _____ (1, 2, 3 in.). Beam _____ (W27 x 177, W30 x 132, or W33 x 240). Plate _____ (PL $\frac{3}{4}$ x 14, PL $\frac{7}{8}$ x 12, or PL 1 x 10).

Comment: As explained in sec. 2-2.2, and elaborated on elsewhere, the concrete slab must be converted to equivalent steel. For this problem, divide only the horizontal dimensions of concrete by 10 and proceed.

    find: I, y_t, and y_b for (a) beam and plate, (b) beam and slab, and/or (c) beam, slab, and plate.

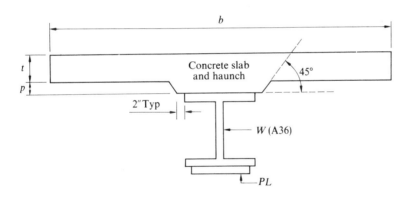

Fig. A-P.1 Composite Section

A-3 given: Girder of _____ (Fig. A-P.2 or Fig. A-P.3).

    find: I, y_t, y_b, Z_t, A_b.

A-4 given: Cross sections of Fig. A-P.4 with b = _____ (10, 11, or 12 in.), t = _____ (1, 1$\frac{1}{2}$, or 2 in.), w = _____ (0.5, 0.75, or 1 in.). Maximum shearing stress = 15 ksi.

    find: (a) Maximum shears. (b) For maximum shear: shearing stress at

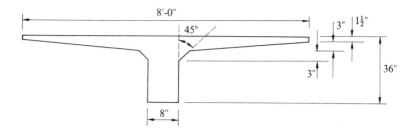

Fig. A-P.2   Single Tee Girder

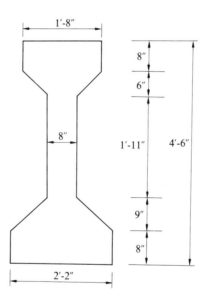

Fig. A-P.3   AASHO Type IV Prestressed Girder

the neutral axis (N/A); 4 in. above N/A; 8 in. above N/A; 12 in. above N/A; at junstion of flange and web (two values for WF only).

plot: Results of (b) using 1 in. $=$ 5 in. vertical, and 1 in. $=$ 5 ksi horizontal.

compute: V/A for rectangle; and V/A_{web} for WF.

conclude: To use kV/A rather than VQ/Ib, what must k be for rectangles and what must k be for WF (approximate).

A-5 given: Simple span beam of Fig. A-P.5 with $L =$ _____ (40, 60, 80, or

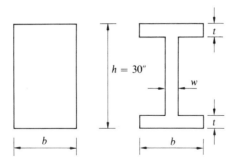

Fig. A-P.4 Rectangle versus WF

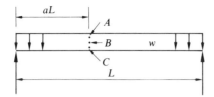

Fig. A-P.5 Simple Span

57.3 ft), $a =$ _____ (0.3, 0.4, or 0.5), $w =$ _____ (1.1, 1.23, or 1.31 k/ft); and points A, B, and C at one-quarter points of depth with _____ (rectangular or WF) cross section of Fig. A-P.4 with dimensions as in problem A-4.

find: Normal and shearing stresses at point _____ (A, B, or C) with planes making angles of multiples of 30° from horizontal.

plot: (a) Shearing stress magnitudes about point. (b) Normal stress magnitudes about point.

BASIC EQUATIONS OF THEORY OF ELASTICITY

The fundamental equations of the theory of elasticity are based on (1) static equilibrium, (2) relationship between strain and displacement, (3) relationship between stress and strain, and (4) compatibility. These equations can be derived using mathematics of tensors, vectors, or geometry. Further, the developments can be broadly general (using, for instance curvilinear coordinates and/or nonlinearities) or be highly restricted.

Because the purpose of this Appendix is to provide necessary background for development of code specifications, simple linearized equations for linear elastic materials are adequate. Such derivation devoid of mathematics of vectors or tensors also facilitates understanding of the physical significance attached to these expressions. For these reasons, the developments here are as simple as possible in leading to three-dimensional equations. As a mnemonic aid and as a very slight introduction, final results also are presented in tensor form.

A complete and detailed treatment of the material summarized in this chapter is available from books by Timoshenko and Goodier[1], Wang[2], Housner and Vreeland[3], and Volterra and Gaines[4].

B-1 STATIC EQUILIBRIUM (NEWTON'S SECOND LAW)

Figure B-1.1 depicts an elemental cube of material, located arbitrarily within a structural member, having faces parallel to an arbitrarily oriented coordinate axes system. In each of the three visible faces, three *typical* stresses (one normal and two tangential) are shown by vector arrows. However, of the nine typical stresses shown, only the three contributing to forces in the x direction are given symbolic representation.

Now, consider the two faces parallel to the X–Y plane. At any point in either face all stresses can be resolved into three resultants, each resultant

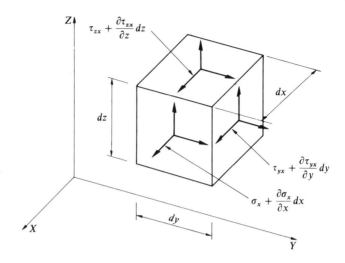

Fig. B-1.1 Elemental Stress Block Equilibrium

being parallel to a coordinate axis. Neglecting higher ordered terms, the magnitude of the stress may be assumed constant over such infinitely small surfaces, but the magnitude of the stress *does* vary between the two parallel faces. For instance, in moving from the lower horizontal face to the upper horizontal face, the shearing stress changes magnitude at the rate of $d\tau_{zx}/dz$, or to indicate that derivatives are obtained with respect to only one of three possible variables, as $\partial\tau_{zx}/\partial z$. Therefore the total shearing *force* in the lower face is

$$\tau_{zx}dxdy \tag{B-1.1}$$

(average stress times area), while the total shearing *force* on the upper face is

$$\left(\tau_{zx} + \frac{\partial\tau_{zx}}{\partial z}dz\right) dxdy \tag{B-1.2}$$

Summation of all forces† in the x direction, including the body force, yields

$$\left(\tau_{zx} + \frac{\partial\tau_{zx}}{\partial z}dz\right) dxdy + \left(\sigma_x + \frac{\partial\sigma_x}{\partial x}dx\right) dydz + \left(\tau_{yx} + \frac{\partial\tau_{yx}}{\partial y}dy\right) dxdz$$
$$- \tau_{zx}dxdy - \sigma_x dydz - \tau_{yx}dxdz + Xdxdydz = 0. \tag{B-1.3}$$

Equation (B-1.3) rapidly reduces to Eq. (B-1.4) below, and Eqs. (B-1.5) and (B-1.6) are derived in a similar manner:

---

† The elemental stress block shown has no surface forces because it is not located at a boundary of the member.

$$\frac{\partial \sigma_x}{\partial x} + \frac{\partial \tau_{yx}}{\partial y} + \frac{\partial \tau_{zx}}{\partial z} + X = 0, \qquad \text{(B-1.4)}$$

$$\frac{\partial \sigma_y}{\partial y} + \frac{\partial \tau_{zy}}{\partial z} + \frac{\partial \tau_{xy}}{\partial x} + Y = 0, \qquad \text{(B-1.5)}$$

$$\frac{\partial \sigma_z}{\partial z} + \frac{\partial \tau_{xz}}{\partial x} + \frac{\partial \tau_{yz}}{\partial y} + Z = 0. \qquad \text{(B-1.6)}$$

It is important to distinguish clearly between body forces and surface forces. In these equations, only body forces appear, and are produced by gravity, centrifugal forces, magnetic attraction, inertia terms from a d'Alembert dynamic formulation, etc. Thus a uniformly loaded simple beam would be subject to (a) body forces to account for the weight of the material in the member itself and (b) surface forces produced by superimposed loads. The latter do not affect these equations, instead they find their way into computations through boundary conditions.

B-1.1 Symbology and Tensor Notation. It is important now to clarify all terms and to introduce tensor notation. First, note that normal stress is subscripted by the direction in which it acts, while shearing stress is subscripted by (1) direction perpendicular to the face on which it acts followed by (2) direction in which it acts. A more uniform symbology uses double subscripting for all stresses, calling them τ and using the form of shearing stress subscripting. Thus, σ_x becomes τ_{xx} while τ_{xy} remains τ_{xy}.

An additional uniformity can be added by replacing x, y, and z with 1, 2, and 3, respectively. Thus σ_x becomes τ_{xx} which is written τ_{11} while τ_{xz} is written τ_{13}. Variables x, y, and z are replaced by x_1, x_2, and x_3.

By use of these symbols, Eqs. (B-1.4) through (B-1.6) can be written:

$$\frac{\partial \tau_{11}}{\partial x_1} + \frac{\partial \tau_{21}}{\partial x_2} + \frac{\partial \tau_{31}}{\partial x_3} + X_1 = 0$$

$$\frac{\partial \tau_{22}}{\partial x_2} + \frac{\partial \tau_{32}}{\partial x_3} + \frac{\partial \tau_{12}}{\partial x_1} + X_2 = 0 \qquad \text{(B-1.7)}$$

$$\frac{\partial \tau_{33}}{\partial x_3} + \frac{\partial \tau_{13}}{\partial x_1} + \frac{\partial \tau_{23}}{\partial x_2} + X_3 = 0$$

Or reordered:

$$\frac{\partial \tau_{11}}{\partial x_1} + \frac{\partial \tau_{21}}{\partial x_2} + \frac{\partial \tau_{31}}{\partial x_3} + X_1 = 0$$

$$\frac{\partial \tau_{12}}{\partial x_1} + \frac{\partial \tau_{22}}{\partial x_2} + \frac{\partial \tau_{32}}{\partial x_3} + X_2 = 0 \qquad \text{(B-1.8)}$$

$$\frac{\partial \tau_{13}}{\partial x_1} + \frac{\partial \tau_{23}}{\partial x_2} + \frac{\partial \tau_{33}}{\partial x_3} + X_3 = 0.$$

Or in summation form:

$$\sum_{j=1}^{j=3} \left[\left(\sum_{i=1}^{i=3} \frac{\partial \tau_{ij}}{\partial x_i} \right) + X_j = 0 \right] \tag{B-1.9}$$

Equation (B-1.9) reduces to the tensor form

$$\tau_{ij,i} + X_j = 0, \tag{B-1.10}$$

where $i = 1 \longrightarrow 3$ and $j = 1 \longrightarrow 3$ and:

1. Repeating a subscript indicates that each value of that subscripted term occurs in each equation.
2. Partial differentiation with respect to a specified variable is indicated by following the preliminary subscripts by a comma and the variable symbol.

B-1.2 Sign Convention. Each elemental cube consists of three sets of two parallel faces, each set being perpendicular to one of the three principal axes. Within each set, the face farthest in the positive direction is called the *near face* and the face farthest in the negative direction is called the *far face*. Thus, the elemental cube of Fig. B-1.1 shows only the three near faces.

In the near faces, stresses acting in the positive direction of the axes are called *positive*. Consequently a positive normal stress implies tension. Figure B-1.2 illustrates positive and negative stresses in the near faces.

By contrast, in the far faces, stresses acting *opposite* to the positive direction of the axes are *positive* stresses. Figure B-1.3 illustrates positive and negative stresses in the far faces.

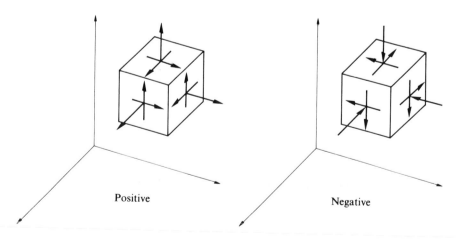

Positive

Negative

Fig. B-1.2   Signs of Stresses in Near Faces

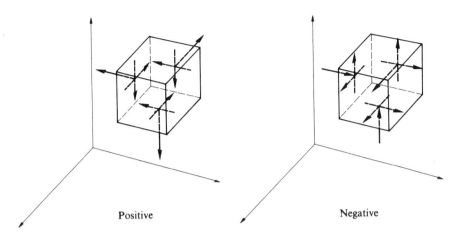

Fig. B-1.3 Signs of Stresses in Far Faces

Note that when this is reduced to a planar case, the theory of elasticity signs for shearing stress are opposite to those used in many structural design texts and periodicals (Fig. B-1.4).

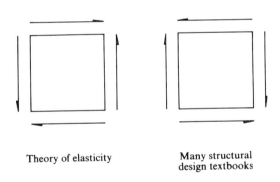

Theory of elasticity

Many structural design textbooks

Fig. B-1.4 Positive Shear (Planar)

B-2 STRAIN DISPLACEMENT

The relationships between stress and strain for a linearly elastic material are adequately illustrated by a planar derivation. Thus, Fig. B-2.4 shows a generalized planar, elemental rectangle before and after displacement.

Before displacement, the lower edge parallel to the X axis has an elemental length dx. In the process of being deformed, this edge AB moves to a new position $A'B'$. This deformation is composed of three components, each of which is discussed independently before using the generalized composite of Fig. B-2.4 to derive the strain displacement equations.

1. Figure B-2.1 depicts rigid body movement which implies that AB maintains both its length and its angular orientation. Displacement is denoted in the x direction by u and displacement in the y direction by v. Only rigid body

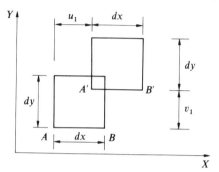

Fig. B-2.1   Rigid Body Movement

movement occurs if both A' and B' displace exactly u_1 in the x direction and v_1 in the y direction. Note that in such a case the elemental square remains square with the lengths of its edges unchanged.

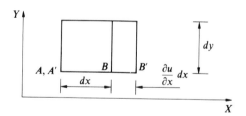

Fig. B-2.2   Axial (Normal) Strain

2. Axial (normal) strain occurs if the edge changes length but does not rotate. In such a case, A and A' are the same point whereas B and B' are different points on a line through A parallel to the X axis. The rate of change of strain in the x direction is a function of x displacement u and x position, or $\partial u/\partial x$. Because strain occurs over the entire length dx, B and B' are separated by the distance $(\partial u/\partial x)dx$ and the length $A'B'$ becomes $dx + (\partial u/\partial x)(dx)$.

3. Shearing (tangential) strain occurs if the edge rotates but does not change length. In such a case, A and A' are the same point whereas B and B' are separate points on a line through B and perpendicular to the X axis. The rate of change in the y direction is a function of the y displacement v related to the x position, or $\partial v/\partial x$. Therefore, B and B' separated by the distance $(\partial v/\partial x)dx$.

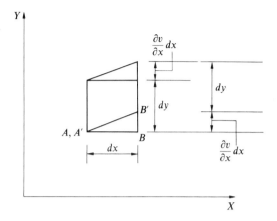

Fig. B-2.3   Shearing (Tangential) Strain

Figure B-2.4 is a generalized composite which depicts rigid body displacement plus axial strain and shearing strain.

B-2.1 Axial Strain. Axial strain in the x direction is denoted ϵ_x and is defined as the quotient of the x component of elongation divided by the original x component of length:

$$\epsilon_x = \frac{A'B' - AB}{AB}. \tag{B-2.1}$$

In more elegant derivations, this equation is linearized by discarding the square of derivatives, considered very small when compared to the derivatives themselves. In this derivation, linearization is obtained by the tacit assumption that $A'B'$ is essentially equal to $A'B''$. Thus

$$\epsilon_x = \frac{A'B' - AB}{AB} = \frac{A'B'' - AB}{AB} = \frac{[dx + (\partial u/\partial x)(dx)] - dx}{dx} = \frac{\partial u}{\partial x} \tag{B-2.2}$$

Similarly,

$$\epsilon_y = \frac{\partial v}{\partial y} \quad \text{and} \quad \epsilon_z = \frac{\partial w}{\partial z}, \tag{B-2.3}$$

where u is displacement in the x direction, v is displacement in the y direction, and w is displacement in the z direction.

B-2.2 Shearing Strain. Shearing strain is defined by two different means: (1) component of (shearing) strain tensor, and (2) engineering (shearing) strain. In the linear theory of elasticity, the first is exactly half of the latter. A com-

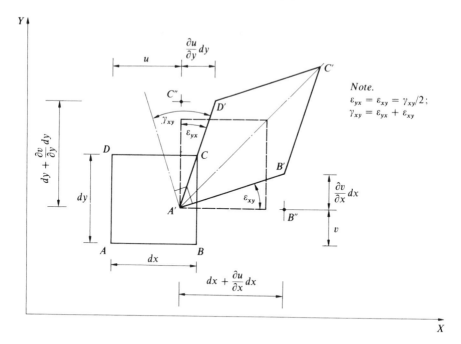

Fig. B-2.4   Strain and Displacement of Planar Element

parison of these two quantities is presented in Fig. B-2.4, ϵ_{yx} and ϵ_{xy} representing *equal* components of the strain tensor and γ_{xy} their sum. It is necessary to use ϵ terms in tensor notation (in order to have a tensor) and desirable to think of γ as the physical distortion produced by pure shear on a rectangular element.

By definition, ϵ is the angular rotation between the unstrained and strained conditions:

$$\epsilon_{xy} = \tan\frac{B'B''}{A'B''} \doteq \frac{B'B''}{A'B''} = \frac{(\partial v/\partial x)\,dx}{\left[dx + (\partial u/\partial x)\,dx\right]} = \frac{\partial v/\partial x}{1 + (\partial u/\partial x)}. \quad \text{(B-2.4)}$$

Because $\partial u/\partial x$ is very much smaller than unity, $1 + \partial u/\partial x \doteq 1$ and Eq. (B-2.4) is readily linearized to

$$\epsilon_{xy} = \frac{\partial v}{\partial x}. \quad \text{(B-2.5)}$$

Engineering strain γ_{xy} is the sum of the two equal components of strain tensor, so

$$\gamma_{xy} = \epsilon_{xy} + \epsilon_{yx} = \frac{(\partial v/\partial x)\left[1 + (\partial v/\partial y)\right] + (\partial u/\partial y)\left[1 + \partial u/\partial x)\right]}{1 + \left[(\partial u/\partial x)(\partial v/\partial y)\right] + (\partial u/\partial x) + (\partial v/\partial y)} \quad \text{(B-2.6)}$$

Equation (B-2.6) linearizes readily to

$$\gamma_{xy} = \frac{\partial v}{\partial x} + \frac{\partial u}{\partial y}.$$ (B-2.7)

In tensor notation, the *strain displacement equation* is

$$\epsilon_{ij} = \frac{1}{2}(u_{i,j} + u_{j,i}),$$ (B-2.8)

where i and j both vary from 1 to 3. Subscripts 1, 2, 3 imply x, y, z, and u_1, u_2, u_3 imply u, v, w. If $i = 1$ and $j = 1$, then

$$\epsilon_{11} = \frac{1}{2}\left(\frac{\partial u_1}{\partial x_1} + \frac{\partial u_1}{\partial x_1}\right) = \frac{\partial u_1}{\partial x_1},$$ (B-2.9)

or without numerical subscripts,

$$\epsilon_{xx} = \frac{\partial u}{\partial x}.$$ [B-2.2]

If $i = 1$ and $j = 3$, then

$$\epsilon_{13} = \frac{1}{2}\left(\frac{\partial u_1}{\partial x_3} + \frac{\partial u_3}{\partial x_1}\right),$$ (B-2.10)

and therefore

$$\epsilon_{xz} = \frac{1}{2}\left(\frac{\partial u}{\partial z} + \frac{\partial w}{\partial x}\right)$$ (B-2.11)

or

$$\gamma_{xz} = \frac{\partial u}{\partial z} + \frac{\partial w}{\partial x}.$$ (B-2.12)

Except when specifically stated or when used in conjunction with tensors, the term shearing strain normally refers to engineering strain γ and not to the component of shearing strain ϵ.

B-3 STRESS–STRAIN

A simple tensile test on a mild steel bar (up to the proportional limit) exhibits the linearly elastic property that stress is proportional to strain, their quotient being defined as (Young's modulus) the *modulus of elasticity*:

$$E = \frac{\sigma_x}{\epsilon_x} \quad \text{or} \quad \epsilon_x = \frac{\sigma_x}{E}.$$ (B-3.1)

However, additional experimental evidence on the same specimen shows that the bar contracts in cross section. The proportionality constant suggesting this contraction is called *Poisson's ratio* and is represented by v. Thus the bar subject to tension along the X axis would exhibit contraction along its Y and Z axes:

$$\epsilon_y = -v\frac{\sigma_x}{E} \quad \text{and} \quad \epsilon_z = -v\frac{\sigma_x}{E}. \tag{B-3.2}$$

Even more careful consideration reveals the possibility that strain in the x direction could be a function of the six stress components, so

$$\epsilon_x = c_1\sigma_x + c_2\sigma_y + c_3\sigma_z + c_4\tau_{xy} + c_5\tau_{yz} + c_6\tau_{xz}. \tag{B-3.3}$$

However, in the linear elastic form, the last three terms are omitted and the constants evaluated so that

$$\epsilon_x = \frac{\sigma_x}{E} - v\frac{\sigma_y}{E} - v\frac{\sigma_z}{E} \tag{B-3.4}$$

or

$$\epsilon_x = \frac{1}{E}\left[\sigma_x - v\left(\sigma_y + \sigma_z\right)\right]. \tag{B-3.5}$$

This equation can be readily verified by superimposing the ϵ_x strains due to tensions, respectively, in the x, y, and z directions.

Furthermore, shearing strain is directly proportional to shearing stress in the linear elastic analysis, so

$$2\,\epsilon_{xy} = \gamma_{xy} = \frac{\tau_{xy}}{G}, \tag{B-3.6}$$

where G is obviously defined as the ratio of shearing stress to shearing strain in a "pure" shear test. Further, it is shown in sec. B-5 that

$$G = \frac{E}{2\left(1 + v\right)}, \tag{B-3.7}$$

so

$$\gamma_{xy} = \frac{2\left(1 + v\right)}{E}\tau_{xy}. \tag{B-3.8}$$

The relationships between linearly elastic stresses and strains can be expressed in tensor form as

$$\epsilon_{ij} = \frac{1 + v}{E}\tau_{ij} - \frac{v}{E}\tau_{kk}\,\delta_{ij}, \tag{B-3.9}$$

where δ_{ij} is the *Kronecker delta*, defined as unity when $i = j$ and as zero when $i \neq j$.

If $i = 1$ and $j = 1$, then Eq. (B-3.9) reduces to

$$\epsilon_{11} = \frac{1 + v}{E}\tau_{11} - \frac{v}{E}\tau_{11}\delta_{11} - \frac{v}{E}\tau_{22}\delta_{11} - \frac{v}{E}\tau_{33}\delta_{11} \qquad (B\text{-}3.10)$$

or

$$\epsilon_{xx} = \frac{1}{E}\left[\tau_{xx} - v(\tau_{yy} + \tau_{zz})\right]. \qquad (B\text{-}3.11)$$

Or if $\tau_{xx} \equiv \sigma_x$, etc., then

$$\epsilon_x = \frac{1}{E}\left[\sigma_x - v(\sigma_y + \sigma_z)\right]. \qquad [B\text{-}3.5]$$

If $i = 2$ and $j = 3$, then

$$\epsilon_{23} = \frac{1 + v}{E}\tau_{23} - \overset{0}{\frac{v}{E}\tau_{11}\phi_{23}} - \overset{0}{\frac{v}{E}\tau_{22}\phi_{23}} - \overset{0}{\frac{v}{E}\tau_{33}\phi_{23}} \qquad (B\text{-}3.12)$$

or

$$\epsilon_{yz} = \frac{1 + v}{E}\tau_{yz}, \qquad (B\text{-}3.13)$$

which corresponds to Eq. (B-3.8) because $\gamma_{yz} = 2\epsilon_{yz}$.

These stress-strain relations for linearly elastic materials are often referred to as *Hooke's law*.

B-4 COMPATIBILITY

Six equations (B-2.8) were developed for strain components in terms of only three displacement components. Obviously, these strain and displacement components must be related to each other so that material remains continuous after deformation. In simplistic terms, (a) the same piece of material cannot be in two different places at the same time and (b) two pieces of adjacent material must remain adjacent, that is, a gap cannot develop between them. These relationships which insure physical compatability are developed by two processes.

B-4.1 Development 1. Three equations are obtained in the following manner: Differentiate an equation for shearing strain with respect to its two subscripts:

$$\gamma_{xz} = \frac{\partial u}{\partial z} + \frac{\partial w}{\partial x} \qquad [B\text{-}2.12]$$

so

$$\frac{\partial^2 \gamma_{xz}}{\partial x\, \partial z} = \frac{\partial^2}{\partial x\, \partial z}\left(\frac{\partial u}{\partial z}\right) + \frac{\partial^2}{\partial x\, \partial z}\left(\frac{\partial w}{\partial x}\right). \qquad (B\text{-}4.1)$$

Now, because u and w must be single-valued continuous functions (that is, elements can exist at only one place at a time and the material must remain continuous) of x and z, Eq. (B-4.1) can be rewritten as

$$\frac{\partial^2 \gamma_{xz}}{\partial x\, \partial z} = \frac{\partial^2}{\partial z^2}\left(\frac{\partial u}{\partial x}\right) + \frac{\partial^2}{\partial x^2}\left(\frac{\partial w}{\partial z}\right) \tag{B-4.2}$$

But from Eq. (B-2.2), $\epsilon_x = \partial u/\partial x$ and $\epsilon_z = \partial w/\partial z$, so that

$$\frac{\partial^2 \gamma_{xz}}{\partial x\, \partial z} = \frac{\partial^2 \epsilon_x}{\partial z^2} + \frac{\partial^2 \epsilon_z}{\partial x^2}. \tag{B-4.3a}$$

However, because $\gamma_{xy} = \epsilon_{xy} + \epsilon_{yx}$, Eq. (B-4.3) is often written as

$$\frac{\partial^2 \epsilon_{xz}}{\partial x\, \partial z} + \frac{\partial^2 \epsilon_{zx}}{\partial x\, \partial z} = \frac{\partial^2 \epsilon_x}{\partial z^2} + \frac{\partial^2 \epsilon_z}{\partial x^2}. \tag{B-4.3b}$$

B-4.2 Development 2. The second set of three equations is obtained by, for instance, summing the partial of γ_{xy} with respect to x and z and of γ_{zx} with respect to y and x, resulting in

$$\frac{\partial^2 \gamma_{xy}}{\partial x\, \partial z} + \frac{\partial^2 \gamma_{zx}}{\partial y\, \partial x} = \frac{\partial^2}{\partial x\, \partial z}\left(\frac{\partial u}{\partial y} + \frac{\partial v}{\partial x}\right) + \frac{\partial^2}{\partial y\, \partial x}\left(\frac{\partial w}{\partial x} + \frac{\partial u}{\partial z}\right) \tag{B-4.4}$$

Again the single-valued continuity of u, v, and w allows the right-hand side of Eq. (B-4.4) to be rewritten as:

$$\frac{\partial^2}{\partial y\, \partial z}\left(\frac{\partial u}{\partial x}\right) + \frac{\partial^2}{\partial x^2}\left(\frac{\partial v}{\partial z}\right) + \frac{\partial^2}{\partial x^2}\left(\frac{\partial w}{\partial y}\right) + \frac{\partial^2}{\partial y\, \partial z}\left(\frac{\partial u}{\partial x}\right) \tag{B-4.5}$$

or

$$2\frac{\partial^2 \epsilon_x}{\partial y\, \partial z} + \frac{\partial^2}{\partial x^2}\left(\frac{\partial v}{\partial z} + \frac{\partial w}{\partial y}\right) = 2\frac{\partial^2 \epsilon_x}{\partial y\, \partial z} + \frac{\partial^2 \gamma_{yz}}{\partial x^2}. \tag{B-4.6}$$

Substitution of Eq. (B-4.6) into the right-hand side of Eq. (B-4.4) and rearrangement of terms yields.

$$2\frac{\partial^2 \epsilon_x}{\partial y\, \partial z} = \frac{\partial}{\partial x}\left(-\frac{\partial \gamma_{yz}}{\partial x} + \frac{\partial \gamma_{xy}}{\partial z} + \frac{\partial \gamma_{zx}}{\partial y}\right) \tag{B-4.7a}$$

If γ values are replaced by ϵ values, then Eq. (B-4.7a) may be written as

$$\frac{\partial^2 \epsilon_x}{\partial y\, \partial z} + \frac{\partial^2 \epsilon_{yz}}{\partial x^2} = \frac{\partial^2 \epsilon_{yx}}{\partial x\, \partial z} + \frac{\partial^2 \epsilon_{xz}}{\partial y\, \partial x}. \tag{B-4.7b}$$

B-4.3 Tensor Form. Both sets of three equations each may be summarized in tensor form as

$$\epsilon_{ij,kl} + \epsilon_{kl,ij} = \epsilon_{kj,il} + \epsilon_{il,kj} \tag{B-4.8}$$

which has only the following *unique* combinations of subscripts (note the simplicity of the table as represented by the groupings indicated).

line	i	j	k	ℓ
1	1	1	2	3
2	2	2	3	1
3	3	3	1	2
4	1	1	2	2
5	2	2	3	3
6	3	3	1	1

$$(B\text{-}4.9)$$

The combination represented by line 1 is written as

$$\epsilon_{11,23} + \epsilon_{23,11} = \epsilon_{21,13} + \epsilon_{13,21} \qquad (B\text{-}4.10)$$

or, using the rules stated at the end of sec. B-1.1, as

$$\frac{\partial^2 \epsilon_x}{\partial y \partial z} + \frac{\partial^2 \epsilon_{yz}}{\partial x^2} = \frac{\partial^2 \epsilon_{yx}}{\partial x \partial z} + \frac{\partial^2 \epsilon_{xz}}{\partial y \partial x} \qquad [B\text{-}4.7b]$$

which was previously derived. Lines 2 and 3 provide coefficients which produce equations of similar form. The combination of subscripts of line 4 results in

$$\epsilon_{11,22} + \epsilon_{22,11} = \epsilon_{21,12} + \epsilon_{12,21} \qquad (B\text{-}4.11)$$

or

$$\frac{\partial^2 \epsilon_x}{\partial y^2} + \frac{\partial^2 \epsilon_y}{\partial x^2} = \frac{\partial^2 \epsilon_{yx}}{\partial x \partial y} + \frac{\partial^2 \epsilon_{xy}}{\partial y \partial x} \qquad (B\text{-}4.12)$$

which is of the form of Eq. (B-4.3b).

B-5 RELATIONSHIP BETWEEN E AND G

Figure B-5.1 represents a right isosceles triangular element before and after deformation. The magnitude of the normal stresses are equal and can be related to τ_{yz} by summation of *forces* perpendicular to the hypotenuse. Thus, $[\sigma_y(1)^2(\sqrt{2}/2)$ is *stress* times *area* times *trigonometric function*] to obtain component perpendicular to hypotenuse:

$$\tau_{yz}(1)(1)\sqrt{2} = \sigma_y(1)^2\frac{\sqrt{2}}{2} + \sigma_z(1)^2\frac{\sqrt{2}}{2}, \qquad (B\text{-}5.1)$$

or since σ_y and σ_z are equal in magnitude,

$$\tau_{yz} = \sigma_z \qquad (B\text{-}5.2)$$

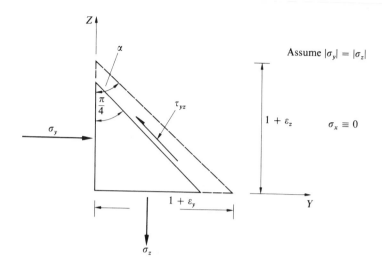

Fig. B-5.1   Isosceles Right Triangular Element

$(\tau_{yz} = \tau_{zy}$ and therefore simply is called τ below).

Further, the tangent of angle α can be expressed

$$\tan \alpha = \frac{1 + \epsilon_y}{1 + \epsilon_z},$$ (B-5.3)

where α is equal to $(\pi/4) - (\gamma_{yz}/2)$. Relate Fig. B-2.4 to Fig. B-5.1 for verification.

By Eq. (B-3.9),

$$\epsilon_y = \frac{1 + v}{E}\sigma_y - \frac{v}{E}\sigma_x - \frac{v}{E}\sigma_y - \frac{v}{E}\sigma_z$$ (B-5.4)

or, because $\sigma_y = -\sigma_z$ and $\sigma_x = 0$,

$$\epsilon_y = -\frac{(1 + v)\sigma_z}{E} = -\frac{(1 + v)\tau}{E}.$$ (B-5.5)

Similarly,

$$\epsilon_z = \frac{(1 + v)\sigma_z}{E} = \frac{(1 + v)\tau}{E}.$$ (B-5.6)

Substitution of these expressions for ϵ_y and ϵ_z into Eq. (B-5.3) yields

$$\tan \alpha = \frac{1 - [(1 + v)\tau/E]}{1 + [(1 + v)\tau/E]}.$$ (B-5.7)

However, for small angles,

$$\tan \alpha = \tan \left(\frac{\pi}{4} - \frac{\gamma_{yz}}{2}\right) = \tan \left(\frac{\pi}{4} - \frac{\gamma}{2}\right)$$

$$= \frac{\tan (\pi/4) - \tan \gamma/2}{1 + [\tan (\pi/4)\tan (\gamma/2)]} = \frac{1 - (\gamma/2)}{1 + (\gamma/2)}. \qquad \text{(B-5.8)}$$

Comparison of the right-hand sides of Eqs. (B-5.7) and (B-5.8) reveals, without computation, that

$$\frac{1 + v}{E}\tau = \frac{\gamma}{2} \qquad \text{(B-5.9)}$$

or

$$\gamma = \frac{2(1 + v)}{E}\tau. \qquad \text{(B-5.10)}$$

But Eq. (B-5.10) is the relationship between shearing stress and shearing strain, and therefore $[2(1 + v)]/E$ is the reciprocal of the proportionality constant called the *modulus of rigidity* of the material, represented by the symbol G. Thus

$$G = \frac{\tau}{\gamma} = \frac{E}{2(1 + v)}. \qquad \text{(B-5.11)}$$

B-6   SUMMARY

Fifteen interrelated equations have been developed for linearly elastic materials that have 15 unknowns: six stresses, σ_x, σ_y, σ_z, τ_{xy}, τ_{yz}, τ_{zx}; six strains ϵ_x, ϵ_y, ϵ_z, γ_{xy}, γ_{yz}, γ_{zx}; and three displacements u, v, and w. The three equations of equilibrium, the six strain displacement equations, and the six compatibility equations are summarized using tensor notation in Table B-6.1 and using expanded form in Table B-6.2. The six stress-strain equations are added for completeness. To solve a particular problem, it also is necessary to utilize boundary conditions. These equation are used to derive the plate equation in Appendix C and to solve torsional problems in Chap. 9.

REFERENCES

1. Timoshenko, S. and J. N. Goodier, *Theory of Elasticity*, 2d ed., McGraw-Hill, New York (1951).

2. Wang, Chi-Teh, *Applied Elasticity*, McGraw-Hill, New York (1953).

3. Housner, George W. and Thad Vreeland, Jr., *The Analysis of Stress and Deformation*, MacMillan, New York (1966).

4. Volterra, Enrico and J. H. Gaines, *Advanced Strength of Materials*, Prentice-Hall, Englewood Cliffs, New Jersey (1971).

Table B-6.1 Summary, Tensor Form

Type	Basis	Form	Number of equations	Number of unknowns			Equation Number
				Stress	Strain	Displacement	
Static equilibrium	Physical law	$\tau_{ij,i} + X_j = 0$	3	6			(B-6.1)
Strain displacement	Geometrical (small displacements)	$\epsilon_{ij} = \frac{1}{2}(u_{ij} + u_{j,i})$	6		6	3	(B-6.2)
Stress-strain	Material Properties	$\epsilon_{ij} = \frac{1+v}{E}\tau_{ij} - \frac{v}{E}\tau_{kk}\delta_{ij}$	6	6	6		(B-6.3)
Compatibility	Stress-strain-displacement consistent with loads and body constraints	$\epsilon_{ij,kl} + \epsilon_{kl,ij} = \epsilon_{kj,il} + \epsilon_{il,kj}$	6		6		(B-6.4)

$$
\begin{array}{cccc}
i & j & k & l \\
1 & 1 & 2 & 3 \\
2 & 2 & 3 & 1 \\
3 & 3 & 1 & 2 \\
\end{array}
$$

Six unique combinations for 81 possibilities

$$
\begin{array}{cccc}
1 & 1 & 2 & 2 \\
2 & 2 & 3 & 3 \\
3 & 3 & 1 & 1 \\
\end{array}
$$

Type	Basis	Form					Equation Number
Supplementary	Kronecker delta $\delta_{ij} = \begin{cases} 0 & \text{when } i \neq j \\ 1 & \text{when } i = j \end{cases}$						(B-6.5)
	Engineering strain versus component of strain (shearing only) $\gamma_{ij} = \epsilon_{ij} + \epsilon_{ji}$ and $\epsilon_{ij} = \epsilon_{ji}$						(B-6.6)
	Modulus of Rigidity $G = \dfrac{E}{2(1 + v)}$						(B-6.7)

Table B-6.2   Summary, Extended Form

Equilibrium

$$\frac{\partial \sigma_x}{\partial x} + \frac{\partial \tau_{yx}}{\partial y} + \frac{\partial \tau_{zx}}{\partial z} + X = 0 \tag{B-6.8}$$

$$\frac{\partial \tau_{xy}}{\partial x} + \frac{\partial \sigma_y}{\partial y} + \frac{\partial \tau_{zy}}{\partial z} + Y = 0 \tag{B-6.9}$$

$$\frac{\partial \tau_{xz}}{\partial x} + \frac{\partial \tau_{yz}}{\partial y} + \frac{\partial \sigma_z}{\partial z} + Z = 0 \tag{B-6.10}$$

Strain Displacement

$$\epsilon_x = \partial u / \partial x \tag{B-6.11}$$
$$\epsilon_y = \partial v / \partial y \tag{B-6.12}$$
$$\epsilon_z = \partial w / \partial z \tag{B-6.13}$$
$$\gamma_{xy} = \partial u / \partial y + \partial v / \partial x \tag{B-6.14}$$
$$\gamma_{yz} = \partial v / \partial z + \partial w / \partial v \tag{B-6.15}$$
$$\gamma_{zx} = \partial w / \partial x + \partial u / \partial z \tag{B-6.16}$$

Compatability

$$\frac{\partial^2 \epsilon_{xx}}{\partial y \partial z} + \frac{\partial^2 \epsilon_{yz}}{\partial x^2} = \frac{\partial^2 \epsilon_{yx}}{\partial x \partial z} + \frac{\partial^2 \epsilon_{xz}}{\partial y \partial x} \tag{B-6.17}$$

$$\frac{\partial^2 \epsilon_{yy}}{\partial z \partial x} + \frac{\partial^2 \epsilon_{zx}}{\partial y^2} = \frac{\partial^2 \epsilon_{zy}}{\partial y \partial x} + \frac{\partial^2 \epsilon_{yx}}{\partial z \partial y} \tag{B-6.18}$$

$$\frac{\partial^2 \epsilon_{zz}}{\partial x \partial y} + \frac{\partial^2 \epsilon_{xy}}{\partial z^2} = \frac{\partial^2 \epsilon_{xz}}{\partial z \partial y} + \frac{\partial^2 \epsilon_{zy}}{\partial x \partial z} \tag{B-6.19}$$

$$\frac{\partial^2 \epsilon_{xx}}{\partial y^2} + \frac{\partial^2 \epsilon_{yy}}{\partial x^2} = \frac{\partial^2 \epsilon_{yx}}{\partial x \partial y} + \frac{\partial^2 \epsilon_{xy}}{\partial y \partial x} \tag{B-6.20}$$

$$\frac{\partial^2 \epsilon_{yy}}{\partial z^2} + \frac{\partial^2 \epsilon_{zz}}{\partial y^2} = \frac{\partial^2 \epsilon_{zy}}{\partial y \partial z} + \frac{\partial^2 \epsilon_{yz}}{\partial z \partial y} \tag{B-6.21}$$

$$\frac{\partial^2 \epsilon_{zz}}{\partial x^2} + \frac{\partial^2 \epsilon_{xx}}{\partial z^2} = \frac{\partial^2 \epsilon_{xz}}{\partial z \partial x} + \frac{\partial^2 \epsilon_{zx}}{\partial x \partial z} \tag{B-6.22}$$

$$2\frac{\partial^2 \epsilon_x}{\partial y \partial z} = -\frac{\partial^2 \gamma_{yx}}{\partial x^2} + \frac{\partial^2 \gamma_{yx}}{\partial x \partial z} + \frac{\partial^2 \gamma_{xy}}{\partial y \partial x} \tag{B-6.17a}$$

$$\frac{\partial^2 \epsilon_x}{\partial y^2} + \frac{\partial^2 \epsilon_y}{\partial x^2} = \frac{\partial^2 \gamma_{xy}}{\partial x \partial y} \tag{B-6.20a}$$

Stress-Strain

$$\epsilon_x = \frac{1}{E} \left[\sigma_x - v(\sigma_y + \sigma_z) \right] \tag{B-6.23}$$

$$\epsilon_y = \frac{1}{E} \left[\sigma_y - v(\sigma_z + \sigma_x) \right] \tag{B-6.24}$$

$$\epsilon_z = \frac{1}{E} \left[\sigma_z - v(\sigma_x + \sigma_y) \right] \tag{B-6.25}$$

$$\gamma_{xy} = 2\epsilon_{xy} = \tau_{xy}/G \tag{B-6.26}$$
$$\gamma_{yz} = \tau_{yz}/G \tag{B-6.27}$$
$$\gamma_{zx} = \tau_{zx}/G \tag{B-6.28}$$
$$G = E/[2(1 + v)] \tag{B-6.29}$$

PLATE THEORY

This development depends upon the linearized equation of the theory of elasticity presented in Appendix B and summarized in Tables B-6.1 and B-6.2. Final results include equations for critical buckling of plates which undergird many code formulas.

A complete and detailed treatment of the material summarized in this chapter is available from books by Timoshenko and Gere [1], Bleich [2], Wang [3], Timoshenko and Goodier [4], Housner and Vreeland [5], and Volterra and Gaines [6].

C-1 THIN PLATES SUBJECT TO GRAVITY LOADS

A fundamental assumption is made that the plate to be analyzed has a constant thickness which is very much less than either its width or length. This implies minimal axial strain in the z direction (perpendicular to the plate). Thus, for practical purposes, it is valid to assume that

$$\epsilon_z = \gamma_{xz} = 2\epsilon_{xz} = \gamma_{yz} = 2\epsilon_{yz} = 0. \tag{C-1.1}$$

This reduces the six strain displacement equations to only three pertinent one (see Table B-6.1):

$$\epsilon_x = \frac{\partial u}{\partial x}, \qquad \epsilon_y = \frac{\partial v}{\partial y}, \qquad \gamma_{xy} = \frac{\partial u}{\partial y} + \frac{\partial v}{\partial x}. \tag{C-1.2}$$

In a similar manner, the six stress-strain equations also reduce to three:

$$\epsilon_x = \frac{(\sigma_x - \nu\sigma_y)}{E}, \qquad \epsilon_y = \frac{(\sigma_y - \nu\sigma_x)}{E}, \qquad \gamma_{xy} = \frac{\tau_{xy}}{G}. \tag{C-1.3}$$

It is advantageous to rewrite the stress-strain equations as strain-stress

489

equations, by simultaneous solution of the first two of Eqs. (C-1.3) and simple inversion of the last. That is,

$$\sigma_x = \frac{E}{1 - \nu^2}(\epsilon_x + \nu\epsilon_y),$$

$$\sigma_y = \frac{E}{1 - \nu^2}(\epsilon_y + \nu\epsilon_x), \qquad \text{(C-1.4)}$$

and

$$\tau_{xy} = \frac{\gamma_{xy}}{G}; \; (\gamma_{xy} = 2\epsilon_{xy}).$$

C-1.1 Stress Displacement Equations. Equations for displacements can be obtained by using tacit assumptions that $\gamma_{xz} = \gamma_{yz} = 0$. For instance,

$$\gamma_{xz} = \frac{\partial w}{\partial x} + \frac{\partial u}{\partial z} = 0. \qquad \text{(C-1.5)}$$

Or multiplication by ∂z and integration yields

$$\int \partial u = \int - \frac{\partial w}{\partial x}\, \partial z \qquad \text{(C-1.6)}$$

so that

$$u = - z\frac{\partial w}{\partial x} + f_1(x, y). \qquad \text{(C-1.7)}$$

The middle surface is defined as a plane mid-way between the top and bottom surfaces of the plate, corresponding to the neutral axis of a rectangular homogeneous beam. The arbitrary function of integration f_1 depends on x and y only, and the middle surface of the plate has no stretch or shear, so $f_1(x,y) = 0$ with $z = 0$. Therefore because f_1 is not a function of z, $f(x, y)$ must be zero for any z, and

$$u = - z\frac{\partial w}{\partial x}, \qquad \text{(C-1.8)}$$

and, similarly,

$$v = - z\frac{\partial w}{\partial y}. \qquad \text{(C-1.9)}$$

These relationships can be visualized in Fig. C-1.1 where an overview of the plate is shown, having a length L in the x direction, width b in the y direction, and a total depth h in the z direction. Adjacent is shown a partial cross section of the plate, parallel to the y–z plane, before and after load. Before deformation, point Q in the middle surface is directly below point R located

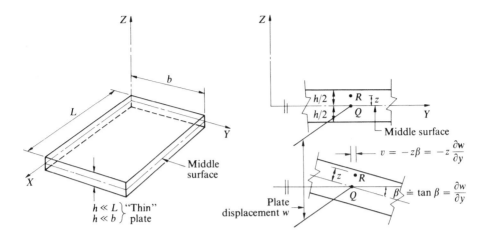

Fig. C-1.1   Thin Plate Geometry

toward the top surface. After loading, R is displaced horizontally a distance v, equal to the product of *vertical distance from the neutral surface* and *angle of rotation*, or

$$v = - z\beta. \tag{C-1.10}$$

The negative sign is necessary to make the displacement v positive (z is positive but β is negative). Angle β is almost equal to $\tan \beta$ (small angles), and $\tan \beta$ is the first derivative of displacement with respect to y, or $\partial w/\partial y$. Substitution in Eq. (C-1.10) gives

$$v = - z \frac{\partial w}{\partial y}, \tag{C-1.9}$$

which concurs with the previous mathematical treatment.

To obtain equations for stress in terms of vertical displacement w, strain displacement Eq. (C-1.2) is substituted into strain-stress Eq. (C-1.4), with $\partial u/\partial x$ replaced by $-z(\partial^2 w/\partial x^2)$,

$$\sigma_x = \frac{-Ez}{1 - v^2} \left(\frac{\partial^2 w}{\partial x^2} + v \frac{\partial^2 w}{\partial y^2} \right) ,$$

$$\sigma_y = \frac{-Ez}{1 - v^2} \left(\frac{\partial^2 w}{\partial y^2} + v \frac{\partial^2 w}{\partial x^2} \right) , \tag{C-1.11}$$

and

$$\tau_{xy} = \frac{-Ez}{1 + v} \frac{\partial^2 w}{\partial x \, \partial y}.$$

With stresses expressed in terms of known vertical displacement, it is desirable to find equations expressing bending moments, twist, and shear also expressed in terms of w. The latter expressions are all *per unit length.*

C-1.2 Plate Bending Moments. The plate bending moment *per unit length* is defined as M_x. In a manner analogous to that for a beam, ΔM_x can be determined for a plate dz deep, having unit width, as

$$\Delta M_x = \sigma_x (dz)(1)z, \tag{C-1.12}$$

where ΔM_x is the differential moment per unit length, $\sigma_x (dz)(1)$ is the differential force, and z is the moment arm. Or,

$$M_x = \int_{-(h/2)}^{h/2} \sigma_x z \, dz, \tag{C-1.13}$$

where $h \equiv$ depth of plate.

By replacement of σ_x by its displacement formulation [Eq. (C-1.11)], Eq. (C-1.13) becomes

$$M_x = \frac{-E}{1 - v^2} \left(\frac{\partial^2 w}{\partial x^2} + v \frac{\partial^2 w}{\partial y^2} \right) \int_{-(h/2)}^{h/2} z^2 \, dz, \tag{C-1.14}$$

where only the single variable z from σ_x has been left under the integral sign. By introduction of a term D, defined as the flexural rigidity of the plate,

$$D = \frac{Eh^3}{(1 - v^2)12}, \tag{C-1.15}$$

the final equation for M_x is obtained after integrating $z^2 \, dz$ (M_y is found in a similar manner):

$$M_x = - D \left(\frac{\partial^2 w}{\partial x^2} + v \frac{\partial^2 w}{\partial y^2} \right),$$
$$M_y = - D \left(\frac{\partial^2 w}{\partial y^2} + v \frac{\partial^2 w}{\partial x^2} \right). \tag{C-1.16}$$

In a somewhat similar process, twisting moments *per unit length* can be determined as

$$M_{xy} = M_{yx} = - D(1 - v) \frac{\partial^2 w}{\partial x \, \partial y}. \tag{C-1.17}$$

C-1.3 Plate Shearing Force. Plate shearing force *per unit length* is called Q_x and Q_y. By the method of the preceding subsection and based on a fundamental assumption that $\tau_{xz} \equiv 0$,

$$Q_x = \int_{-(h/2)}^{h/2} \tau_{xz} \, dz = 0 \; (?). \tag{C-1.18}$$

This anomaly is overcome by summation of moments about the edge parallel to and farthest from the X axis (Fig. C-1.2) and neglecting higher ordered terms,

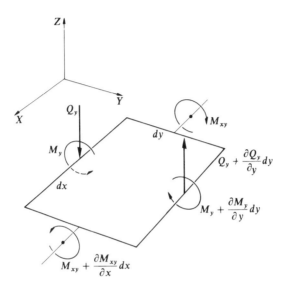

Fig. C-1.2   Moments with Respect to the X Axis

$$+ \left(\frac{\partial M_{xy}}{\partial x} dx \right) (dy) \quad - \left(\frac{\partial M_y}{\partial y} dy \right) (dx) \quad + Q_y \ dy \ dx = 0. \tag{C-1.19}$$

Moment Length   Moment Length Force Arm
per unit   per unit   per
length   length   unit
                                  length

Moment   Length
per
unit
length

Signs are based on use of the negative sense for all forces on edges nearest the coordinate axes, application of moments to act on the bottom of the plate, and application of twisting moments on edges nearest the coordinate axes acting to produce negative vertical displacements. Then, rearranging and dividing Eq. (C-1.19) by $dxdy$, we have

$$Q_y = \frac{\partial M_y}{\partial y} - \frac{\partial M_{xy}}{\partial x}. \tag{C-1.20}$$

Likewise,

$$Q_x = \frac{\partial M_x}{\partial x} - \frac{\partial M_{yx}}{\partial y}.$$

C-1.4 The Plate Equation. By use of the terms above, it is possible to develop the plate equation. Initial formulation is provided by summing forces in the vertical z direction, with the unit load normal to the plate called q:

$$\frac{\partial Q_x}{\partial x}\, dxdy + \frac{\partial Q_y}{\partial y}\, dxdy + q\, dxdy = 0. \tag{C-1.21}$$

Differential area $dxdy$ vanishes and partials of Q can be obtained by differentiating Eq. (C-1.20). Substitution of these partial derivatives into (C-1.21) yields

$$\frac{\partial^2 M_x}{\partial x^2} - 2\frac{\partial^2 M_{xy}}{\partial x \partial y} + \frac{\partial^2 M_y}{\partial y^2} = -q. \tag{C-1.22}$$

By taking second partial derivatives of Eqs. (C-1.16) and (C-1.17), we can develop the substitute expressions:

$$\frac{\partial^2 M_x}{\partial x^2} = -D\left(\frac{\partial^4 w}{\partial x^4} + v\frac{\partial^4 w}{\partial x^2 \partial y^2}\right),$$

$$\frac{\partial^2 M_y}{\partial y^2} = -D\left(\frac{\partial^4 w}{\partial y^4} + v\frac{\partial^4 w}{\partial x^2 \partial y^2}\right),$$

$$-2\frac{\partial^2 M_{xy}}{\partial x \partial y} = -D\left(2\frac{\partial^4 w}{\partial x^2 \partial y^2} - 2v\frac{\partial^4 w}{\partial x^2 \partial y^2}\right).$$

Using these relations, Eq. (C-1.22) reduces to:

$$\frac{\partial^4 w}{\partial x^4} + 2\frac{\partial^4 w}{\partial x^2 \partial y^2} + \frac{\partial^4 w}{\partial y^4} = \frac{q}{D}. \tag{C-1.23}$$

Equation (C-1.23) is essentially complete, but normally is simplified in appearance by introducing the *del-fourth* operator:

$$\nabla^4 = \left(\frac{\partial^4}{\partial x^4} + 2\frac{\partial^4}{\partial x^2 \partial y^2} + \frac{\partial^4}{\partial y^4}\right) \tag{C-1.24}$$

so that

$$\nabla^4 w = \frac{q}{D}. \tag{C-1.25}$$

When loads act perpendicular to the plate face, Eq. (C-1.25) is adequate for bending analysis. Ability to consider loads acting parallel to a plate face, and subsequently problems of buckling, requires the more general equations derived in the next section.

C-2 EXTENSION OF THE THIN PLATE EQUATION

To be useful for stability analysis, the plate equation must be generalized to handle forces applied along its edges. Two types of edge loadings are possible: (1) loads normal to the edge of the plate, and (2) forces parallel to the edge of the plate, commonly called tractions. In Fig. C-2.1, three views are shown for a

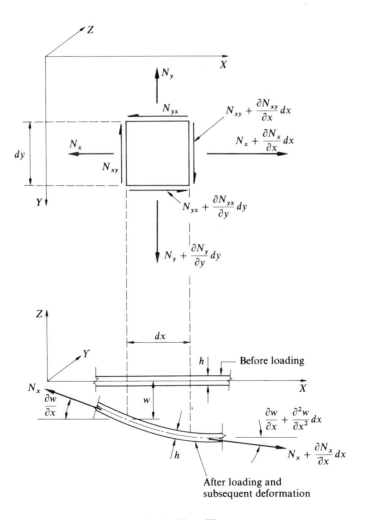

Fig. C-2.1 Plate Element

plate element originally parallel to the neutral surface (the x–y plane): a top view with stresses indicated, a sectional view before loading, and a sectional view after deformation has occurred.

Static equilibrium in the x direction and the y direction, respectively, indicate that

$$\frac{\partial N_x}{\partial x} + \frac{\partial N_{yx}}{\partial y} = 0/(dx)(dy)$$

and

$$(\text{C-2.1})$$

$$\frac{\partial N_y}{\partial y} + \frac{\partial N_{xy}}{\partial x} = 0/(dx)(dy),$$

where N_x is the normal force *per unit length* in the x direction ($\sigma_x h$); N_y is the normal force *per unit length* in the y direction ($\sigma_y h$); N_{xy} is the tangential force *per unit length* in the y direction ($\tau_{xy} h$); and N_{yx} is the tangential force *per unit length* in the x direction ($\tau_{yx} h$).

In the sectional view, only normal forces and their angles are shown. Leftward of the section the force $N_x dy$ acts at an angle $\partial w/\partial x$, obtained as the first derivative of the deflection w. At the other end, the force is incremented to $N_x + (\partial N_x/\partial x)dx$ which acts at an angle changing from $\partial w/\partial x$ at the rate $\partial(\partial w/\partial x)/\partial x$ over a length dx, becoming $\partial w/\partial x + (\partial^2 w/\partial x^2)dx$.

Vertical forces in the z direction are more easily visualized by considering only forces in either the x–z or the y–z plane, and by considering normal and tangential forces independently. Summation of *only* normal forces with respect to z in the x–z plane *only* produces

$$N_x \left(\frac{\partial w}{\partial x}\right) dy - \left(N_x + \frac{\partial N_x}{\partial x} dx\right)\left(\frac{\partial w}{\partial x} + \frac{\partial^2 w}{\partial x^2} dx\right) dy \qquad (\text{C-2.2})$$

or, neglecting higher ordered terms involving $(dx)^2$,

$$-\left(N_x \frac{\partial^2 w}{\partial x^2} dxdy + \frac{\partial N_x}{\partial x}\frac{\partial w}{\partial x} dxdy\right). \qquad (\text{C-2.3})$$

For tangential (tractive) forces, angles are measured on either side of the section shown in Fig. C-2.1, rather than at the ends as for normal forces. Determination of an average angle for the far side as the first derivative of w yields $\partial w/\partial x$ as the average angle through which N_{yx} acts. The average angle in the near face is changed from $\partial w/\partial x$ at the rate $\partial(\partial w/\partial x)/\partial y$, for a distance dy. Summing vertical components of tractive forces in the x–y plane only, we get

$$(N_{yx}) dx \left(\frac{\partial w}{\partial x}\right) - \left(N_{yx} + \frac{\partial N_{yx}}{\partial y} dy\right) dx \left(\frac{\partial w}{\partial x} + \frac{\partial^2 w}{\partial x \partial y} dy\right)$$

which reduces to

$$- \left(N_{yx} \frac{\partial^2 w}{\partial x \partial y} dxdy + \frac{\partial N_{yx}}{\partial y} \frac{\partial w}{\partial x} dxdy \right).$$

Summation of internal forces in the y–z plane provides four additional terms which represent total force in the vertical z direction:

$$- N_x \frac{\partial^2 w}{\partial x^2} dxdy - \frac{\partial N_x}{\partial x} \frac{\partial w}{\partial x} dxdy - N_y \frac{\partial^2 w}{\partial y^2} dxdy - \frac{\partial N_y}{\partial y} \frac{\partial w}{\partial y} dxdy$$

$$- 2N_{xy} \frac{\partial^2 w}{\partial x \partial y} dxdy - \frac{\partial N_{xy}}{\partial x} \frac{\partial w}{\partial y} dxdy - \frac{\partial N_{yx}}{\partial y} \frac{\partial w}{\partial x} dxdy - q \, dxdy. \qquad \text{(C-2.4)}$$

Rewritten, this becomes:

$$\left[\left(\frac{\partial N_x}{\partial x} + \frac{\partial N_{yx}}{\partial y} \right) \frac{\partial w}{\partial x} + \left(\frac{\partial N_y}{\partial y} + \frac{\partial N_{xy}}{\partial x} \right) \frac{\partial w}{\partial y} + \right.$$

$$\left. N_x \frac{\partial^2 w}{\partial x^2} + N_y \frac{\partial^2 w}{\partial y^2} + 2N_{xy} \frac{\partial^2 w}{\partial x \partial y} + q \right] dxdy. \qquad \text{(C-2.5)}$$

Because of relationships expressed by Eq. (C-2.1), the first two compound terms vanish. Comparison of Eq. (C-2.5) and plate equation (C-1.25) suggests a complete plate equation (linearized and not including body forces):

$$\nabla^4 w = \frac{1}{D} \left(q + N_x \frac{\partial^2 w}{\partial x^2} + N_y \frac{\partial^2 w}{\partial y^2} + 2N_{xy} \frac{\partial^2 w}{\partial x \partial y} \right). \qquad \text{(C-2.6)}$$

or, in terms of stresses,

$$\nabla^4 w = \frac{1}{D} \left(q + \sigma_x h \frac{\partial^2 w}{\partial x^2} + \sigma_y h \frac{\partial^2 w}{\partial y^2} + 2\tau_{xy} h \frac{\partial^2 w}{\partial x \partial y} \right). \qquad \text{(C-2.7)}$$

C-3   MODIFIED PLATE EQUATION

To make the plate equation applicable in both the inelastic and elastic range, it is convenient to define two terms:

1. $E_t \equiv$ tangent modulus of material at critical stress.
2. $\alpha \equiv$ ratio of E_t to E, that is, $\alpha = E_t/E$. These alterations were suggested by Bleich [2] who showed that the resulting solutions agree well with experimental results.

In summary, the equations are:

$$D \left(\alpha \frac{\partial^4 w}{\partial x^4} + 2\sqrt{\alpha} \frac{\partial^4 w}{\partial x^2 \partial y^2} + \frac{\partial^4 w}{\partial y^4} \right) = q + \sigma_x h \frac{\partial^2 w}{\partial x^2} + \sigma_y h \frac{\partial^2 w}{\partial y^2} +$$

$$2\tau_{xy} h \frac{\partial^2 w}{\partial x \partial y}, \qquad \text{(C-3.1)}$$

$$M_x = -D \left(\alpha \frac{\partial^2 w}{\partial x^2} + v \sqrt{\alpha} \frac{\partial^2 w}{\partial y^2} \right), \tag{C-3.2}$$

$$M_y = -D \left(v \sqrt{\alpha} \frac{\partial^2 w}{\partial x^2} + \frac{\partial^2 w}{\partial y^2} \right), \tag{C-3.3}$$

$$M_{xy} = M_{yx} = -D \sqrt{\alpha} \, (1 - v) \frac{\partial^2 w}{\partial x \partial y}, \tag{C-3.4}$$

$$Q_x = -D \frac{\partial}{\partial x} \left(\alpha \frac{\partial^2 w}{\partial x^2} + \sqrt{\alpha} \frac{\partial^2 w}{\partial y^2} \right), \tag{C-3.5}$$

$$Q_y = -D \frac{\partial}{\partial y} \left(\sqrt{\alpha} \frac{\partial^2 w}{\partial x^2} + \frac{\partial^2 w}{\partial y^2} \right), \tag{C-3.6}$$

and

$$D = \frac{Eh^3}{(1 - v^2)12}. \tag{C-1.15}$$

If α is defined as 1, these reduce to the usual linearly elastic equations.

C-4 SOLUTION OF PLATE EQUATION FOR BUCKLING (ASSUMED MODE TECHNIQUE)

Two similar approaches to solving the plate equation for critical buckling stress are demonstrated in this section. To clarify concepts, a discussion of results for an elastic plate, simply supported on all four sides, is used as a vehicle to interpret the results physically.

C-4.1 Solution Type I. Critical buckling load for a rectangular plate supported on four sides, loaded through the middle surface by opposing axial compression forces in only two opposite faces which are simply supported, can be determined by assuming a displacement form, such as †

$$w = Y(y) \sin \frac{n\pi x}{L}, \tag{C-4.1}$$

where Y is an undetermined function of y only and it can be shown that sin $[(n\pi x)/L]$ satisfies the boundary conditions for simple support in the x direction for all values of n. With x equal to either 0 or L, sin $[(n\pi x)/L] = 0$

---

†It may appear surprising that no summation sign is used with sin $(n\pi x/L)$; however, one term is adequate to show a buckled configuration. For gravity loads, the summation sign would be mandatory.

signifying that $w = 0$, satisfying the first requirement for a simple support, that is, there is no deflection. The additional requirement for zero moment is fulfilled when the edge is straight, $(\partial^2 w/\partial y^2) = 0$, and when it is shown that the second derivative of Eq. (C-4.1) is zero when $x = 0$ or L, so

$$M_x = -D \left(\alpha \frac{\partial^2 w}{\partial x^2} + v\sqrt{\alpha}\, \frac{\partial^2 w}{\partial y^2} \right) = 0.$$

Thus the boundary conditions are satisfied by $w = 0$ and $(\partial^2 w)/(\partial x^2) = 0$ at $x = 0$ and L.

Differentiation of the assumed displacement function Eq. (C-4.1) with respect to x and y as required by Eq. (C-3.1) (with $\sigma_x h = -N_x$ and all other loads zero) results in a linear differential equation. Note that the load N_x is compressive, so the sign is negative. Subsequent differentiation of w twice also produces a negative sign; thus the load term is positive when to the right of the equal sign:

$$\frac{d^4 Y}{dy^4} - 2\sqrt{\alpha}\, \frac{n^2\pi^2}{L^2}\, \frac{d^2 Y}{dy^2} + \left[\alpha \left(\frac{n^4\pi^4}{L^4} \right) - \left(\frac{N_x}{D} \right) \frac{n^2\pi^2}{L^2} \right] Y = 0$$

$$(\text{C-4.2})$$

Then if we define

$$\phi_1 = \frac{n\pi}{L} \sqrt[4]{\alpha} \sqrt{\mu + 1} \quad \text{and} \quad \phi_2 = \frac{n\pi}{L} \sqrt[4]{\alpha} \sqrt{\mu - 1}, \quad (\text{C-4.3})$$

where

$$\mu = \sqrt{[N_x/(D_\alpha)]\, [L/(n\pi)]^2}, \quad (\text{C-4.4})$$

a solution in terms of w is

$$w = \sin \frac{n\pi x}{L}(C_1 \cosh \phi_1 y + C_2 \sinh \phi_1 y + C_3 \cos \phi_2 y + C_4 \sin \phi_2 y). \quad (\text{C-4.5})$$

By utilization of boundary conditions along the supports, $y = 0$ and $y = L$, constants are evaluated to obtain a final result. Equation (C-4.5) may be used for plates which are simply supported on two opposite sides with the other two support conditions accounted for by proper evaluation of constants in accordance with such boundary conditions. This method is used to find solution for the various k coefficients presented in Chap. 11. However, for plates simply supported on all four sides, the procedure given next is more direct.

C-4.2 Solution Type II. A displacement function for a plate simply supported on all four sides, which satisfies boundary conditions w and $M = 0$ at $x = 0$ and L, $y = 0$ and b, is

$$w = A_{mn} \sin \frac{m\pi x}{L} \sin \frac{n\pi y}{b}, \tag{C-4.6}$$

where A_{mn} is an undertermined coefficient and m and n are integers. By successive differentiation, expressions are determined for the displacement derivatives of Eq. (C-3.1):

$$\frac{\partial w}{\partial x} = A_{mn} \frac{m\pi}{L} \cos \frac{m\pi x}{L} \sin \frac{n\pi y}{b},$$

$$\frac{\partial^2 w}{\partial x^2} = -A_{mn} \left(\frac{m\pi}{L}\right)^2 \sin \frac{m\pi x}{L} \sin \frac{n\pi y}{b},$$

$$\frac{\partial^3 w}{\partial x^3} = -A_{mn} \left(\frac{m\pi}{L}\right)^3 \cos \frac{m\pi x}{L} \sin \frac{n\pi y}{b},$$

$$\frac{\partial^4 w}{\partial x^4} = A_{mn} \left(\frac{m\pi}{L}\right)^4 \sin \frac{m\pi x}{L} \sin \frac{n\pi y}{b}.$$

Likewise

$$\frac{\partial^4 w}{\partial y^4} = A_{mn} \left(\frac{n\pi}{b}\right)^4 \sin \frac{m\pi x}{L} \sin \frac{n\pi y}{b} \tag{C-4.7}$$

and

$$\frac{\partial^4 w}{\partial x^2 \partial y^2} = A_{mn} \left(\frac{n\pi}{b}\right)^2 \left(\frac{m\pi}{L}\right)^2 \sin \frac{m\pi x}{L} \sin \frac{n\pi y}{b}.$$

Substitution of Eqs. (C-4.7) into Eq. (C-3.1), after suitable rearrangement produces

$$A_{mn} D \sin \frac{m\pi x}{L} \sin \frac{n\pi y}{b} \left[\alpha \left(\frac{m\pi}{L}\right)^4 + 2\sqrt{\alpha} \left(\frac{m\pi}{L}\right)^2 \left(\frac{n\pi}{b}\right)^2 + \left(\frac{n\pi}{b}\right)^4 \right]$$

$$= \sigma_x h A_{mn} \left(\frac{m\pi}{L}\right)^2 \sin \frac{m\pi x}{L} \sin \frac{n\pi y}{b}, \tag{C-4.8}$$

where the term to the right of the equal sign is positive because σ_x is negative, that is, compressive. Equation (C-4.8) reduces to

$$D \left[\sqrt{\alpha} \left(\frac{m\pi}{L}\right)^2 + \left(\frac{n\pi}{b}\right)^2 \right]^2 = \sigma_x h \left(\frac{m\pi}{L}\right)^2 \tag{C-4.9}$$

where $(\sigma_x h)$ represents a compressive force on two opposite edges.

By replacing σ_x by σ_{CR}, we obtain an equation for the critical stress:

$$\sigma_{CR} = \frac{D\pi^2 \left[\sqrt{\alpha}\,(m/L)^2 + (n/b)^2\right]^2}{h\,(m/L)^2}. \tag{C-4.10}$$

The smallest nontrivial solution of σ_{CR} is when $n = 1$. Dividing numerator and denominator by $(m/L)^2$ yields

$$\sigma_{CR} = \frac{D\pi^2}{h} \left(\sqrt{\alpha}\, \frac{m}{L} + \frac{L}{mb^2} \right)^2,$$

where n, having been set to 1, is no longer evident. If $1/b^2$ is factored from this equation, the result is:

$$\sigma_{CR} = \frac{D\pi^2}{hb^2} \left[\sqrt{\alpha}\, m \left(\frac{b}{L} \right) + \frac{1}{m} \left(\frac{L}{b} \right) \right]^2.$$

Expanding D [Eq. (C-1.15)], we obtain

$$\sigma_{CR} = \left[\sqrt{\alpha}\, m \left(\frac{b}{L} \right) + \frac{1}{m} \left(\frac{L}{b} \right) \right]^2 \frac{\pi^2 E}{12(1 - \nu^2)} \left(\frac{h}{b} \right)^2, \qquad \text{(C-4.11)}$$

where the composite term contains the only variable. To minimize σ_{CR}, it is necessary to minimize the coefficient terms [which is usually expressed as buckling coefficient k, see Eq. (C-4.18)] with respect to m. That is, set the partial of the coefficient with respect to m equal to zero; thus

$$\frac{\partial}{\partial m} \left[\sqrt{\alpha}\, m \frac{b}{L} + \left(\frac{1}{m} \right) \frac{L}{b} \right]^2$$

$$= 2 \left[\sqrt{\alpha}\, m \frac{b}{L} + \left(\frac{1}{m} \right) \frac{L}{b} \right] \left[\sqrt{\alpha} \left(\frac{b}{L} \right) - \frac{L}{m^2 b} \right] = 0. \qquad \text{(C-4.12)}$$

Using the last term, because the first term is always greater than zero for a positive m, we have

$$\sqrt{\alpha}\, \frac{b}{L} - \frac{L}{m^2 b} = 0, \qquad m^2 = \frac{L^2}{b^2 \sqrt{\alpha}}, \qquad \text{(C-4.13)}$$

or

$$m = \frac{L}{b \sqrt[4]{\alpha}}. \qquad \text{(C-4.14)}$$

Substitution of this value of m into the coefficient terms of Eq. (C-4.11) provides the minimum buckling coefficient k as

$$k = \left[\sqrt{\alpha} \left(\frac{L}{b \sqrt[4]{\alpha}} \right) \frac{b}{L} + \left(\frac{b \sqrt[4]{\alpha}}{L} \right) \frac{L}{b} \right]^2 \qquad \text{(C-4.15)}$$

$$= \left[\frac{\sqrt{\alpha}}{\sqrt[4]{\alpha}} + \sqrt[4]{\alpha} \right]^2 = \frac{\alpha}{\sqrt{\alpha}} + 2\sqrt{\alpha} + \sqrt{\alpha} = 4\sqrt{\alpha}. \qquad \text{(C-4.16)}$$

An important anomaly occurs in this determination of k for the special case of a rectangular plate simply supported on four sides. As used in Eq. (C-4.6), m must be an integer in order to satisfy boundary conditions. In Eq. (C-4.12), m is treated as a real number as can be clearly seen in Eq. (C-4.14). Thus the minimum value of k established by using a noninteger m represents an absolute minimum which is only reached when m is an integer.

A general critical stress equation is obtained by replacing the coefficient term of Eq. (C-4.11) by $k\sqrt{\alpha}$ and α by E_T/E:

$$\sigma_{CR} = k \frac{\pi^2 \sqrt{EE_T}}{12(1 - v^2)} \left(\frac{h}{b}\right)^2. \tag{C-4.17}$$

Or, with $\alpha = 1$,

$$\sigma_{CR} = k \frac{\pi^2 E}{12(1 - v^2)} \left(\frac{h}{b}\right)^2. \tag{C-4.18}$$

Equations (C-4.17) and (C-4.18) become specific for any particular end conditions when properly corresponding numerical values are used for the coefficient k. For a rectangular plate simply supported on four sides with $\alpha = 1$, k is 4. Values for several different support conditions are tabulated in Chap. 11.

C-4.3 Physical Significance. A rectangular elastic plate with $\alpha = E_T/E = 1$, having all four sides simply supported has a minimum k of 4 [Eq. (C-4.16)]. This coefficient k can be written in general form by eliminating α from Eq. (C-4.11);

$$k = \left(\frac{mb}{L} + \frac{L}{mb}\right)^2 = \left(\frac{mb}{L}\right)^2 + 2 + \left(\frac{L}{mb}\right)^2. \tag{C-4.19}$$

By giving m successive integer values of 1, 2, 3, etc., we can compute corresponding values of k for various ratios of L/b.

Figure C-4.1 has a tabular computation which shows k, with m changing by columns and L/b by rows. Below this computation, k is plotted versus L/b for each value of m. As expected, the minimum k value for each m is 4.0, *but* in each case corresponding to different ratios of L/b. The portion of each curve which is below every other curve is drawn solid, representing the lower bound. Remaining portions of the lines are drawn broken. Curves for $m = 1$ and $m = 2$ intersect at a cusp. Location of cusps for adjacent curves can be found by equating $\sqrt{k}$ for any m to $\sqrt{k}$ for the next higher m, defined as $m + 1$, thus

$$\frac{mb}{L} + \frac{L}{mb} = (m + 1)\frac{b}{L} + \frac{L}{(m + 1)b} \tag{C-4.20}$$

$$k = \left(\frac{mb}{L}\right)^2 + 2 + \left(\frac{L}{mb}\right)^2 \qquad \text{[Eq. (C-4.19)]}$$

	m = 1			m = 2			m = 3		
L/b	$(b/L)^2$	$(L/b)^2$	k_1	$(2b/L)^2$	$(L/2b)^2$	k_2	$(3b/L)^2$	$(L/3b)^2$	k_3
0.5	4.00	0.25	6.25	16.00	0.06	18.06	36.00	0.03	38.03
1.0	1.00	1.00	4.00	4.00	0.25	6.25	9.00	0.11	11.11
$\sqrt{2}$	0.50	2.00	4.50	2.00	0.50	4.50	4.50	0.22	6.72
2.0	0.25	4.00	6.25	1.00	1.00	4.00	2.25	0.44	4.69
$\sqrt{6}$	0.17	6.00	8.17	0.67	1.50	4.17	1.50	0.67	4.17
3.0	0.11	9.00	11.11	0.44	2.25	4.69	1.00	1.00	4.00

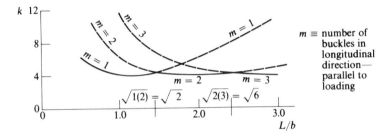

Fig. C-4.1 Plate Buckling; k versus L/b

so that

$$\frac{L}{b} = \sqrt{m(m + 1)}. \qquad (\text{C-4.21})$$

By substitution, when $m = 1$, $L/b = \sqrt{1(2)}$; when $m = 2$, $L/b = \sqrt{2(3)}$; when $m = 3$, $L/b = \sqrt{3(4)}$; etc. In Fig. C-4.1, it can be seen that these high points of k occur at L/b ratios of $\sqrt{2}$, $\sqrt{6}$, $\sqrt{12}$, etc.

For any m, when the L/b ratio is integer, k is 4.0. In every such case, the plate will have exactly m buckles, similar to those shown at the left-hand side of Fig. C-4.2 for $m = 3$. This "feels" reasonable because physically each line of contraflexture is acting very much as a simple support.

By contrast, when the ratio of L/b is at a cusp, say $L/b = \sqrt{6}$, with m either 2 or 3, while n is 1, and $k = 4.1\overline{6}$, the situation is different.

$$k_1 = \left(\frac{1\sqrt{6}}{2}\right)^2 + 2 + \left(\frac{2}{1\sqrt{6}}\right)^2 = 1.5 + 2 + \frac{2}{3} = 4.1\overline{6},$$

$$k_1 = \left(\frac{1\sqrt{6}}{3}\right)^2 + 2 + \left(\frac{3}{1\sqrt{6}}\right)^2 = \frac{2}{3} + 2 + 1.5 = 4.1\overline{6}.$$

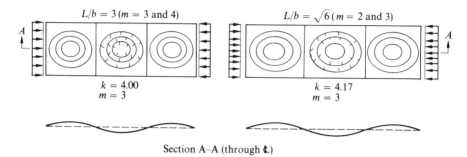

Both plates simply supported on four sides

Section A–A (through ℄)

Fig. C-4.2   Buckling Mode and k as Function of L/b

Thus curvature between nodal lines is identical and buckling is still symmetrical or anti-symmetrical, with the buckling portion between nodal lines no longer square. This is illustrated at the right-hand side of Fig. C-4.2 for a plate having $k = 4.1\overline{6}$ at a cusp with L/b ratio of 3. Critical load is higher for nonsquare sections when $m = 1$, and correspondingly, critical load of plates having m greater than 1 is also higher; buckling configurations are no longer square.

Study of the tabular calculation of Fig. C-4.1 shows that k never exceeds the value obtained at the first cusp when $L/b = \sqrt{2}$ and $k = 4.5$. Such a plate can carry $(4.5)/4$ or 112.5% more load without buckling than one having an integer L/b ratio. (As L/b gets smaller and smaller, the critical buckling coefficient does exceed 4.5, which, corresponding to columns, indicates that failure is then more likely to be from insufficient material strength rather than insufficient stiffness.) As the L/b ratio increases, maximum values of k occur at successive cusps and approach 4 as a limiting value. Thus, for plates simply supported on four sides, k is usually fixed at 4, which represents both a safe and not overly conservative value.

REFERENCES

1. Timoshenko, Stephen P. and James M. Gere, *Theory of Elastic Stability*, 2d ed. McGraw-Hill, New York (1961).

2. Bleich, Friedrich, *Buckling Strength of Metal Structures*, McGraw-Hill, New York (1952).

3. Wang, Chi-Teh, *Applied Elasticity*, McGraw-Hill, New York (1953).

4. Timoshenko, S. and J. N. Goodier, *Theory of Elasticity*, 2d ed., McGraw-Hill, New York (1951).

5. Housner, George W. and Thad Vreeland, Jr., *The Analysis of Stress and Deformation*, MacMillan, New York (1966).

6. Volterra, Enrico and J. H. Gaines, *Advanced Strength of Materials*, Prentice-Hall, Englewood Cliffs, New Jersey (1971).

DESIGN AIDS

d	F	d	F	d	F	d	F	d	F	d	F
$b = 5''$		$b = 7''$		$b = 9''$		$b = 11\frac{1}{2}''$		$b = 15''$		$b = 19''$	
8	0.027	12	0.084	16	0.192	21	0.423	28	0.98	36	2.05
$8\frac{1}{2}$	0.030	$12\frac{1}{2}$	0.091	$16\frac{1}{2}$	0.204	$b = 12''$		29	1.05	38	2.29
9	0.034	$b = 7\frac{1}{2}''$		$b = 9\frac{1}{2}$		22	0.484	30	1.13	$b = 21''$	
$9\frac{1}{2}$	0.038	13	0.106	17	0.229	23	0.529	31	1.20	40	2.80
$b = 6''$		$13\frac{1}{2}$	0.114	$17\frac{1}{2}$	0.242	$b = 13''$		$b = 17''$			
10	0.050	$b = 8''$		$b = 10''$		24	0.624	32	1.45		
$10\frac{1}{2}$	0.055	14	0.131	18	0.270	25	0.677	33	1.54		
11	0.061	$14\frac{1}{2}$	0.140	$18\frac{1}{2}$	0.285	26	0.732	34	1.64		
$11\frac{1}{2}$	0.066	15	0.150	19	0.301	27	0.790				
		$15\frac{1}{2}$	0.160	20	0.333						

$$F \equiv \frac{bd^2}{12,000}$$

D-1 Restricted F Table

Spacing	#4	#5	#6
@ 3''	0.80	1.24	1.76
@ 4''	0.60	0.93	1.32
@ 6''	0.40	0.62	0.88
@ 9''	0.27	0.41	0.59

D-2 Restricted Slab Bar List

Restricted Bar List

Bar #	2	3	4	5	6	7	8	9	10	11	14	18
Wt. (lb/ft)	0.167	0.376	0.668	1.043	1.502	2.044	2.670	3.400	4.303	5.313	7.65	13.60
Diam. (in.)	0.25	0.375	0.50	0.625	0.750	0.875	1.00	1.128	1.270	1.410	1.693	2.257
Area (in.²)	0.05	0.11	0.20	0.31	0.44	0.60	0.79	1.00	1.27	1.56	2.25	4.00
Perim. (in.)	0.786	1.178	1.571	1.963	2.356	2.749	3.142	3.544	3.990	4.430	5.32	7.09

Number of Bars versus Area/Minimum Web Width

(Each cell: Area (in.²) / Minimum Web Width (in.))

No. of Bars	4	5	6	7	8	9	10	11
1	0.20	0.31	0.44	0.60	0.79	1.0	1.27	1.56
2	0.40 / 6½	0.62 / 6½	0.88 / 7	1.20 / 7	1.58 / 7	2.0 / 7½	2.54 / 8	3.12 / 8
3	0.60 / 8	0.93 / 8	1.32 / 8½	1.80 / 9	2.37 / 9	3.0 / 9½	3.81 / 10½	4.68 / 11
4	0.80 / 9½	1.24 / 9½	1.76 / 10	2.40 / 10½	3.16 / 11	4.0 / 12	5.08 / 13	6.24 / 14
5	1.00 / 11	1.55 / 11½	2.20 / 12	3.00 / 12½	3.95 / 13	5.0 / 14	6.35 / 15½	7.80 / 16½
6	1.20 / 12½	1.86 / 13	2.64 / 13½	3.60 / 14½	4.74 / 15	6.0 / 16½	7.62 / 18	9.36 / 19½
7	1.40 / 14	2.17 / 14½	3.08 / 15½	4.20 / 16½	5.53 / 17	7.0 / 19	8.89 / 20½	10.92 / 22½
8	1.60 / 15½	2.48 / 16	3.52 / 17	4.80 / 18	6.32 / 19	8.0 / 21	10.16 / 23	12.48 / 25
9	1.80 / 17	2.79 / 18	3.96 / 19	5.40 / 20	7.11 / 21	9.0 / 23½	11.43 / 25½	14.04 / 28
10	2.00 / 18½	3.10 / 19½	4.40 / 20½	6.00 / 22	7.90 / 23	10.0 / 25½	12.70 / 28	15.60 / 31

D-3   Restricted Bar List

Nominal n	n	f_c	f_c'	Values of K f_s								f_v†
				16	18	20	22	24	27	30	33	
10	10.1	1125	2500	201	190	179	170	161	150	140	132	55
9	9.2	1350	3000	252	238	226	214	204	190	178	168	60
8	8.0	1800	4000	359	341	324	309	295	277	260	246	70
7	7.1	2250	5000	468	446	426	407	390	366	346	327	78

D-4   WSD K

Note. Values of K computed by using $\frac{1}{2}f_c\,kj$, where $k = nf_c/(nf_c + f_s)$ with n *not* rounded and $j = 1 - k/3$.

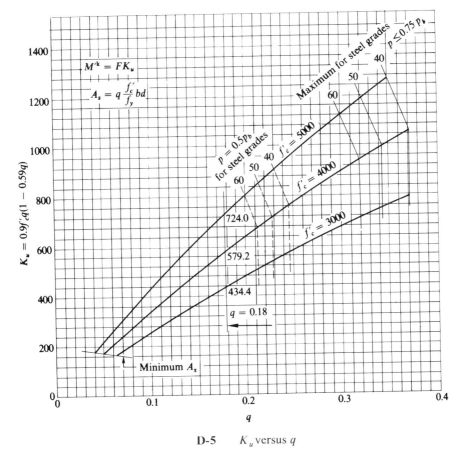

D-5 K_u versus q

---

† f_v is allowable shear stress permitted without using stirrups.

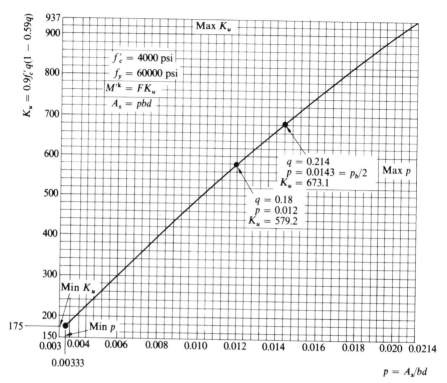

D-6 K_u versus p ($f'_c = 4000$, Grade 60 Steel)

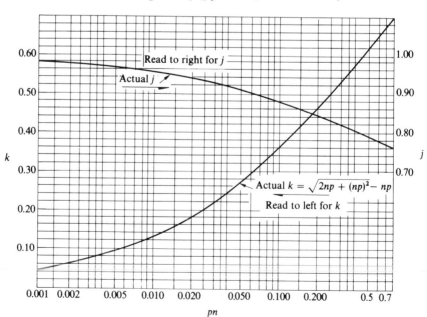

D-7 k and j versus pn

SUPPLEMENTARY FACTS

E-1 COORDINATE SYSTEMS

Right-hand rule for establishing coordinate systems usually is taught in mechanics courses. Typically the rule states that when the thumb, forefinger, and middle finger of the right hand are placed into a mutually perpendicular position, the thumb points along the X axis, the forefinger along the Y axis, and the middle finger along the Z axis. Any finger can be used to represent the X axis so long as the other axes are named in a counterclockwise manner, with the palm of the hand visible.

Although most authors use right-handed coordinates, some use only one orientation. Figure E-1.1 illustrates various right-handed coordinate systems as oriented in various literature sources.

E-2 SIGN CONVENTIONS

The philosophical attitude toward sign conventions used in this text is that they are developed and used to facilitate efficient, error-free computation. For this reason, sign conventions are not thought of as "fashionable," but rather in the light of their usefulness.

E-2.1 Beam Sign Convention. This sign convention was used almost exclusively, until recent years, for problems in statics and strength of materials. A colloquial expression for this sign convention is: *If it holds water, bending moment is positive.* Positive and negative moments produce curvatures as shown in Fig. E-2.1. Note that the sign of the quotient of these end moments producing single curvature is positive. (Positive axes are assumed upward and to the right; some frame analysis uses a sign convention assuming positive axes upward and to the left.)

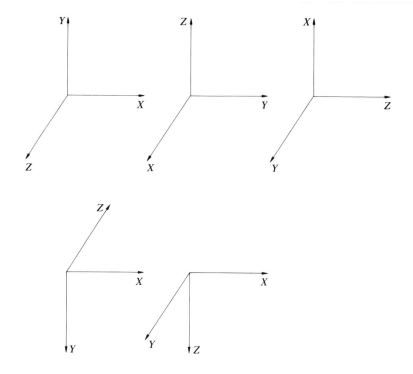

Fig. E-1.1   "Right-Handed" Coordinate Systems

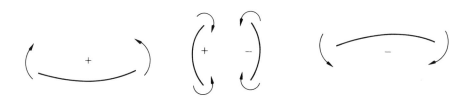

Fig. E-2.1   Beam Sign Convention

Using beam sign convention, both fixed end moments for gravity loads and carry-over factors are negative, as indicated in Fig. E-2.2.

E-2.2 Frame Sign Convention. This convention often is used when analyzing frames so that no reference axes are needed. *Clockwise resisting moments are considered as positive.* The results are shown in Fig. E-2.3. Note that single curvature results in a negative sign for the quotient of end moments.

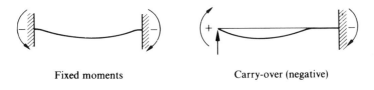

Fixed moments   Carry-over (negative)

Fig. E-2.2   Beam Signs for Moment Distribution

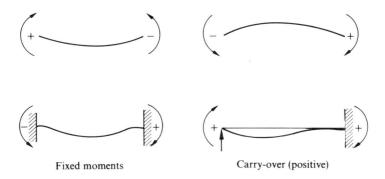

Fixed moments   Carry-over (positive)

Fig. E-2.3   Frame Sign Convention

E-2.3 Elasticity Sign Convention. This convention is used extensively, sometimes even in simple statics courses. It has universal appeal for theoretical development and is used in this text in conjunction with all developments which depend upon theory of elasticity. However, it has drawbacks when applied to "linear" structures, particularly with regard to signs for shearing stresses. See Appendix B.

E-2.4 Right-hand Rule. Moments may be represented by vectors when referred to right-hand coordinate systems. The direction of moment is determined by figuratively "wrapping" one's right hand around the vector, thumb pointing in arrow direction. Fingers then indicate direction of revolution as imposed by moment.

E-3   THEORIES OF FAILURE AND VON MISES' YIELD CRITERIA

Seely and Smith [1] present an extensive discussion and analysis of various "Theories of Failure" which are summarized in Table E-3.1. One of the more

Table E-3.1 Theories of Failure

	states that inelastic action at a point in a body at which any state of stress exists begins *ONLY* when the ...	at some point reaches a value equal to the ...	occurring in a simple tensile specimen when yielding starts.
Maximum principal stress theory (Rankine's theory)	maximum principal stress	tensile or compressive stress	
Maximum shearing stress theory (Coulomb's theory; Guest's law)	maximum shearing stress	shear stress	
Maximum strain theory (St. Venant's theory)	maximum strain	strain	
Total energy theory (Beltrami and Haigh)	energy per unit volume absorbed	energy per unit volume absorbed	
Energy of distortion theory (Huber; Von Mises; Hencky; Bridgman)	strain energy per volume	strain energy per unit volume	
Octahedral shearing stress theory	octahedral shearing stress	equals 0.47 of the tensile elastic strength	

important is the theory of distortion energy often attributed to Von Mises. Basis of this theory is that yielding begins when the distortion energy (strain energy per unit volume) equals the distortion energy at yield for an identical material in a simple tension test.

In terms of principal stresses [2], this theory is expressed by the equation

$$\tfrac{1}{2}[(\sigma_1 - \sigma_2)^2 + (\sigma_2 - \sigma_3)^2 + (\sigma_3 - \sigma_1)^2] = \sigma_y^2, \qquad \text{(E-3.1)}$$

where σ_1, σ_2, and σ_3 are principal stresses, and σ_y is the yield stress.

For the case of plane stress, the principal stress σ_3 is zero and Eq. (E-3.1) reduces to

$$\sigma_1^2 - \sigma_1\sigma_2 + \sigma_2^2 = \sigma_y^2 \qquad \text{(E-3.2)}$$

which is plotted as an ellipse in Fig. E-3.1.

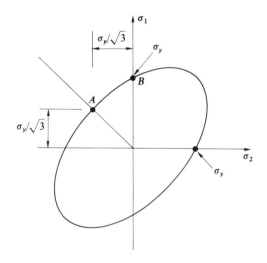

Fig. E-3.1   Von Mises' Yield Condition for Plane Stress

For the case of pure shear $\sigma_1 = -\sigma_2 = \tau$ (visualize Mohr's circle) and Eq. (E-3.2) can be written as $\sigma_1^2 - (\sigma_1)(-\sigma_1) + (-\sigma_1)^2 = 3\,(\sigma_1)^2 = \sigma_y^2$. Replacement of σ_1 by τ_y yields

$$\tau_y = \sigma_y/\sqrt{3}, \qquad \text{(E-3.3)}$$

where $\tau_y \equiv$ yield stress in pure shear.

Coordinates with $|\sigma_1| = |-\sigma_2| = \sigma_y/3$ are plotted at point A on the figure.

A straight line equation which approximates the curve between points A and B of Fig. E-3.1 is obtained from the geometry shown in Fig. E-3.2. Substitution in the general equation for a straight line, $y = mx + b$, yields

$$\sigma_1 = \frac{\sigma_y - (\sigma_y/\sqrt{3})}{\sigma_y/\sqrt{3}}\sigma_2 + \sigma_y,$$

which reduces to

$$\sigma_1 = \sigma_y + (\sqrt{3} - 1)\sigma_2. \tag{E-3.4}$$

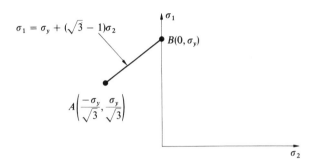

Fig. E-3.2   Approximation to Von Mises' Yield Condition

E-4   TAYLOR'S SERIES

Taylor's series expansions for functions are often used in problem formulations related to structures to markedly reduce complexity without introducing serious error. Although important, recollection of details of application tend to become hazy because the technique is not usually used in design and analysis computations. For these reasons, this short review is given.

If any function can be expressed using x as the only variable, say $f(x)$, and if successive higher ordered derivatives exist in an interval which includes a point where $x = a$, then, in that interval, the function can be represented by Taylor's series:

$$f(x) = f(a) + f'(a)(x - a) + \frac{f''(a)}{2!}(x - a)^2 + \cdots + \frac{f^n(a)}{n!}(x - a)^n + \cdots \tag{E-4.1}$$

for those values of x for which

$$\lim_{n \to \infty} \frac{f^{(n)}(\varepsilon)}{n!}(x - a)^n = 0 \quad \text{with} \quad a < \varepsilon < x. \tag{E-4.2}$$

Here $f'(a)$ signifies the first derivative of $f(x)$ evaluated at point a, and $f^n(a)$ signifies the nth derivative of $f(x)$ also evaluated at a. Another form of the equation is obtained when $x - a$ is represented by h:

$$f(a + h) = f(a) + f'(a)h + \frac{f''(a)h^2}{2!} + \frac{f'''(a)h^3}{3!} + \cdots, \qquad \text{(E-4.3)}$$

where

$f(a) \equiv$ function evaluated by replacing x by a,

$f'(a) \equiv$ first derivative of function evaluated by replacing x by a,

$f''(a) \equiv$ second derivative of function evaluated by replacing x by a, etc.

As an example, consider the expansion of $\sin x$ about some point a. Then

$$\begin{aligned}
f(x) &= \sin x, & f(a) &= \sin a, \\
f'(x) &= \cos x, & f'(a) &= \cos a, \\
f''(x) &= -\sin x, & f''(a) &= -\sin a, \\
f'''(x) &= -\cos x, & f'''(a) &= -\cos a, \ldots
\end{aligned}$$

so, using Eq. (E-4.1), we get,

$$f(x) = \sin x = \sin a + (\cos a)(x - a) + \frac{(-\sin a)}{2!}(x - a)^2$$

$$+ \frac{(-\cos a)}{3!}(x - a)^3 + \frac{(\sin a)}{4!}(x - a)^4$$

$$+ \frac{(\cos a)}{5!}(x - a)^5 + \cdots$$

or letting $a = 0$, we have

$$\sin x = x - \frac{x^3}{3!} + \frac{x^5}{5!} - \frac{x^7}{7!} + \cdots.$$

By use of Eq. (E-4.3) with $h = x - a$, where $a = 0$,

$$\begin{aligned}
f(x) &= \sin x, & f(0) &= 0, \\
f'(x) &= \cos x, & f'(0) &= 1, \\
f''(x) &= -\sin x, & f''(0) &= 0, \\
f'''(x) &= -\cos x, & f'''(0) &= -1,
\end{aligned}$$

$$f(x) = \sin x = 0 + 1(x) + 0 + \frac{(-1)x^3}{3!} + 0 + \frac{(1)x^5}{5!}, \text{ etc.}$$

A more important application in structures is to expand $(\partial x/\partial z)_L$ about a point dz distance away from point L by letting $f(z) \equiv (\partial x/\partial z)_L$ and $x - a = dz$. Then

$$f(z) = (\partial x/\partial z)_L, \qquad f(z + dz) = (\partial x/\partial z)_R,$$
$$f'(z) = (\partial^2 x/\partial z^2)_L, \qquad f'(z + dz) = (\partial^2 x/\partial z^2)_R,$$
$$f''(z) = (\partial^3 x/\partial z^3)_L, \qquad f''(z + dz) = (\partial^3 x/\partial z^3)_R,$$

and by use of Eq. (E-4.1):

$$f(z) = \left(\frac{\partial x}{\partial z}\right)_L = \left(\frac{\partial x}{\partial z}\right)_R + \left(\frac{\partial^2 x}{\partial z^2}\right)_R (-dz) + \left(\frac{\partial^3 x}{\partial z^3}\right)_R \frac{dz^2}{2!} + \cdots,$$

where L signifies the left-hand side of the element of length dz, and R the right-hand side.

E-5 dx VERSUS Δx

The novice often overlooks the very important basic difference in problem formulations using numerical methods which involve dx versus Δx. Figure E-5.1 shows both cases.

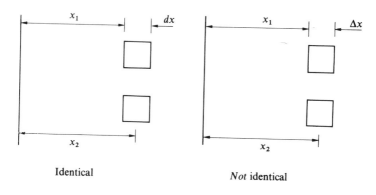

Fig. E-5.1 dx versus Δx

When dx is used for integration in closed form, x_1 measured to the edge of an element may be considered as identical in length to x_2 which is measured to the middle, because dx is infinitely small.

However, when using Δx, care must be taken in establishing the limits of x. Because Δx is finite, $x_2 = x_1 + \Delta x/2$ is not equal to x_1.

REFERENCES

1. Seely, Fred B. and James O. Smith, *Advanced Mechanics of Materials*, 2d ed., Wiley, New York (1962).

2. Mendelson, Alexander, *Plasticity: Theory and Application*, Macmillan, New York (1968).

3. Basler, Konrad, "Strength of Plate Girders in Shear," *Journal of Structural Division*, American Society of Civil Engineers, V. 87, ST 7 (October 1961), pp. 151–180.

INDEX

Particular care has been used to make this index truly valuable to both the student and practitioner. In keeping with the spirit of an integrated engineering-design approach, specific references key portions of ACI-63, ACI-71, and AISC to this text. Further, it is anticipated that the captions **Aids**, **Assumptions**, **Buckling**, **Definition**, **Design** (particularly the subheading Procedures), **Examples**, and **Stress** will be particularly helpful.

A

AASHO, 80, 255, 256, 411

ACI-63

918(h)   Standard hooks, **149**

Formula 12-1   Shear stress, **205**

1301   Bond stress, **157**

Formula 13-1   Bond stress, **157**

Formula 14-1   Spirally reinforced columns, **119**

Formula 17-1   Ultimate shear stress, **205**

1801   Ultimate bond stress, **157**

Formula 18-1   Ultimate bond stress, **157**

Formula 19-1 P_u, bending and axial load, **376, 402, 403**

Formula 19-2 M_u, bending and axial load, **402, 403**

Formula 19-3 M_b, balanced load moment, **403**

Formula 19-4 P_u, column tension failure, general, **406**

Formula 19-5 P_u, column tension failure, symmetrical reinforcement, **372, 406**

Formula 19-6 P_u, column tension failure, no compressive steel, **406**

Formula 19-7 P_o, column, pure compression, **118, 370, 407**

Formula 19-8 P_u, column, ultimate compression strength, **409**

Formula 19-9 P_u, column, ultimate compression strength, alternate form, **408**

Formula 19-10 P_u, column, compressive failure, general, **119, 124, 372, 407**

ACI-71

3.3.2   Nominal maximum size aggregate, **180**

4.3.3   Average concrete strength level, **75**

Chapter 7   Details of reinforcement, **185**

521

Y